AF521918

Selected Titles in This Series

44 **Laura F. Landweber and Eric B. Baum, Editors,** DNA Based Computers II

43 **Panos Pardalos, Sanguthevar Rajasekaran, and José Rolim, Editors,** Randomization Methods in Algorithm Design

42 **Ding-Zhu Du and Frank K. Hwang, Editors,** Advances in Switching Networks

41 **David Aldous and James Propp, Editors,** Microsurveys in Discrete Probability

40 **Panos M. Pardalos and Dingzhu Du, Editors,** Network Design: Connectivity and Facilities Location

39 **Paul W. Beame and Samuel R Buss, Editors,** Proof Complexity and Feasible Arithmetics

38 **Rebecca N. Wright and Peter G. Neumann, Editors,** Network Threats

37 **Boris Mirkin, F. R. McMorris, Fred S. Roberts, and Andrey Rzhetsky, Editors,** Mathematical Hierarchies and Biology

36 **Joseph G. Rosenstein, Deborah S. Franzblau, and Fred S. Roberts, Editors,** Discrete Mathematics in the Schools

35 **Dingzhu Du, Jun Gu, and Panos M. Pardalos, Editors,** Satisfiability Problem: Theory and Applications

34 **Nathaniel Dean, Editor,** African Americans in Mathematics

33 **Ravi B. Boppana and James F. Lynch, Editors,** Logic and Random Structures

32 **Jean-Charles Grégoire, Gerard J. Holzmann, and Doron A. Peled, Editors,** The SPIN Verification System

31 **Neil Immerman and Phokion G. Kolaitis, Editors,** Descriptive Complexity and Finite Models

30 **Sandeep N. Bhatt, Editor,** Parallel Algorithms: Third DIMACS Implementation Challenge

29 **Doron A. Peled, Vaughan R. Pratt, and Gerard J. Holzmann, Editors,** Partial Order Methods in Verification

28 **Larry Finkelstein and William M. Kantor, Editors,** Groups and Computation II

27 **Richard J. Lipton and Eric B. Baum, Editors,** DNA Based Computers

26 **David S. Johnson and Michael A. Trick, Editors,** Cliques, Coloring, and Satisfiability: Second DIMACS Implementation Challenge

25 **Gilbert Baumslag, David Epstein, Robert Gilman, Hamish Short, and Charles Sims, Editors,** Geometric and Computational Perspectives on Infinite Groups

24 **Louis J. Billera, Curtis Greene, Rodica Simion, and Richard P. Stanley, Editors,** Formal Power Series and Algebraic Combinatorics/Séries Formelles et Combinatoire Algébrique, 1994

23 **Panos M. Pardalos, David I. Shalloway, and Guoliang Xue, Editors,** Global Minimization of Nonconvex Energy Functions: Molecular Conformation and Protein Folding

22 **Panos M. Pardalos, Mauricio G. C. Resende, and K. G. Ramakrishnan, Editors,** Parallel Processing of Discrete Optimization Problems

21 **D. Frank Hsu, Arnold L. Rosenberg, and Dominique Sotteau, Editors,** Interconnection Networks and Mapping and Scheduling Parallel Computations

20 **William Cook, László Lovász, and Paul Seymour, Editors,** Combinatorial Optimization

19 **Ingemar J. Cox, Pierre Hansen, and Bela Julesz, Editors,** Partitioning Data Sets

18 **Guy E. Blelloch, K. Mani Chandy, and Suresh Jagannathan, Editors,** Specification of Parallel Algorithms

17 **Eric Sven Ristad, Editor,** Language Computations

16 **Panos M. Pardalos and Henry Wolkowicz, Editors,** Quadratic Assignment and Related Problems

DIMACS

Series in Discrete Mathematics
and Theoretical Computer Science

Volume 44

DNA Based Computers II

DIMACS Workshop
June 10–12, 1996

Laura F. Landweber
Eric B. Baum
Editors

NSF Science and Technology Center
in Discrete Mathematics and Theoretical Computer Science
A consortium of Rutgers University, Princeton University,
AT&T Labs, Bell Labs, and Bellcore

American Mathematical Society

This conference was published in October 1998 and carries the proceedings of a conference held June 10–12, 1996.

This DIMACS volume contains papers from the Second DIMACS Workshop on DNA Based Computers. The Workshop was held on June 10–12, 1996, at Princeton University, Princeton, N. J.

1991 *Mathematics Subject Classification.* Primary 92–06, 92B05, 92C40, 68Q05.

Library of Congress Cataloging-in-Publication Data

DNA based computers II : DIMACS workshop, June 10–12, 1996 / Laura F. Landweber, Eric B. Baum, editors.

p. cm. — (DIMACS series in discrete mathematics and theoretical computer science ; v. 44)

"NSF Science and Technology Center in Discrete Mathematics and Theoretical Computer Science, a consortium of Rutgers University, Princeton University, AT&T Labs, Bell Labs, and Bellcore."

Includes bibliographical references.

ISBN 0-8218-0756-0

1. Molecular computers—Congresses. I. Landweber, Laura F. (Laura Faye), 1967– . II. Baum, Eric B., 1957– . III. NSF Science and Technology Center in Discrete Mathematics and Theoretical Computer Science. IV. DIMACS (Group) V. Series.

QA76.887.D52 1998

511.3—dc21 98-35278

CIP

---

10 9 8 7 6 5 4 3 2 1 04 03 02 01 00 99

Contents

Foreword

The Second Annual Workshop on DNA Based Computers was held on June 10–12, 1996 at Princeton University, N. J. We would like to express our appreciation to Eric Baum, Dan Boneh, Peter Kaplan, Richard Lipton, John Reif, and Nadrian Seeman for their efforts to organize and plan this successful workshop.

The Workshop was part of the broader Special Year on DNA Computing. We extend our thanks to Richard Lipton and Laura Landweber for their work over many months as Special Year organizers.

The Workshop was a forum for bringing together researchers working on all areas that relate directly to computing with DNA, including algorithms, applications, techniques, architectures, computational processes in vivo, and relevant ideas regarding biological evolution.

DIMACS gratefully acknowledges the generous support that makes these programs possible. The National Science Foundation, through its Science and Technology Center program, the New Jersey Commission on Science and Technology, and DIMACS' partners at Rutgers, Princeton, AT&T Labs Research, Bell Labs, and Bellcore generously supported the Special Year.

Fred S. Roberts
Director

Bernard Chazelle
Co-Director for Princeton

Introduction

DNA Computing has fueled a tremendous amount of excitement by offering a fresh paradigm for performing and viewing computations: this is biological mathematics, while also mathematical biology. Encoding of data in DNA strings coupled with the power of molecular biology now allows the execution of computational operations on single strands of DNA in the laboratory. As a small drop of solution can house several quadrillion DNA molecules, DNA computers use much less energy and space than traditional computers. Consequently, they offer the tantalizing potential of one day outperforming electronic computers. Computing with DNA is also massively parallel, with billions and trillions of molecules undergoing simultaneous chemical reactions and performing computations almost spontaneously.

The product of a radically different marriage of computer science to molecular biology, the fledgling field of DNA computers began in 1994, when Leonard Adleman surprised the scientific community by using DNA molecules, protein enzymes, and chemicals to solve an instance of a hard computational problem. These proceedings from the Second DIMACS Workshop on DNA Based Computers at Princeton University – held only one year and a half after the publication of Adleman's seminal experiment – contain a set of refereed papers that highlight some of the exciting progress in this field and together build a strong foundation for the current theory of molecular computation.

This volume introduces such important advances as "mark and destroy" DNA algorithms and the use of surface chemistry in DNA computing. It also includes a healthy discussion of the myriad possibilities as well as potential pitfalls, means of error correction, and general practicality of building DNA based computers. Several articles draw on the analogy between DNA computing and other cutting edge fields of biology such as directed evolution. In principle, both types of experiments search for a single or unique class of molecules in a random or heterogeneous mixture. Such techniques have been enormously successful in discovering biological molecules with novel capacities for performing chemical reactions. As these molecules can even encode the solution to a mathematical problem, nucleic acid based computers may be viewed as a natural extension of test-tube evolution experiments that search a defined space of mathematical solutions. Indeed, they are "solutions of solutions!"

Laura F. Landweber, Princeton University

Acknowledgements

We gratefully acknowledge the generous support from our sponsors: DIMACS, NSF, and the NEC Research Institute. We also thank the other members of the program committee – Dan Boneh, Peter Kaplan, Richard Lipton, John Reif, and Nadrian Seeman – for helping to organize the conference. Lastly, we especially thank Tom Hagedorn for helping to produce this volume.

Laura Landweber, Princeton University

Eric Baum, NEC

DIMACS Series in Discrete Mathematics
and Theoretical Computer Science
Volume **44**, 1999

A Sticker Based Model for DNA Computation

Sam Roweis, Erik Winfree, Richard Burgoyne, Nickolas V. Chelyapov,
Myron F. Goodman, Paul W. K. Rothemund, and Leonard M. Adleman†

ABSTRACT. We introduce a new model of molecular computation that we call the *sticker model*. Like many previous proposals it makes use of DNA strands as the physical substrate in which information is represented and of separation by hybridization as a central mechanism. However, unlike previous models, the stickers model has a random access memory that requires no strand extension, uses no enzymes, and (at least in theory) its materials are reusable.

The paper describes computation under the stickers model and discusses possible means for physically implementing each operation. We go on to propose a specific machine architecture for implementing the stickers model as a microprocessor-controlled parallel robotic workstation. Finally, we discuss several methods for achieving acceptable overall error rates for a computation using basic operations that are error prone.

In the course of this development a number of previous general concerns about molecular computation [**36, 20, 24**] are addressed. First, it is clear that general-purpose algorithms can be implemented by DNA-based computers, potentially solving a wide class of search problems. Second, we find that there are challenging problems, for which only modest volumes of DNA should suffice. Third, we demonstrate that the formation and breaking of covalent bonds is not intrinsic to DNA-based computation. This means that costly and short-lived materials such as enzymes are not necessary, nor are energetically costly processes such as PCR. Fourth, we show that a single essential biotechnology, sequence-specific separation, suffices for constructing a general-purpose molecular computer. Fifth, we illustrate that separation errors can theoretically be reduced to tolerable levels by invoking a trade-off between time, space, and error rates at the level of algorithm design; we also outline several specific ways in which this can be done and present numerical calculations of their performance.

Despite these encouraging theoretical advances, we emphasize that substantial engineering challenges remain at almost all stages and that the ultimate success or failure of DNA computing will certainly depend on whether these challenges can be met in laboratory investigations.

S. Roweis is supported in part by the Center for Neuromorphic Systems Engineering as a part of the National Science Foundation Engineering Research Center Program under grant EEC-9402726 and by the Natural Sciences and Engineering Research Council of Canada.

E. Winfree is supported in part by National Institute for Mental Health (NIMH) Training Grant # 5 T32 MH 19138-06; also by General Motors' Technology Research Partnerships program.

L. Adleman, N. Chelyapov, and P. Rothemund are supported in part by grants from the National Science Foundation (CCR-9403662) and Sloan Foundation.

† To whom correspondence should be addressed. Reprint requests to *roweis@cns.caltech.edu*.

1. Introduction

Much of the recent interest in molecular computation has been fueled by the hope that it might some day provide the means for constructing a massively parallel computational platform capable of attacking problems which have been resistant to solution with conventional architectures. Model architectures have been proposed which suggest that DNA based computers may be flexible enough to tackle a wide range of problems [**1, 2, 4, 25, 8, 6, 34**], although fundamental issues such as the volumetric scale of materials and fidelity of various laboratory procedures remain largely unanswered.

In this paper we introduce a new model of molecular computation that we call the *sticker model.* Like many previous proposals it makes use of DNA strands as the physical substrate in which information is represented and of separation by hybridization as a central mechanism. However, unlike previous models, the stickers model has a random access memory that requires no strand extension, uses no enzymes, and (at least in theory) its materials are reusable.

The paper begins by introducing a new way of representing information in DNA, followed by an abstract description of the basic operations possible under this representation. Possible means for physically implementing each operation are discussed. We go on to propose a specific machine architecture for implementing the stickers model as a microprocessor-controlled parallel robotic workstation, employing only technologies which exist today. Finally, we discuss methods for achieving acceptable error rates from imperfect separation units.

2. The Stickers Model

2.1. Representation of Information. The stickers model employs two basic groups of single stranded DNA molecules in its representation of a bit string. Consider a *memory strand* N bases in length subdivided into K non-overlapping regions each M bases long (thus $N \geq MK$). Each region is identified with exactly one bit position (or equivalently one boolean variable) during the course of the computation. We also design K different *sticker strands* or simply *stickers.* Each sticker is M bases long and is complementary to one and only one of the K memory regions. If a sticker is annealed to its matching region on a given memory strand then the bit corresponding that particular region is *on* for that strand. If no sticker is annealed to a region then that region's bit is *off.* Figure 1 illustrates this representation scheme.

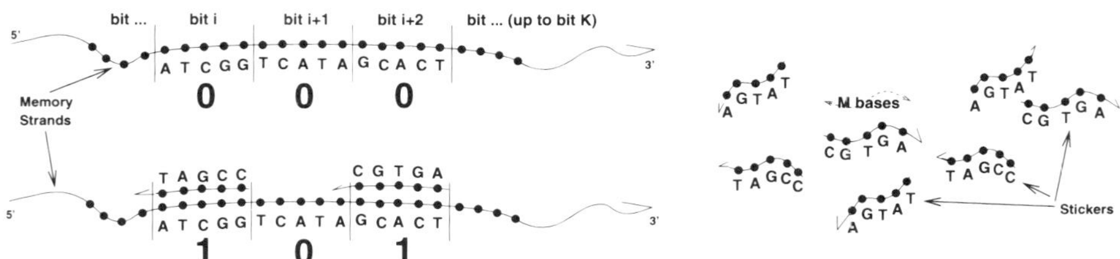

FIGURE 1. A memory strand and associated stickers (together called a *memory complex*) represent a bit string. The top complex on the left has all three bits *off*; the bottom complex has two annealed stickers and thus two bits *on*.

Each memory strand *along with* its annealed stickers (if any) represents one bit string. Such partial duplexes are called *memory complexes.* A large set of bit strings is represented by a large number of *identical* memory strands each of which has stickers annealed only at the required bit positions. We call such a collection of memory complexes a *tube.* This differs from previous representations of information using DNA in which the presence or absence of a particular subsequence in a strand corresponded to a particular bit being on or off (e.g. see [**1, 25**]). In this new model, each possible bit string is represented by a unique *association* of memory strands and stickers whereas previously each bit string was represented by a unique molecule.

To give a feel for the numbers involved, a reasonable size problem (for example breaking DES as discussed in [**3**]), might use memory strands of roughly 12000 bases (N) which represent 580 binary variables (K) using 20 base regions (M).

The information density in this storage scheme is $(1/M)$ bits/base, directly comparable to the density of previous schemes [**1, 8, 25**]. We remark that while information storage in DNA has a theoretical maximum value of 2 bits/base, exploiting such high values in a separation based molecular computer would require the ability to reliably separate strands using only single base mismatches. Instead we choose to sacrifice information density in order to make the experimental difficulties less severe.

2.2. Operations on Sets of Strings. We now introduce several possible operations on sets of bit strings which together turn out to be quite flexible for implementing general algorithms. The four principle operations are *combination* of two sets of strings into one new set, *separation* of one set of strings into two new sets and *setting* or *clearing the* k^{th} *bit* of every string in a set. Each of these logical set operations has a corresponding interpretation in terms of the DNA representation introduced above. Figure 2 summarizes these required DNA interactions.

- The most basic operation is to *combine* two sets of bit strings into one. This produces a new set containing the multi-set union of all the strings in the two input sets. In DNA, this corresponds to producing a new tube containing all the memory complexes (with their annealed stickers undisturbed) from both input tubes.
- A set of strings may be *separated* into two new sets, one containing all the original strings having a particular bit *on* and the other all those with the bit *off.* This corresponds to isolating from the set's tube exactly those complexes with a sticker annealed to the given bit's region. The original input set (tube) is destroyed.
- To *set* (turn on) a particular bit in every string of a set, the sticker for that bit is annealed to the appropriate region on every complex in the set's tube (or left in place if already annealed).
- Finally, to *clear* (turn off) a bit in every string of a set, the sticker for that bit must be removed (if present) from every memory complex in the set's tube.

Computations in this model consist of a sequence of combination, separation, and bit setting/clearing operations. This sequence must begin with some initial set of bit strings and must ultimately produce one (possibly null) set of strings deemed to be "the answers". We call the tube containing the initial set of bit strings the *mother tube* for a computation. Thus, to complete our theoretical description of

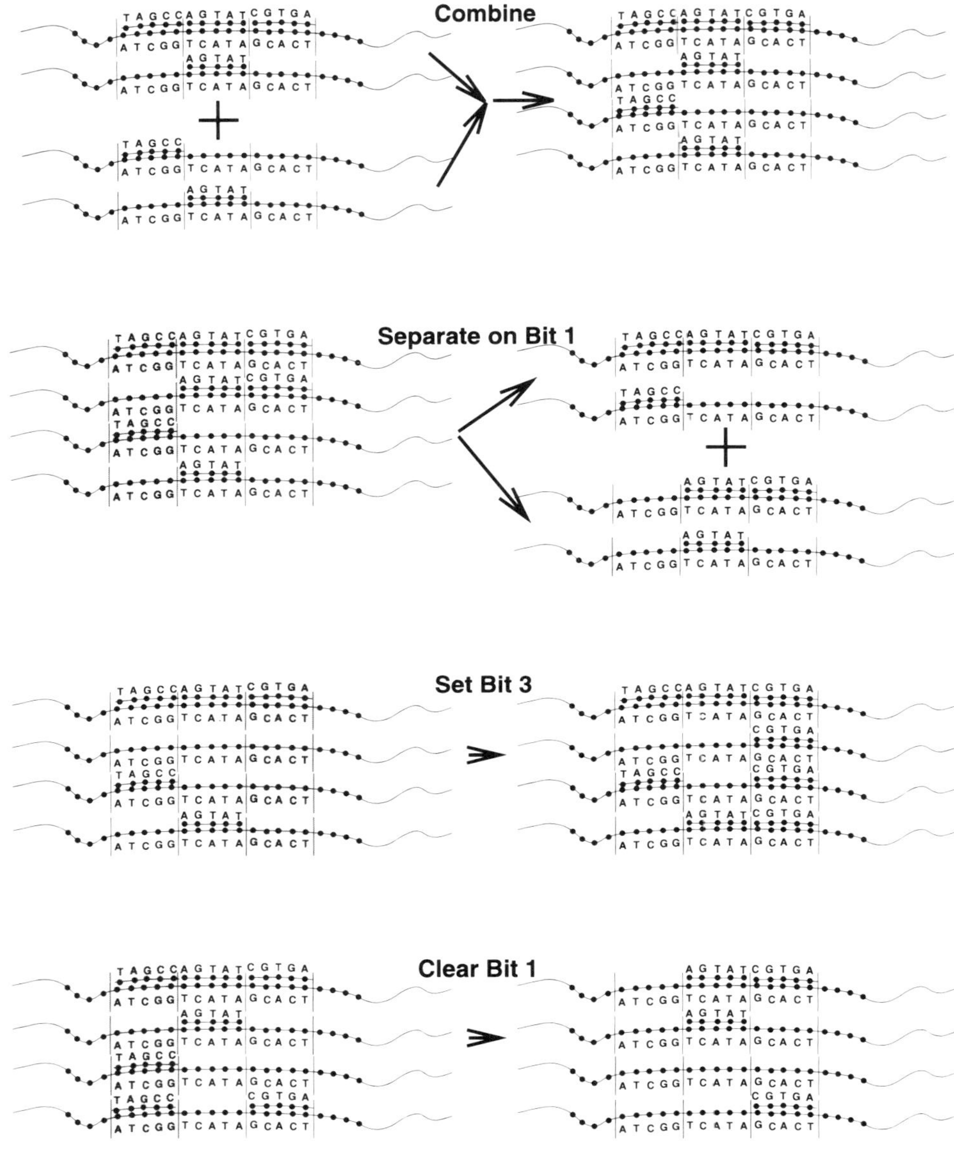

FIGURE 2. DNA manipulations required for the four operations of the stickers model.

how to compute with the stickers model, we must describe how to create a mother tube of memory complexes and also how to read out at least one bit string from a (possibly empty) final tube of answers (or recognize that the tube contains no strands). We consider creation of the mother tube first:

- It will suffice for our purposes to consider creating a mother tube which corresponds to the (K, L) *library set* of strings. A (K, L) *library set* contains strings of length K generated by taking the set of all possible bit strings of length L followed by $K - L$ zeros. There are thus 2^L length K strings in the set[1].

Our paradigm of computation will generally be to cast hard problems as large combinatorial searches over inputs of length L. We search for the few rare "answer" strings by processing all 2^L possible inputs in parallel and eliminating those that fail the search criteria. It is important that the memory strand we design may have more than L bit regions. The first L bits represent the encoding of the input and are the random portion of the initial library. The remaining $K - L$ bits are used for intermediate storage and answer encoding and are initially off on all complexes. All bits can be written to and read from later in the computation as needed. In this way creating a mother tube which is a (K, L) *library set* corresponds to generating all possible inputs (of length L) and zeroing the workspace (length $K - L$).

Lastly, we indicate how to obtain a solution at the end of the computation:

- To read a string from the final "answer" set, one memory complex must be isolated from the answer tube and its annealed stickers (if any) determined. Alternately, it must be reported that the answer tube contains no strands.

2.3. Example Problem. To illustrate the power of the operations defined above we work through the solution of the NP-Complete[2] *Minimal Set Cover* problem [**19**] within the stickers model. Informally, assume we are given a collection of B bags each containing some objects. The objects come in A types. The problem is to find the smallest subset of the bags which between them contains at least one object of every type. Formally the problem is as follows:
Given a collection $C = \{C_1, \ldots, C_B\}$ *of subsets of* $\{1, \ldots, A\}$ *what is the smallest subset* I *of* $\{1, \ldots, B\}$ *such that* $\bigcup_{i \in I} C_i = \{1, \ldots, A\}$*?*
The solution of the problem in our model is straightforward. We create memory complexes representing all possible 2^B choices of bags. We mark all those which include bag i as containing every type appearing in the subset C_i. Then we separate out those complexes which have been marked as containing all A types and read out the one(s) which uses the fewest bags. Formally, the sticker algorithm for minimal set cover is:

- `Design a memory strand with` $K = B + A$ `bit regions.`
  Bits $1 \ldots B$ *represent which bags are chosen, bits* $B + 1 \ldots B + A$ *represent which object types are present.*
- `Initialize a` (K, B) `library set in a tube called` T_0.
- `for i=1 to B`
  `Separate` T_0 `into` T_{on} `and` T_{off} `based on bit` i
  `for j=1 to` $|C_i|$
  `Set bit` $N + C_i[j]$ `in` T_{on}
  `Combine` T_{on} `and` T_{off} `into` T_0

  Mark the final A *positions of each complex to record which object types it contains.*
- `for i=B+1 to B+A`
  `Separate` T_0 `into` T_0 `and` T_{bad} `based on bit` i

[1]For example, the (7,3) *library set* is the set {0000000, 0010000, 0100000, 0110000, 1000000, 1010000, 1100000, 1110000}.

[2]Technically the NP-Complete version of this problem is the binary decision version in which we ask if there exists a collection of a *particular* size that covers the set, not for the collection of the smallest size.

```
        Discard T_bad
```

Get rid of ones which do not have all A types.

```
• for i=0 to B-1
        for j=i down to 0
            Separate T_j into T_(j+1)' and T_j based on bit i+1
            Combine T_(j+1) and T_(j+1)' into T_(j+1)
```

Count how many bags were used. At the end of the outer loop, tube T_i contains all complexes which used exactly i bags.

```
•                                     Read T_1;
  else if it was empty then           Read T_2;
  else if it was empty then           Read T_3;
                                      ...
```

where above $|C_i|$ is the number of items in subset C_i and $C_i[j]$ is the jth item in subset C_i. Note that the above algorithm takes $O(AB)$ steps, and the input is $O(AB)$ bits.

We point out that, as we will envision a robotic system performing the experiments automatically, we allow arbitrary sequential algorithms for controlling the molecular operations. However, these operations must be performed "blind"; the only interface to molecular parallelism is via *initialize, combine, separate, set, clear,* and *read.* Thus the electronic algorithms are responsible for "experiment design" i.e. compiling higher-level problem specifications into concise sequences of molecular operations but they cannot get any feedback from the DNA during the course of the experiment.

As a final comment we note that the stickers model is capable of simulating (in parallel) independent universal machines, one per memory complex, under the usual theoretical assumption of an unbounded number of sticker regions[3]. It should be noted that the stickers model is universal, in the sense discussed, even in the absence of the *clear* operation, although more compact algorithms are possible using *clear*.

3. Physical Implementation of the Model

Each logical operation in our model has a corresponding interpretation (which we gave as we introduced the operations) in terms of what must happen to the DNA memory strands and associated stickers when that operation is carried out. In what follows we examine various physical procedures which are candidates for implementing these requirements for all the operations described above. We speak in terms of *tubes* instead of *sets*; recall that a *tube* consists of the collection of memory complexes that represents a *set* of bit strings.

[3]This can be seen as the consequence of two observations. First, a memory complex in the stickers model can simulate a feedforward circuit, in the spirit of [**8**]. Using the *clear* operation, a clocked feedback circuit can also be simulated. Second, allowing the circuit to grow with each clock cycle, we can simulate a universal machine. The electronic algorithm is responsible for designing the new gates to fit into the circuit; each new gate will require a new bit and hence a new sticker region in the memory strand. For concreteness, a feedforward circuit C_t can be automatically designed which computes the instantaneous description of a TM at time step t from the description at $t-1$. Thus, the stickers model can simulate in parallel the execution of a TM on all 2^L length L inputs.

Often there are several possible implementations of a given operation; each has its own assumed strengths and weaknesses on which we speculate. However, which implementations, if any, turn out to be viable will ultimately have to be decided by laboratory experiments.

3.1. Combination. *Combination* of two tubes can be performed by rehydrating the tube contents (if not already in solution) and then combining the fluids together (by pouring or pumping for example) to form a new tube. It should be noted that even this seemingly straightforward operation is plagued by constraints: if DNA is not handled gently the shear forces from pouring and mixing it will fragment it into ≈ 15 kilobase sections [**23**].

Also of concern for this operation and indeed for all others is the amount of DNA which remains stuck to the walls of tubes, pumps, pipette tips, etc. and thus is "lost" from the computation. Even if this "lost" DNA is a minute fraction of the total (which would be unimportant to molecular biologists) it is problematic for computation because we are working with relatively few copies of each relevant molecule.

3.2. Separation. The ultimate goal of the separation operation is to physically isolate those complexes in a tube that have a sticker annealed to some position from those that do not without disturbing any annealed stickers. The mechanism of DNA hybridization will be central to any proposal. In general, separation by hybridization is is performed by bringing the solution containing the original set of memory complexes into contact with many identical single stranded *probes*. In our case, each bit position has a particular type of probe (with a unique nucleotide sequence) that is used when separation on that bit is performed. The probe sequence is designed such that probes hybridize only to the region of the memory strand corresponding to their bit and nowhere else. During separation, the original complexes with the key bit *off* will be captured on the probes while all those with the bit *on* will remain unbound in solution because the region is covered by a sticker. Next, the unbound ("on") complexes are physically isolated, for example by conjugating the probes to magnetic beads or affixing the probes to solid support and then washing. Lastly, the "off" memory complexes are recovered from the probes that bound them by elution (say by heating and washing). The result is two new tubes, one containing the memory complexes for each of the output sets of the operation.

Notice that if heating is used to achieve the final step of elution this must be done without also removing all of the stickers from the memory strands. This necessitates that the probes have a lower binding affinity for their corresponding regions than do the stickers. This might be achieved by making the probe sequences not exactly complementary to their regions on the memory strands (or merely shorter) to create a differential between the temperature of probe-strand and sticker-strand dissociation. An alternative is to use perfectly complementary sequences for both the probes and stickers but to make the stickers out of an alternate backbone material (such as PNA or DNG [**14, 13**]) which would exhibit stronger and more specific binding to the DNA memory strand than DNA probes[4]. PNA and DNG offer the additional advantage that decreasing salt concentration causes PNA/DNA and DNG/DNA to bind more strongly while the opposite is true for DNA/DNA

---

[4]PNA "clamps" [**15**] have been shown to form $(\text{PNA})_2$/DNA triplexes with remarkable affinity and specificity. These clamps could also be used as stickers.

binding. Thus the final elution step might be achieved by washing in a zero salt solution rather than by heating. There are other possibilities for creating differential affinity between the stickers and probes[5].

3.3. Setting and Clearing. To *set* a bit in every string of a set the most obvious choice is direct annealing. An excess amount of the sticker corresponding to the bit is added to the tube containing the set's memory complexes. One sticker should anneal to every complex that does not already have one, always in the position opposite the region corresponding to the bit being set. Subsequently the excess (unused) stickers are removed, perhaps by filtration or by *separating* out all the memory complexes. This latter proposal could be achieved by having a *universal region* on every memory strand (say at the very beginning or end) that is never covered by a sticker and designing a probe for that region as described in the separation operation above. Such a universal region is a generally useful idea for recovering all memory complexes from a given solution which may contain other species.

To *clear* a bit in every string of a set requires removing the stickers for only that bit from every complex in a tube. Simple heating will obviously not work since *all* stickers from *all* bit regions will come off. One possibility is to designate certain bit regions as *weak* regions. These regions have weak stickers which dissociate more easily from the memory strand than regular stickers. By heating to some intermediate temperature *all the weak stickers* can be made to dissociate at once, keeping all of the regular stickers in place.

In order to implement the *clear* operation in full generality, it may be possible to use the phenomenon of PNA strand invasion by triple helix formation [**29**]. It has been shown that under appropriate conditions, two single stranded oligos of all-pyrimidine PNA will "invade" an existing complementary DNA/DNA duplex to form a $(\text{PNA})_2/\text{DNA}$ triple helix, displacing the pyrimidine DNA strand. This process is most efficient with PNA "clamps"[**15**] which contain both the Watson-Crick and Hoogsteen PNA strands in a single molecule. We suggest that if for example 21 nucleotide DNA stickers are used, then a 14 base PNA clamp could be designed which forms a triple helix with the central 7 nucleotides of the DNA sticker. By mixing PNA clamps specific to a particular bit with a tube of memory complexes, and heating, the PNA clamps should form triple helices with the targeted sticker, destabilizing and thus "prying" it off at a temperature lower than the dissociation temperature for the unaffected stickers. The specificity and reliability of this operation are not yet known experimentally; indeed the mechanism of triplex formation[**12**] may be incompatible with the requirement that non-targeted stickers remain in place. In terms of physical implementation prospects, *clear* seems to be the most problematic of our operations. Recall, however, that it can be eliminated without significantly sacrificing the computational power of the model.

3.4. Initialization and Final Output. To make a combinatorial library containing roughly one copy of every possible bit string of length L followed by $K - L$ zeros, it is first necessary to synthesize roughly 2^L identical copies of a properly designed memory strand with $K \geq L$ regions. Stickers must then be added

[5]For example crosslinking techniques might be used to covalently bond the stickers to the memory strands so that they could not come off during elution, although this confounds the *clear* operation and does not keep with the reusable spirit of the model.

"randomly" to these strands in positions $1 \ldots L$. One procedure that achieves this is outlined below. Note that the method requires only a single step.

The strands are split into two equal volumes. To one volume is added an excess of stickers for all bits $1..L$; this results in all bits $1..L$ being set on all strands. The unused stickers are then removed, for example by filtration or by separating on a universal region of the memory strand. The two volumes are then recombined and heated causing all stickers to dissociate. Finally the mixture is cooled again, causing the stickers to randomly anneal to the memory strands. Since each bit position has only one sticker for every two strands, the resulting memory complexes have any given bit set with probability one half (very nearly independently). Under this model, the odds that any particular bit string is *not* present in the final library is $(1 - 1/2^L)^{2^L}$ which for the L of interest is almost exactly $1/e$. In other words each string is created at least once with probability roughly 63%. This percentage can obviously be increased by synthesizing more than 2^L strands initially. Notice that this procedure is relatively robust to errors in stoichiometry: For example, if the original strands are split into volumes whose ratio is not 1 but 1.5 then (for say $L = 56$) a randomly chosen string is created with probability 37%, still not vanishingly small[6].

To obtain an output string it is necessary to be able to *detect* the presence or absence of memory complexes in a solution. If any are present, we also need to be able to isolate at least one memory complex and then identify which stickers (if any) are annealed to it.

Detection of complexes might be accomplished by fluorescent labeling of each memory strand. Single molecule detection can then be performed by running the solution through a fine capillary tube. Such detection has already been achieved experimentally, see for example [**11**]. This technique may also be effective for isolating a single complex if the time between detection events is large enough. In addition to the capillary tube method mentioned above, other proposals (e.g. based on PCR) for complex detection are possible.

The final step of identifying annealed stickers may be possible by direct imaging – since we know the order of bit regions we could imagine just *looking* and reading off the answer string (perhaps using electron microscopy). Alternately once a complex is isolated its stickers may be eluted and poured over a detection hybridization grid [**26**] to determine which ones were present. While these possibilities are intriguing, more practical approaches based on PCR are more likely to work in the near term [**3**]. However, we show below that detection alone is sufficient to obtain an output string. The approach is to use binary tree decoding:

Begin with the solution containing all putative answer complexes (of which there may be none). Detect complexes in it. If there are none, then no answer has been found. If there are some then separate them based on the first bit of the answer string[7]. Detect complexes in each of the resulting solutions and retain the one which is not empty. If neither is empty then there is more than one answer and either can be retained. Repeat this separation and detection for all the bits of the answer string.

---

[6]The probability that a random bit string is created is $1 - \sum_{k=0}^{L} \frac{\binom{L}{k}}{2^L}[1 - r^k(1+r)^{-L}]^{2^L}$ where r is the ratio of the volumes into which we split initially.

[7]The answer string which we are interested in reading out may be a substring of the entire string encoded by the memory strand in which case separation only needs to be done for those bits.

3.5. Memory Strand and Sticker Design. At several points in the above discussion it was necessary to design the sequence of the memory strand or stickers to have certain properties. In this section we summarize those requirements and explore possibilities for achieving them.

The most fundamental requirement of sequence design is to achieve sticker specificity. It is critical that the stickers only anneal to the memory strands when opposite their assigned region and not in any other position. Thus the memory strand sequence must be designed so that any region's complementary sticker is only complementary to that one region and has much reduced affinity at all other alignments along the strand. As a first approximation to this we will require a certain minimum number of base mismatches at all other alignments. Notice that this is a much stronger requirement than simply requiring each sticker to mismatch all bit regions but its own. It must mismatch every other M long window (possibly spanning two bit regions) on the strand. Mathematically, we wish to design a sequence of length N such that there exist K non-overlapping subsequences of length M each (call them "regions") with the following property: For each region, its complement has at least D_1 mismatches with *every* other subsequence of length M in the entire sequence. The quantity D_1 is the minimum number of mismatches needed for a sticker M bases long not to anneal.

It is also important to eliminate secondary structure in the memory strand itself. We must prevent the memory strand from annealing to itself and creating a hairpin structure, as this makes regions inaccessible for proper use in the system. Fulfilling this requirement can be loosely modeled by the combinatorial problem of designing a N long sequence such that the complement of *every* subsequence of length M has at least D_2 mismatches with every other subsequence of length M. The quantity D_2 is the minimum number of mismatches to prevent the memory strand from self-annealing.

Finally, we must design separation probes such that they stick specifically to the appropriate region and they have sufficiently lower affinity there than the stickers. This ensures that there exists a wash temperature (and salinity) for which the probes will dissociate while the stickers will remain in place. Again, as a first approximation we require that the probes have at least D_3 mismatches within their region and at least $D_4 > D_3$ mismatches everywhere else.

These criteria may seem daunting. However, there are some ways to make this task potentially easier. Notice that in general we may leave portions the memory strand unused; that is we may not identify those portions with any regions so that the product of K and M does not always equal N (but certainly still $KM \leq N$). In other words, we leave "gaps" between the bit regions on the memory strand. In order to avoid the secondary structure problem, it has been suggested that the memory strand be composed of only pyrimidines (or purines) and the stickers of only purines (or pyrimidines)[**27**]. The applied mathematics literature on "comma free codes" and on "de Bruijn sequences" (when $D = 1$) contains detailed discussions of many of the important issues (see [**28**] and [**17**] for reasonable introductions). Also, [**36, 5**] have discussed sequence design in the context of DNA computation.

Finally, D_1 would be reduced if higher-affinity PNA or DNG stickers were used; furthermore, D_3 would possibly be reduced to zero. Other variables other than or in addition to temperature could be manipulated, such as salt concentration and chemical solvent, in order to achieve the relative affinities required for each operation. It is worth speculating about the possibility of using naturally occurring sequences (e.g. plasmids) for the memory strands because of the obvious ease of their mass production. However it remains to be seen if natural sequences can be found which meet the above restrictions.

We emphasize that the criteria outlined above are for illustration only; a more sophisticated approach would have to take into consideration the sequence-dependent thermodynamic parameters for oligonucleotide hybridization. There are several data sets available for calculating ΔH and ΔS for DNA/DNA hybridization [**35, 10, 31**], and similar data could be obtained for PNA and/or DNG interactions. Allowances would also have to be made for potential bubble mismatches at incorrect sticker hybridization sites, and secondary structure due to triple helix formation must be prevented. The *clear* operation, if used, would introduce additional constraints. Although such sophisticated design approaches could suggest potentially useful memory strand, sticker, and probe sequences, correct operation will have to be tested experimentally.

Our conclusion is that although design of the memory strand and the stickers may be difficult, the design space is large; and once a strand with K regions is found, it can be used and reused in the stickers model for any problem requiring K or fewer bits of memory. Since the stickers model uses only a single type of memory strand, in contrast to the 2^K different molecules required in the representation of [**8**], the design process is simplified and the functionality of the strand can be tested experimentally once and for all.

3.6. Experimental Feasibility. The above stickers model presents many challenging requirements for strand design and experimental implementation. Several objections might be raised to the effect that it is unreasonable to expect that these requirements can be met. We attempt to briefly address some of these issues here.

Objection: No matter what methods are proposed, DNA based techniques will suffer from strands being misprocessed. What error rates would be required in order to still accomplish useful computation?

Response: For many search problems, including DES and NP-complete problems, probabilistic algorithms have practical value. Answers suggested by the molecular computer, so long as there aren't too many, can be verified electronically. To ensure that a complex carrying the solution to the problem has a 90% chance of ending up in the "answers" tube after a 1000-step computation, separation error probabilities of less than 0.01% are required. To eliminate false-positive distractors, it may be necessary to refine the "answers" tube by repeating the steps of the computation [**2, 22**]. This and other related error-handling strategies are discussed in Section 5.

Objection: Purity and yield of 90% for purification of DNA are considered excellent in molecular biology. The conditions imposed for separation of memory complexes are much more challenging, since long strands may be used, stickers must not be knocked off, and both supernate and eluant are required. Yet DNA computation requires much lower error rates, both for purity and yield.

Response: Isolation of particular target DNA in complicated cDNA libraries is a routine task in molecular biology. 10^5-fold enrichment of target DNA, with 80% recovery, has been reported using, for example, triplex affinity capture [**21**]. The use of PNA probes also shows some promise: 99% purification with 50% yield using PNA 15-mers has been reported [**30**]. However, current techniques do not meet our requirements for the separation operator. We do not believe that this is due to a fundamental limit. So long as yield is extremely high (i.e. memory complexes don't get "lost"), our calculations (see Section 5) suggest that a poor separation can be improved dramatically by automated processing. Furthermore, we have the opportunity to design our own sequences that can be effectively separated, for example by ensuring that the memory strand has no secondary structure. We recognize that attaining high step yield may be a major challenge, however.

Objection: Even without trying to process them at all, stickers will be falling off their memory strands at some rate k_d. Once a sticker dissociates, it may then hybridize to and thus corrupt some other complex. During operations such as separate, when memory complexes must be melted from probes, k_d surely increases. By the time the computation is complete, the contents of the memory complexes may be completely scrambled.

Response: Suppose we would like to ensure that fewer than 0.01% of stickers fall off during the course of a 1000 hour computation. This would require a k_d of less than $0.3 \times 10^{-9}/sec$. A generic DNA 20-mer can be estimated to have the required k_d at 42°C in 1 M [Na^+] [**39**]. PNA and DNG stickers would be expected to have an even lower dissociation rate, especially at low salt. High wash temperatures may be avoided by using DNA probes and PNA or DNG stickers, and washing in low salt. Additionally, we must be careful not to encourage other circumstances, such as rough physical handling, which might induce sticker dissociation.

Objection: If DNA is subjected to high temperatures for a significant portion of a 1000 hour computation, it may be damaged by deamination, depurination, or strand breakage by hydrolysis, thus rendering it non-functional. (Such objections are discussed briefly in for example [**36**].)

Response: Under physiological conditions of salinity, pH, and temperature, the depurination half-life of a base is 1,000,000 hours, and the hydrolysis half-life of a depurinated base is 400 hours [**18**]. Thus, after 1000 hours, approximately 0.1% of bases will be damaged, and 10% of 2500-mer strands will remain unbroken. This last figure is very dependent on the length of the strands; only 0.1% of 5000-mers would survive. While not good, this indicates that for "short" strands, errors due to damage can be compensated for by a mild increase in the volume of DNA used in a computation. Additionally, improved rates may be possible by carefully adjusting solvent salinity, pH, and composition – and again minimizing rough physical handling.

In summary, although there are many serious engineering challenges, we do not see any as being clearly insurmountable.

4. A Stickers Machine Proposal

This section describes the details of one possible machine that implements computation using the stickers model. The machine is a sort of "parallel robotic workstation for molecular computation" in which various robotic and fluid flow apparatuses are controlled by a central programmable electronic computer. It contains of a rack of many test tubes, a small amount of robotics, some fluid pumps and heaters/coolers and some conventional microelectronics. For each of the operations in the model, we have made a specific choice of physical procedures to implement it. Thus the machine represents one particular realization of many possible variations on the ideas discussed above. The proposal is meant to provoke thought about the

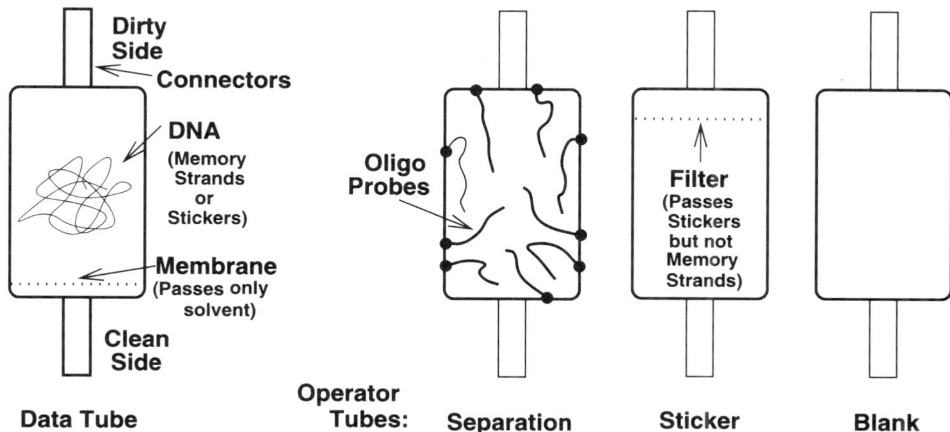

FIGURE 3. Data and Operator Tubes in the stickers machine.

engineering issues involved in eventually constructing a molecular computer and not as a serious or viable construction plan.

The workstation stores all DNA which represents information during the computation in so called *data tubes*. Each data tube is a closed cylinder with a nipple connector in either end that allows fluid to flow in or out. Near one end on the inside is a permanent membrane which passes solvent but not stickers or memory strands. This membrane gives a polarity to the data tube: the connector on the end closest to the membrane is the "clean" side while the opposite connector is "dirty". No DNA is ever present on the clean side or in the clean connector. When a data tube is not in use it is held clean side down with all of the DNA in the tube resting on the membrane.

The data tubes (which may be empty) hold either sets of memory complexes or supplies of unbound stickers. Specifically, each set of bit strings has associated with it a data tube which holds the memory strands and annealed stickers representing those strings. Also each bit has associated with it a data tube which contains a supply of stickers corresponding to that bit.

Whenever a new set of complexes is created (e.g. from a separation operation) it is placed in a new data tube. Whenever a set of complexes is destroyed (e.g. from a combination operation) the data tube that used to contain it is discarded (or perhaps vigorously washed and sterilized for reuse).

In addition to data tubes there also exist *operator tubes* of similar external construction but with different internal contents. A "blank" operator tube is merely an empty tube with nipple connectors on each end. A "sticker" operator tube is identical except for a permanent filter on its inside which passes stickers but not memory strands. A "separation" operator tube contains many identical copies of one bit's oligo probe. (There is a different separation operator tube for each bit.) It is designed so that the probes cannot escape from the tube but unbound memory complexes can. For example, the probes might be fastened to solid support (by biotinylating them and using a biotin binding matrix) or to large beads with filters that pass memory strands but not beads. For all of the operator tubes, both ends are considered "dirty". Figure 3 illustrates the data and operator tubes.

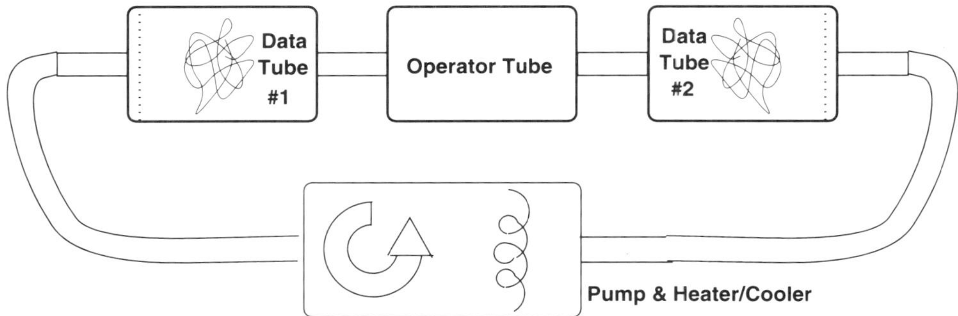

FIGURE 4. Setup for a generic operation in the stickers machine.

At any time during the operation of the machine, some tubes are in use and other are not. All tubes that are not in use are stored on a large rack or carousel. Any single operation takes place as follows: under control of the electronic computer two data tubes are selected and removed from the rack by a robot. One operator tube is also selected and removed. The dirty sides of the data tubes are connected to the operator tube, one data tube at each end of the operator. The clean sides of the data tubes are joined by a pump. Solution is cycled through all three tubes. The direction of flow may be towards the first data tube, or vice versa, or both intermingled. The temperature, salinity, direction, and duration of the flow is controlled by the electronic computer. Once the flow stops, one or more of the tubes is disconnected and replaced on the rack (or discarded). New tubes then come in from the rack until there are once again two data tubes and one operator tube and the next operation begins. Notice that in general clean connectors never touch dirty ones and only clean connectors contact the pumping system. This setup for a generic operation is shown in Figure 4. We will now review how each of our conceptual operations can be performed as outlined generically above. The descriptions below are summarized graphically in Figure 5.

- To combine two sets of complexes simply select the two data tubes and a blank operator tube. Cycle cold solution towards (say) the first data tube. This catches all the memory complexes in the first data tube. The second data tube and the blank operator are discarded.
- To separate a set of complexes based on the value of some bit, select the data tube containing the complexes to be separated and also an empty data tube. Select the separation operator tube for the bit in question. Cycle cold solution in both directions for some time; this allows the probes to bind those complexes that have the bit in question off. Next cycle cold solution towards the empty data tube, forcing all the unbound memory complexes into it. Detach this (originally empty) tube and return it to the rack; it holds the complexes with the bit in question on. Replace it with another empty data tube. Cycle hot solution (or perhaps low salinity solution) towards this new data tube. This releases the memory complexes bound to the probes and forces them into the new data tube. Detach this tube and return it to the rack also; it contains complexes with the bit off. Discard the original data tube (now empty) and return the operator tube to the rack.

- To set a bit (add a sticker to a set of complexes), select the data tube containing the complexes and also the data tube containing the sticker supply for the sticker to be added. Using the sticker operator tube cycle cold solution in both directions for some time. This washes the stickers over the memory complexes allowing them to anneal. Now cycle cold solution towards the sticker data tube. This returns the unused stickers and leaves all the memory complexes caught on the filter in the operator tube. Disconnect the sticker data tube and return it to the rack. Replace it with an empty data tube. Cycle cold solution towards the memory complex data tube. This expels the memory complexes from the operator tube and returns them to their data tube. Return the memory complex data tube to the rack and discard the operator tube and empty data tube.

Additional parallelism can be added in many places. For example, setting or clearing bits might be applied to many data tubes at once by stacking all of them after the operator tube. Also, many copies of the robotics might be included to allow several operations to be performed simultaneously (this would also require multiple copies of, for example, the separator operator and sticker operator tubes).

As we have described it, the stickers machine requires relatively rudimentary robotics and electronics. Simple fluid pumps and heaters/coolers are also necessary. It can be stocked with a generic supply of empty data tubes, blank operator tubes, sticker operator tubes, and salt solutions of various concentrations. It contains data tubes containing both the original sets of memory strands and the sticker supplies for each bit. It also needs to be loaded with the separation operator tubes for each bit. An important feature is that these tubes are reusable from problem to problem, so long as the number of bits required does not exceed the number of regions on the designed memory strand. For a problem of reasonable size, a few thousand tubes might be required (for example DES as described in [**3**]). With each data tube being a few $m\ell$ in size and operator tubes perhaps a hundred times this size it is not inconceivable that such a machine might fit on a desktop or lab bench. This example directly addresses the concern that any useful or hard computation will require an enormous volume of DNA by demonstrating both a specific problem and a specific machine proposal for which this seems far from true.

5. Reducing Error Rates: A Refinery Model[8]

In this section we introduce a second possible implementation of the stickers model. In contrast to the "stickers machine" discussed above, the "stickers refinery" addresses the issue of how to perform reliable computation using a very unreliable separation operator. The refinery model also illustrates the principle of pipelining, whereby a large volume of memory complexes can be processed by small capacity operators with minimal slowdown. These advantages come at the cost of a time-space trade-off which we find reasonable.

5.1. An Error Framework. There are three fundamental types of errors that might be made by any molecular computer which attempts to sort a huge library of initial candidate solution complexes into those which encode a solution to a problem and those which do not. It may give some *false positives*, namely some of the complexes that it classifies as solving the problem actually may not. It may also have *false negatives* which occur when complexes that are classified

[8]The MATLAB code which was used to generate all of the figures in this section is available by request from *roweis@cns.caltech.edu*.

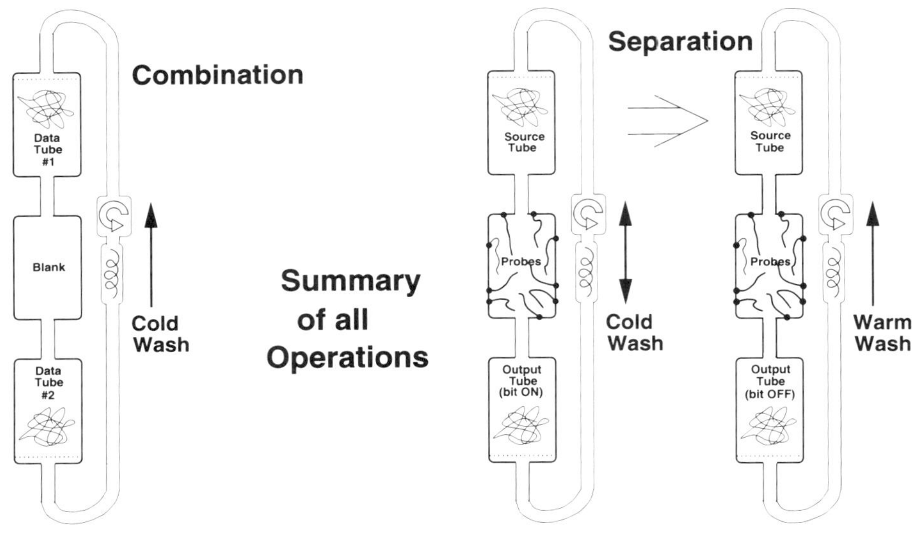

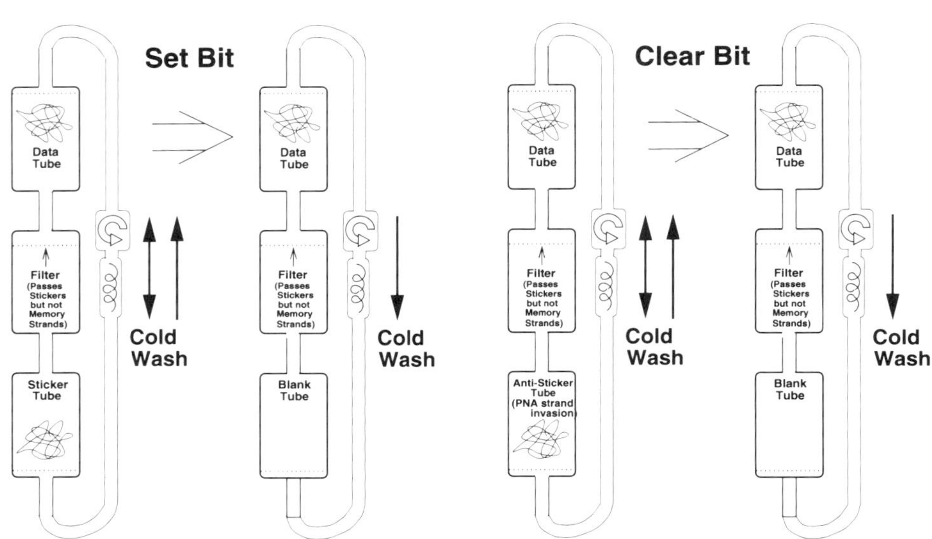

FIGURE 5. Graphical synopsis of all operations in the stickers machine.

as not solving the problem actually do solve it. Finally, the machine may incur some *strand losses* – some of the complexes which were present in the input may not appear in the output at all: they may simply get lost somewhere inside the machine. What are the error requirements to do useful computation? It is clear that we want low false positive and false negative rates and few strand losses, but how low do they need to be?

Our model of a molecular computer is a machine that takes as input a tube encoding a large number of potential solutions to some problem and produces as output two tubes, one labeled *Yes* and the other *No*. In the *Yes* tube are all those complexes which the machine has decided encode solutions to the problem, in the *No* tube are all those complexes which it has decided do not encode solutions. Call a *good* complex one which *actually does* encode a solution and a *bad* complex one which *actually does not*. Because the machine is not perfect, there may be some *good* complexes in the *No* output, some *bad* complexes in the *Yes* output, as well as some losses.

Now we are in a position to state our requirements for error rates: We want two things to be true with high probability (say $1 - \epsilon$) each time we run the molecular computer: there is at least one *good* complex in the *Yes* tube *and* the *ratio of good to bad complexes* in the *Yes* tube is reasonable (say ≥ 1). Informally, when we get the answer tube, we will fish around in it, pull out a random complex (if there are any), and read the solution that it encodes. We will be disappointed if either *(a)* we do not find any complexes in the answer tube or *(b)* the complex we read does not actually encode a solution. Our goal is to be disappointed with low probability.

We would like to be able to answer the question: "How good do individual operations have to be for disappointment to be rare?" Unfortunately, it is very complicated to express the above requirements in terms of conditions on the fidelity of the individual operations such as *separate*. In fact, even for reasonably simple error models, the answers are extremely dependent on the particular architecture of the molecular computer and on the problem being solved[9]. Instead we will work with a model which allows us to characterize the *fraction of complexes not yet correctly processed* (denoted simply δ) at some time T after we begin the computation. This quantity can be easily understood as follows: we turn on our molecular computer at time 0 and feed it its input. It works away, placing some complexes in the *Yes* tube and some in the *No* tube. At time T we stop the machine and collect the *Yes* and *No* tubes. At this point, original input complexes fall into three categories: *(1)* those which have been correctly placed[10] into either *Yes* or *No*, *(2)* those which have been incorrectly placed into *Yes* or *No*, and *(3)* those which were either lost or were still being processed by the machine when we turned it off. The fraction of complexes not yet correctly processed (δ) is the fraction of the original input complexes which fall into either categories *(2)* or *(3)* above at time T. We would like δ to be very near zero. Below we develop a model which allows us to compute δ for various machine architectures and also various time and

[9]For example, one could imagine a simple model of errors which is characterized by only three numbers (each between 0 and 1): a false positive rate R_{fp}, a false negative rate R_{fn} and a loss rate R_{loss}. Any given complex is "lost" with probability R_{loss}. If not lost, *good* complexes go to the *Yes* tube with probability $(1 - R_{fn})$ and *bad* complexes go to the *No* tube with probability $(1 - R_{fp})$ regardless of the specific bit string they encode. Under such a model, if our input tube contains G *good* complexes and B *bad* complexes (typically $G << B$) then we require a false negative rate R_{fn} which is less than some function $f_1(G, B, \epsilon)$, a false positive rate R_{fp} which is less than $f_2(G, B, \epsilon)$, and a loss rate R_{loss} which is less than $f_3(G, B, \epsilon)$, where ϵ is the fraction of runs of the experiment that will result in disappointment. However, it turns out that even when f_1,f_2, and f_3 have been determined, the conversion from these three numbers to a requirement on the fidelity of individual operations is highly architecture dependent; compare for example the simple *OR* of all bits in a bit string with the simple *AND*.

[10]Note that *good* complexes can be incorrectly processed at some step(s), yet still end up in the "Yes" tube; similarly *bad* complexes can end up in "No" after incorrect processing. We still count these cases as incorrect.

space tradeoff factors in terms of only the fidelity of the atomic operations which are used by the machine, independent of the problem being solved.

5.2. Computing δ. We will consider a very simple mathematical model of a molecular computer as a series of exactly S identical separation operations. The separation operation is used because it is a fundamental operation in the stickers model; both the *set bit* and *clear bit* operations can be described in terms of only separations (see Section 5.7). This model assumes that the algorithm used to process complexes has the effect of passing each one though at most S separations (an assumption which is true for all algorithms that terminate within a known time)[11]. It further assumes that complexes do not interfere with one another, nor do different bit positions on a single strand. For the moment, let us also assume that there are no strand losses; we will return to this crucial issue later.

Assume that (regardless of which bit is being used to separate and of the values of any other bits) each separation operation takes one unit of time to complete and has a probability p of correctly processing each complex in its input[12]. Notice that we expect p to be near unity. In every separation, we assume that each complex ends up in one or the other of the output tubes; no strands are physically lost. Now any computation will take S units of time and when it is done, the fraction of complexes not yet correctly processed will be a depressingly high $\delta = (1 - p^S)$. (For example if $p = 0.9$ and $S = 100$ then $\delta = 0.99997$.) The main point of this section is that *without changing* p (i.e. without improving the basic biotechnology used to implement operations) and without reducing S (i.e. without moving to easier problems) the fraction δ can be made much smaller using intelligent space and time tradeoffs.

Imagine that you have in hand enough hardware (i.e. units that perform separations and test tubes) to perform a given computation. A *space tradeoff of factor* H involves obtaining $H - 1$ extra identical copies of that hardware, which we may use in parallel. A *time slowdown of factor* M involves taking $M \cdot S$ units of time instead of merely S to perform the computation. How can these factors be used to reduce errors? Given any algorithm A for performing a computation and factors H and M we would like to investigate algorithm *transformations* which give us a new algorithm A' (that runs in no more than $M \cdot S$ time and requires no more than H copies of the hardware) that has a smaller δ than the original A.

[11]Recall that since "answer readout" and "strand detection" are not permitted during the course of the computation, the algorithm which controls the processing cannot get any feedback and so cannot do any "if then else" type branching. To see that the model assumption is not as restrictive as it may seem, consider architectures that are of the form of feedforward layered circuits with S layers. Each layer receives some number of input tubes from the previous layer and produces some (possibly different) number of output tubes which it passes to the next layer. No tube may go through more than one separation per layer. In this way, for any individual complex such architectures look like a series of S identical separation operations, although different complexes may take different paths through the circuit. The first layer receives as its input the single tube which was the input to the entire problem. The final layer (S) produces as its output the final output tubes for the problem. Any (terminating) algorithm for doing a stickers computation can be converted into a feedforward circuit of this kind.

[12]In practice, operations like *separation* have a much higher probability of correctly processing some inputs than others. For example if hybridization is used, it is much harder for probes to erroneously capture complexes than it is for them to let through complexes which they should capture. All of the mathematics which follows can easily be done for the assymetric probability case although it is somewhat more complicated.

5.3. Repeating the Computation. A basic transformation, *repeating* was proposed by Adleman in [**2**]. It makes use of a slowdown factor of M by proposing A' as follows:

- Repeat M times:
  Run A on input I, producing tubes Y and N.
  Discard tube N and rename tube Y to tube I.
- Return tube I as the ''Yes'' tube and an empty tube as ''No''.

This approach is of value when the original algorithm A very reliably places *good* complexes into its *Yes* output (i.e. low false negatives) even if it often also places *bad* complexes into *Yes* (i.e. high false positives). Note that if the original algorithm instead has high false negatives and low false positives then the following version of *repeating* can be used:

- Make an empty tube Z.
- Repeat M times:
  Run A on input I, producing tubes Y and N.
  Combine tube Y into tube Z, destroying Y.
  Rename tube N to tube I.
- Return tube Z as the ''Yes'' tube and tube I as ''No''.

By how much does *repeating* reduce δ ? The performance of this transformation is bounded by the performance of an imaginary transformation called *repeating with an oracle* which makes use of a new *oracle* operation. The *oracle* takes as input two tubes Y and N and produces as output three tubes: Y',N', and X. In Y' are all the *good* complexes that were in the input tube Y, in N' are all the *bad* complexes that were in the input tube N, and in X are all the *bad* complexes from Y along with all the *good* complexes from N. In other words, the *oracle* "fixes-up" Y and N by putting any incorrectly processed complexes into X. Using this magical operation, *repeating with an oracle* transforms A into the following A':

- Make an empty tube Z.
- Repeat M times:
  Run A on input I, producing tubes Y and N.
  Run the oracle on Y and N, producing Y',N', and X.
  Discard tube N' and combine tube Y' into tube Z, destroying Y'.
  Rename tube X to tube I.
- Return tube Z as the ''Yes'' tube and tube I as ''No''.

This transformation improves δ from $1-p^S$ to $(1-p^S)^M$. The vanilla *repeating* transformations can approach but never exceed this improvement. The reason that plain *repeating* works well at all is that for very disparate false positive and negative rates, one can approximate the action of the oracle easily. While these transformations do yield some reduction in δ they require enormous slowdowns to improve even modest sized problems. For larger problems, these transformations require enormous slowdowns. Figure 6 shows the slowdown factors required to achieve various performance levels for the case in which $p = 0.9$ and $S = 100$ or $S = 1000$.

5.4. A New Operation: Compound Separation. It is possible to make much better use of space and time tradeoffs than the above transformations do. Shortly, we will develop new transformations which do this, but first we must introduce a new operation which they employ known as *compound separation*.

The central observation is that the following algorithm, analogous to "countercurrent cascade stages" in chemical engineering [**38**], will exponentially improve upon the accuracy of the *Separation* step:

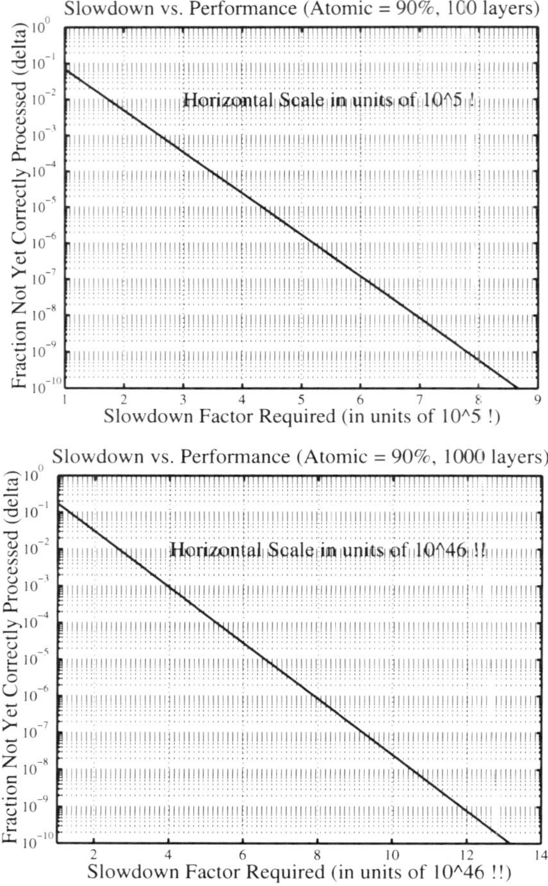

FIGURE 6. Performance vs. Slowdown for *repetition with an oracle*

- Begin with a tube T_0 whose contents we wish to separate based on bit k.
- Begin also with $2N$ extra tubes called $T_{-N}, \ldots, T_{-1}$ and $T_1, \ldots, T_N$, initially empty.
- for t=1 to Q

    for j=-N+1 to N-1 s.t. t+j $\equiv$ 1 (mod 2)

        Separate T_j into T_{on} and T_{off} based on bit k
        Combine T_{on} and T_{j+1} into T_{j+1}
        Combine T_{off} and T_{j-1} into T_{j-1}

(Notice that for odd `t`, odd numbered tubes start off empty and for even `t`, even numbered tubes start off empty.)

Thus each complex will perform a biased random walk in tubes T_{-N} through T_N, with absorption at the boundaries. Most memory complexes which have bit k on will end up in T_N, while most memory complexes which have bit k off will end up in T_{-N}. A graphical illustration of the process is shown in Figure 7. The statistics of such processes have been thoroughly worked out (see the "Gambler's Ruin" problem [**16**]). Let p be the probability that a separation step *correctly* moves a complex into T_{on} or T_{off}. At the end of the algorithm, we would like to know the

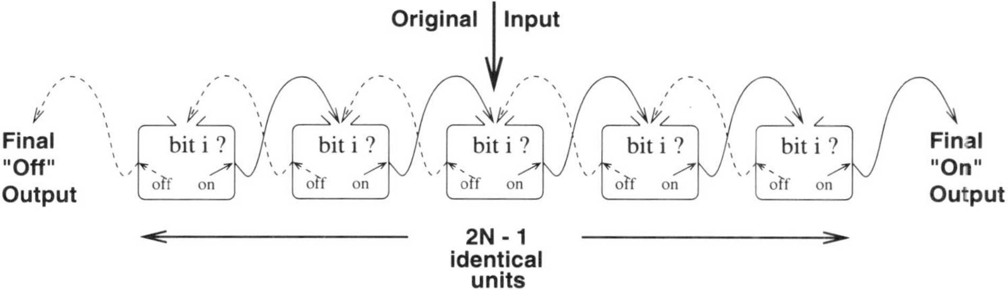

FIGURE 7. A Compound Separator

probabilities that a complex with bit k on (or off) will either be in tube T_{-N}, T_N, or still stuck in some other tube. Let us first consider the case $Q = \infty$, i.e. each complex continues to be processed until it is absorbed at either T_{-N} or T_N. Then a complex has probability p_∞ of being correctly processed, where

$$p_\infty = \frac{1}{1 + (\frac{1-p}{p})^N}.$$

For example, if $p = 0.9$, we choose $N = 5$, and then $p_\infty \approx 1 - 10^{-5}$. It is critical to this argument that no memory complexes are lost in the woodwork. However, it is not crucial that Q be ∞. The expected time $t_{compound}$ for a complex to arrive in either T_{-N} or T_N is

$$\langle t_{compound} \rangle = \frac{N}{2p-1}\left(1 - 2\frac{1-r^N}{1-r^{2N}}\right)$$

where $r = \frac{p}{1-p}$. In the example, $\langle t_{compound} \rangle \approx 6.25$. In fact, in this example, $Q = 20$ ensures that fewer than 10^{-4} of the complexes are not correctly processed. Figure 8 shows the performance (δ) of *compound separation* as a function of number of steps (Q) for various chain lengths (N).

We have shown that, by applying the *compound separation* algorithm above, we can achieve excellent error rates even when the fundamental separation operation is not reliable. This comes at the cost of a small linear slowdown (and a few extra tubes). In general we need to perform $Q \cdot N$ separations instead of one.

Notice that this algorithm can be easily parallelized: if N "atomic" separator units are available instead of just one then the slowdown factor can be reduced to Q by performing all the separations simultaneously (i.e. do all iterations of the inner `for j ...` loop in parallel). We will call this parallelized algorithm *parallel compound separation.*

Although the basic mathematics are not new, to our knowledge the first application of this idea to molecular computation appeared in [**22**]. Their "Super Extract" operation is very similar (although not identical to) the compound separation we have proposed above. We refer the reader to the excellent discussion and detailed analysis (including some interesting bounds) contained therein.

5.5. Better Transformations: The Refinery Idea. What if we were to replace every *separate* operation in our original algorithm A with a *compound separation*? This would incur a slowdown factor of $M = Q \cdot N$ but would give an

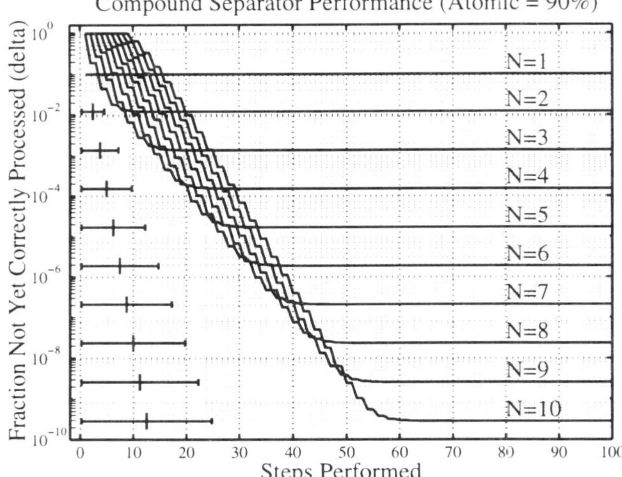

FIGURE 8. Compound Separation Performance. The bars on the left show the mean time $\pm$ one standard deviation for complexes to be absorbed at a boundary.

enormous reduction in δ since the fidelity of each separation has improved exponentially. This is exactly the idea behind the *serial refinery* transformation which exploits a slowdown factor of M. It proposes A' to be:

- Run A on input I, replacing each *separation* operation with a *compound separation* operation of chain size N and duration Q where $Q \cdot N \leq M$.

Notice that if a space tradeoff factor of H is also available then we can employ the *one layer refinery* transformation which makes use of the available parallelism H and slowdown M by specifying A' to be:

- Run A on input I, replacing each *separation* operation with a *parallel compound separation* operation of chain size N and duration Q where $Q \cdot N \leq M \cdot H$.

The one layer refinery is so named because if A originally processed one layer in parallel before moving on to the next layer, with sufficient parallelism A' may now process each layer in parallel for Q steps, and then move on to the next layer.

For the moment we defer the issue of how to decide on the optimal factorization of M or $M \cdot H$ into $Q \cdot N$ although we return to it shortly. (The obvious choice is to choose $N = H$ and $Q = M$.) First let us find out how much improvement in δ this transformation buys us. The exact expression for δ is complicated[13] but easily computable. The plots in Figure 9 show the performance (δ) of the *one layer refinery* transformation as a function of slowdown factor (M) for various compound separator chain lengths (N) and for $S = 100$ and $S = 1000$. The plots assume that we have chosen $Q = M$ and $N = H$.

[13]For the afficionado, $\delta = 1 - [\hat{p}(N,Q)]^S$ where $\hat{p}$ is the probability of getting absorbed at the correct boundary in Q steps or less in a biased random walk (bias probability $= p$) with absorption at boundaries N and $-N$. In turn, $\hat{p}(N,Q) = \sum_{i=0}^{Q}\left[\frac{2^i}{2N}(1-p)^{(i-N)/2}p^{(i+N)/2}\sum_{v=1}^{2N-1}\cos^{i-1}\frac{\pi v}{2N}\sin\frac{\pi v}{2N}\sin\frac{\pi v}{2}\right]$ where the expression in square brackets is the probability of absorption in exactly i steps. All of the mathematics can be extended to the case when the random walk bias is different in each direction; see [**16**].

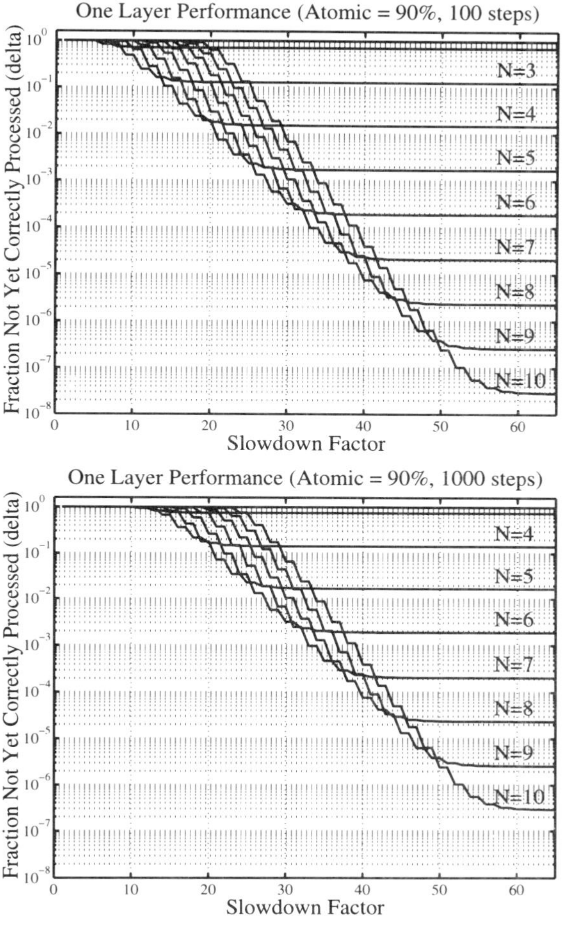

FIGURE 9. One Layer Refinery performance for S=100 and S=1000. Plots assume that we have chosen $Q = M$ and $N = H$ (see text).

5.6. A Fully Parallel Refinery Architecture. In the remainder of this section we show how, by exploiting the ideas above, a new machine architecture called the "stickers refinery" which achieves the same low error rates as the one layer refinery and greater speed-up by continuously processing all steps in the computation – at the cost, of course, of additional space. The refinery architecture may have other advantages as well, which we will comment on below.

As shown in Figure 8, the mean time for a complex to get through a single compound separation chain is considerably less than the Q required to obtain maximal performance, typically by a factor of about 4. Most of the time during a computation is spent waiting for a few straggling complexes to come out of a separator chain. We can avoid this wasted time by proceeding to process complexes as soon as they are absorbed in T_{-N} or T_N. The *parallel refinery* transformation creates A' by replacing each *separation operation* in A by a *parallel compound separation* of chain length N, and then iteratively processing the entire computation in parallel for T iterations.

Specifically, suppose A has $S \cdot W$ separations (S feedforward layers, at most W per layer) and uses tubes $T^0 \dots T^J$, where separation i separates $T^{j_{in,i}}$ into $T^{j_{on,i}}$ and $T^{j_{off,i}}$ based on bit k_i. Then the *parallel refinery* transformation given as A' which is defined as:

- Begin with $3 \cdot S \cdot W \cdot (2N-1)$ tubes T_n^j, and $T_{on,n}^j$ and $T_{off,n}^j$ for $-N+1 \leq n \leq N-1$.
- Initially, T_0^1 contains the mother tube complexes.
- for t=1 to T

```
    for j=1 to S·W (do all j in parallel)
        for n=-N+1 to N-1 (do all n in parallel)
            Separate T^j_n into T^j_{on,n} and T^j_{off,n} based on bit k_j
    for j=1 to S·W (do all j in parallel)
        for n=-N+2 to N-2 (do all n in parallel)
            Combine T^j_{on,n-1} and T^j_{off,n+1} into T^j_n
        Combine T^j_{off,-N+1} into T_0^{j_{off,j}}
        Combine T^j_{on,N-1} into T_0^{j_{on,j}}
```

Compared to the original A, the fully parallel refinery requires a space tradeoff factor of $H = (2N-1) \cdot S$ (since every separation is expanded) and a slowdown factor (not necessarily integer) of $M = \frac{T}{S}$. The question is, what parallelism H and slowdown M are required to obtain a desired performance δ? We answer this question by calculating δ given N and T, as before. First we note that the probability that a given complex is correctly processed after T steps can be decomposed into the probability $p_{done}(N,T)$ that it is in either the "Yes" tube or the "No" tube after T steps (i.e. not still in the machine when we stop) and the probability $p_{correct}(N)$ that a complex arriving in a final tube has been correctly processed[14]. Recall that a complex has probability $p_\infty = 1/\left(1+\left(\frac{1-p}{p}\right)^N\right)$ of having been correctly separated every time it leaves a compound separator, so $p_{correct}(N) = (p_\infty)^S$. The distribution of emergence times can be obtained by convolving the distribution for a single compound separation, thus numerically calculating $p_{done}(N,T)$. Then $\delta = 1 - p_{done}p_{correct}$. The result of doing such a computation for $p = 0.9$, $N = 1 \dots 10$ and $S = 100$ and $S = 1000$ are shown in Figure 10.

5.7. Advantages of the Full Refinery. With the fully parallel refinery, we can obtain the same target error performance and a roughly 4-fold smaller slowdown factor then the one layer refinery at the cost of S-fold more space and parallelism. This may not seem like a beneficial trade-off since S can be potentially large and 4 is small. In fact, it turns out that the 4-fold speedup can be achieved with an extra space tradeoff of much less than S times[15]. However, the fully parallel refinery affords a number of interesting possibilities. For example, suppose our fundamental separation units can handle limited volume, but we need to process a 10000-fold larger volume of DNA. We can "pipeline" the computation by inputting small aliquots of the mother tube at each step, and waiting until the last aliquot

[14] This second probability is independent of when the complex emerges.

[15] If we consider where the complexes are at some time t, we see that the vast majority of them are near layer $t/\langle t_{compound}\rangle$, leaving the rest of the machine empty – a waste. This observation leads to an intermediate class of refinery algorithms in which a moving window of $L < S$ layers of the circuit are being continually processed as in the full parallel refinery algorithm. Since the distribution of complexes is fairly thin, L can be small, thus requiring less space while achieving nearly identical performance.

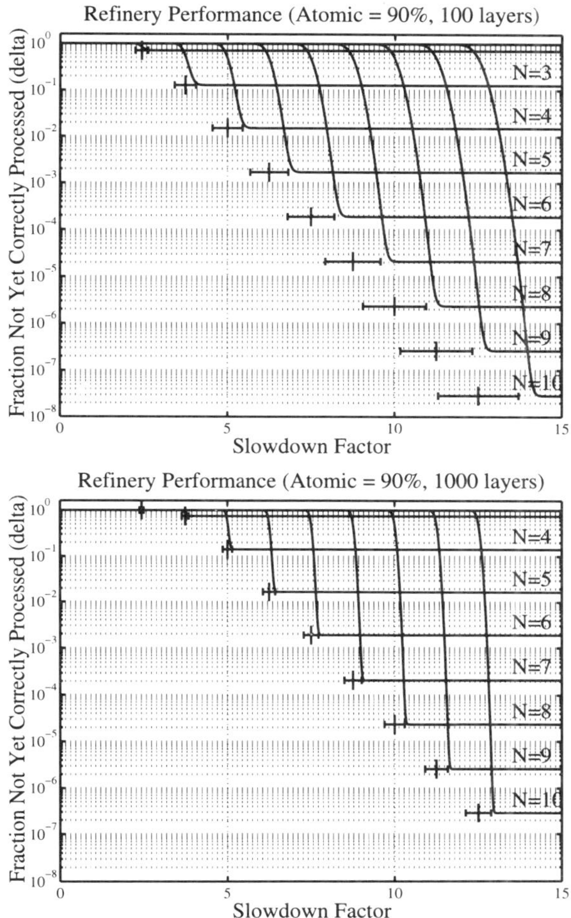

FIGURE 10. Fully Parallel Refinery performance for S=100 and S=1000. The bars on the left show the mean time ± one standard deviation for complexes to emerge from the entire refinery.

gets out. Now most of the machine is being utilized most of the time instead of idly pumping solution around. If the non-pipelined parallel refinery would have taken 10000 steps, then after about 20000 steps the entire computation will be finished, performing a 10000-fold larger search than the non-pipelined version while taking only twice the time. In other words, we are now exploiting for computation all of the additional parallelism and time employed beyond that used by the naive algorithm, while gaining vastly improved error rates *for free.*

The parallel refinery model does not require re-use of any separation unit to serve at multiple points in the algorithm, and thus a general purpose robotic workstation (such as the stickers machine) is unnecessary. We envision a special-purpose refinery system being assembled, from standard units, for each problem to be solved. A separation unit consists of a reservoir into which complexes are received, an affinity column with DNA probes on solid support, pumps and heaters for the wash and elution, and two exit channels (labeled "on" and "off") which lead permanently

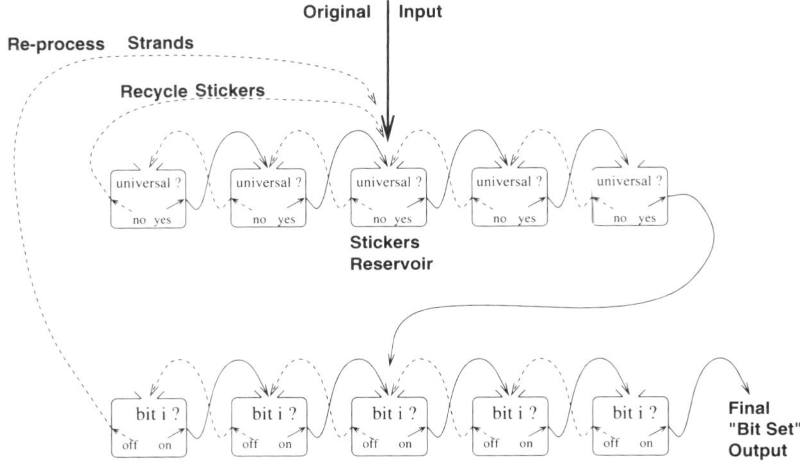

FIGURE 11. A Reliable *set* Operation

(through piping or tubing) to the reservoirs of other separation units. We refer to such a machine as embodying the "stickers refinery architecture". It is our hope that a refinery architecture will alleviate the problem of "lost strands", because the physical permanence of all connections allows temporarily stuck strands to eventually become unstuck and still complete the computation.

Note that the performance of the operations *set* and *clear* can also be improved using these ideas. A *set* operation can be implemented by two compound separations, the first separation based on the universal tag, and the second separation based on the bit being set, as diagrammed in Figure 11. The starting tube is seeded at the beginning of the computation with an excess of stickers, which the universal separation recycles. Complexes which failed to acquire the sticker are returned to the starting tube, where they have another chance to hybridize with a sticker. A similar technique could be used for *clear*, adding a step to purify stickers from PNA clamps.

5.8. Using refineries. It is illustrative to consider using the refinery to solve a particular problem. We will consider breaking DES, for which the naive algorithm A has $S = 6500$ and $W = 32$. Let's suppose $p = 0.9$. Using the one layer refinery algorithm and $N = 10$, we incur a space factor of 19 and a slowdown of ≈ 60 (no further slowdown helps); this achieves $\delta \approx 1.9 \times 10^{-6}$. We started with 2^{56} keys, exactly one of which is good. We can be sure (except for 1.9 in a million) that the good key will end up in the "Yes" tube, but $2^{56}\delta \approx 1.4 \times 10^{11}$ bad keys will be incorrectly processed. Will the incorrectly processed complexes also end up in the "Yes" tube as distractors? In the case of the DES algorithm, we argue that they won't end up in the "Yes" tube [**3**]. However, we cannot make the same argument for generic algorithms, and so we consider the worst case scenario in which all of the incorrectly processed complexes are distractors. In this case, we need to achieve $\delta \approx 10^{-17}$ to get the number of distractors below 1. With the one layer refinery, this could either be realized by increasing the space factor to 43 ($N = 22$) and the slowdown to ≈ 125, or by simply re-running the $N = 10$ version mentioned above three times in a row[16] (giving a space factor of 19 and a slowdown of ≈ 180). This last approach is an interesting example of what can be further achieved by *composing* the various algorithm transformations we discussed above.

6. Conclusions

In this paper we have tried to visualize a practical molecular computer. A number of previous concerns [**36, 20, 24**] have been addressed. First, it is now clear, from our own work and that of others, that general-purpose algorithms can be implemented by DNA-based computers, potentially solving a wide class of search problems. Second, we now understand that there are challenging problems, such as breaking DES, for which only modest volumes of DNA (e.g. 2 grams) should suffice. Third, we demonstrated that the formation and breaking of covalent bonds is not intrinsic to DNA-based computation. This means that costly and short-lived materials such as enzymes are not necessary, nor are energetically costly processes such as PCR. All the materials in the stickers model are potentially reusable from one computation to the next. Fourth, we have shown that a single essential biotechnology, sequence-specific separation, suffices for constructing a general-purpose molecular computer. Fifth, we now know that separation errors can theoretically be reduced to tolerable levels by invoking a trade-off between time, space, and error rates at the level of algorithm design; we have also illustrated several specific ways in which this can be done and presented encouraging numerical calculations of their performance.

That several major roadblocks have been overcome at a theoretical level suggests that real applications of molecular computation may be feasible in the future. Nonetheless, we emphasize that substantial engineering challenges remain at almost all stages and that the ultimate success or failure of DNA computing will certainly depend on whether these challenges can be met in laboratory investigations.

7. Acknowledgments

The authors would like to express their appreciation to Professor John Baldeschwieler for his contributions to this paper through early discussions of this work. Sam Roweis and Erik Winfree are also grateful to their advisor, Professor John Hopfield for his perpetual wisdom and long term advice.

References

[1] L. Adleman, Molecular Computation of Solutions to Combinatorial Problems. *Science* 266: 1021–1024 (Nov. 11) 1994.

[2] L. Adleman, On Constructing a Molecular Computer. In R. J. Lipton and E. B. Baum, editors, *DNA Computers: Proceeding of a DIMACS Workshop,* DIMACS Series in Discrete Mathematics and Theoretical Computer Science, American Mathematical Sociey, 1-21, 1996.

[3] L. Adleman, P. W. K Rothemund, S. Roweis, E. Winfree. On Applying Molecular Computation to the Data Encryption Standard. In L. Landweber and E. Baum, editors, *DNA Computers: Proceeding of the Second DIMACS Workshop,* DIMACS Series in Discrete Mathematics and Theoretical Computer Science, American Mathematical Sociey, 31-44, 1998.

[4] M. Amos, A. Gibbons, and D. Hodgson. Error-resistant Implementation of DNA Computations. In L. Landweber and E. Baum, editors, *DNA Computers: Proceeding of the Second DIMACS Workshop,* DIMACS Series in Discrete Mathematics and Theoretical Computer Science, American Mathematical Society, 151-161, 1998.

[16] Thus the expected number of distractors will be 1.4×10^{11} (first run), 2.7×10^{5} (second run), 0.5 (third run).

[5] E. B. Baum. DNA Sequences Useful for Computation. In L. Landweber and E. Baum, editors, *DNA Computers: Proceeding of the Second DIMACS Workshop,* DIMACS Series in Discrete Mathematics and Theoretical Computer Science, American Mathematical Society, 235-241, 1998.
[6] D. Beaver. Molecular Computing. Penn State University Tech Report CSE-95-001
[7] D. Boneh, Christopher Dunworth, and Richard J. Lipton. Breaking DES Using a Molecular Computer. Technical Report CS-TR-489-95, Princeton University, 1995.
[8] D. Boneh, C. Dunworth, R. Lipton, and J. Sgall. On the Computational Power of DNA. Draft 1995.
[9] D. Boneh, C. Dunworth, R. Lipton, and J. Sgall. Making DNA Computers Error Resistant. In L. Landweber and E. Baum, editors, *DNA Computers: Proceeding of the Second DIMACS Workshop,* DIMACS Series in Discrete Mathematics and Theoretical Computer Science, American Mathematical Society, 163-170, 1998.
[10] K. J. Breslauer, R. Frank, H. Blöcker, L. A. Marky. Predicting DNA duplex stability from the base sequence. *Proc. Natl. Acad. Sci. USA*, 83: 3746–3750, June 1986.
[11] A. Castro and E. B. Shera. Single-molecule Detection: Applications to Ultrasensitive Biochemical Analysis. *Applied Optics*, 34 (18): 3218–3222, June 20, 1995.
[12] V. V. Demidov, M. V. Yavnilovich, B. P. Belotserkovskii, M. D. Frank-Kamenetskii, and P. E. Nielsen. Kinetics and Mechanism of Polyamide ("peptide") nucleic acid binding to duplex DNA. *Proc Natl. Acad. Sci. USA* 92: 2637–2641, 1995.
[13] R. O. Dempcy, K. A. Browne, and T. C. Bruice. Synthesis of a thymidyl pentamer of deoxyribonucleic guanidine and binding studies with DNA homopolynucleotides. *Proc. Natl. Acad. Sci. USA*, 92: 6097–6101, June 1995.
[14] M. Egholm, O. Buchardt, L. Christensen, C. Behrens, S. M. Freier, D. A. Driver, R. H. Berg, S. K. Kim, B. Norden, and P. E. Nielsen. PNA hybridizes to complememntary oligonucleotides obeying the Watson-Crick hydrogen bonding rules. *Nature*, 365: 566–568 (7 Oct.), 1993.
[15] M. Egholm, L. Christensen, K. L. Dueholm, O. Buchardt, J. Coull, and P. E. Nielsen. Efficient pH-independent sequence-specific DNA binding by pseudoisocytosine-containing bis-PNA. *Nucleic Acids Research*, 23(2): 217–222, 1995.
[16] W. Feller. *An Introduction to Probability Theory and Its Applications, Volume I, 3rd edition.* John Wiley & Sons. 1968.
[17] H. Fredricksen. A Survey of Full Length Nonlinear Shift Register Cycle Algorithms. *SIAM Review*, 24(2):195-221, 1982.
[18] E. C. Friedberg. DNA Repair. W. H. Freeman and Co., 1985.
[19] M. R. Garey and D. S. Johnson. Computers and Intractability: A Guide to the Theory of NP-Completeness. W. H. Freeman and Co., 1979.
[20] J. Hartmanis. On the Weight of Computations. *Bulletin of the European Association for Theoretical Computer Science*, 55: 136–138, 1995.
[21] T. Ito, C. L. Smith, C. R. Cantor. Sequence-specific DNA purification by triplex affinity capture. *Proc. Natl. Acad. Sci. USA*, 89: 495–498, January 1992.
[22] R. Karp, C. Kenyon and O. Waarts. Error Resilient DNA Computation. *Laboratoire de l'Informatique du Parallelisme, Ecole Normale Superieure de Lyon*, Research Report Number 95-20 Sept. 1995.
[23] A. Kornberg. *DNA Replication, 2nd edition.* W.H. Freman and Company, New York, New York, 1992.
[24] M. Linial, N.Linial, Y.M.D. Lo, K.F.C. Yiu, S.L.Wong, B. Bunow, L. Adleman, On the Potential of Molecular Computing, *Science*, 268: 481–484, 1995.
[25] R. Lipton. DNA Solution of Hard Computational Problems. *Science* 268: 542–545 (Apr. 28) 1995.
[26] D. Patersen, Electric Genes. *Scientific American,* 272:33–34, May 1995.
[27] Kalim Mir, personal communication.
[28] B. Nevelin. Comma-free and Synchronizable Codes. *J. Theor. Biol.*, 144:209–212, 1990.
[29] P. E. Nielsen, M. Egholm, R. H. Berg, and O. Buchardt. Sequence-Selective Recognition of DNA by Strand Displacement with a Thymine-Substituted Polyamide. *Science*, 254: 1497–1500, 1991.
[30] H. Orum, P. E. Nielsen, M. Jorgensen, C. Larsson, C. Stanley, and T. Koch. Sequence-specific Purification of Nucleic Acids by PNA-Controlled Hybrid Selection. *BioTechniques* 19: 472–480, 1995.

[31] J. Pestruska and M. F. Goodman. Enthalpy-Entropy Compensation in DNA Melting Thermodynamics. *Journal of Biological Chemistry* 270(2): 746–750, January 1995.

[32] Parallel Molecular Computation: Models and Simulations. Seventh Annual ACM Symposium on Parallel Algorithms and Architectures (SPAA95), Santa Barbara, June, 1995.

[33] On the Power of Bio-Computers. Draft Feb. 28, 1995

[34] P. W. K. Rothemund. A DNA and Restriction Enzyme Implementation of Turing Machines. In R. J. Lipton and E. B. Baum, editors, *DNA Computers: Proceeding of a DIMACS Workshop,* DIMACS Series in Discrete Mathematics and Theoretical Computer Science, American Mathematical Sociey, 75-119, 1996.

[35] J. Santalucia, H. T. Allawi, A. Seneviratne. Improved nearest-neighbor parameters for predicting DNA duplex stability. *Biochemistry* 35(11): 3555-3562, 1996.

[36] W. D. Smith and A. Schweitzer. DNA Computers in Vitro and in Vivo. NECI Technical Report, March 20, 1995.

[37] W. P. C. Stemmer. The Evolution of Molecular Computation. *Science*, 270: 1510–1510, 1995.

[38] P. C. Wankat. *Separations in Chemical Engineering: Equilibrium Staged Separations.* Elsevier Science Publishing Co., Inc., New York, New York, 1988.

[39] J. G. Wetmur. DNA Probes: Applications of the Principles of Nucleic Acid Hybridization. *Critical Reviews in Biochemistry and Molecular Biology* 36 (3/4) : 227–259, 1991.

Laboratory for Molecular Science, University of Southern California

and

(S. Roweis and E. Winfree) Computation and Neural Systems Option, California Institute of Technology

(R. Burgoyne) Department of Biomedical Engineering, U.S.C.

(N. Chelyapov, P. Rothemund, and L. Adleman) Department of Computer Science, U.S.C.

(M. Goodman) Department of Biological Sciences, U.S.C.

DIMACS Series in Discrete Mathematics
and Theoretical Computer Science
Volume **44**, 1999

On Applying Molecular Computation to the Data Encryption Standard

Leonard M. Adleman, Paul W. K. Rothemund, Sam Roweis, Erik Winfree

ABSTRACT. Recently, Boneh, Dunworth, and Lipton [**2**] described the potential use of molecular computation in attacking the United States Data Encryption Standard (DES). Here, we provide a description of such an attack using the *sticker model* of molecular computation. Our analysis suggests that such an attack might be mounted on a table-top machine, using approximately a gram of DNA and might succeed even in the presence of a large number of errors.

1. Introduction

With their work on DES, Boneh, Dunworth, and Lipton [**2**] provided the first example of a "practical" problem which might be susceptible to molecular computation. DES is one of the most widely used cryptographic systems. It produces a 64-bit ciphertext from a 64-bit plaintext under the control of a 56-bit key. While it has been argued that special purpose electronic hardware [**7**] or massively parallel supercomputers might be used to break DES in a reasonable amount of time, it appears that today's most powerful sequential machines would be unable to accomplish the task. We continue the work of [**2**] by considering the difficulty of breaking DES on the recently proposed *sticker model* of molecular computation [**5**]. While our results are encouraging, it must be stressed that the feasibility of such an attack will ultimately be decided in the laboratory.

In this paper we consider the so called plaintext-ciphertext attack. Here the cryptanalyst obtains a plaintext and its corresponding ciphertext and wishes to determine the key used to perform the encryption. The most naive approach to this problem is to try all 2^{56} keys, encrypting the plaintext under each key until a key that produces the ciphertext is found. Remarkably, a significantly more

---

Key words and phrases. Data encryption standard, molecular computation, DNA computation, sticker model.

L. M. Adleman and P. W. K. Rothemund are supported in part by the National Science Foundation under grant CCR-9403662 and the Sloan Foundation. S. Roweis is supported in part by the Center for Neuromorphic Systems Engineering as a part of the National Science Foundation Engineering Research Center Program under grant EEC-9402726 and by the Natural Sciences and Engineering Research Council of Canada. E. Winfree is supported in part by National Institute for Mental Health (NIMH) Training Grant # 5 T32 MH 19138-06; also by General Motors' Technology Research Partnerships program.

efficient attack is not known and consequently this brute force approach will be the one considered here.

We begin by describing the algorithm to implement a plaintext-ciphertext attack for breaking DES at a logical level. This allows us to identify the fundamental operations we need to implement on a *stickers machine*, and serves as a roadmap for what follows.

2. The Molecular Algorithm

Start with approximately 2^{56} identical ssDNA *memory strands* [**5**] each 11580 nucleotides long. We think of each memory strand as containing 579 contiguous blocks (referred to as regions in [**5**]) $B_0, B_1, B_2, ..., B_{578}$ each 20 nucleotides long. As is appropriate in the sticker model there are 579 stickers $S_0, S_1, ..., S_{578}$—one complementary to each block. (We refer to memory strands with annealed stickers as *memory complexes*.) Hence, we consider that each strand represents a 579-bit memory, and we sometimes use B_i to refer to the bit which B_i represents. Block B_0 is never set and is used later in the implementation of the algorithm (Subsection 3.1). Blocks B_1 though B_{56} of the memory strands are used to store a key, the next 64 blocks, $B_{57}, ..., B_{120}$, will eventually encode the corresponding ciphertext, and the remainder of blocks are used for intermediate results during the computation. The *stickers machine* which processes the memory strands to compute the ciphertexts does so under the control of a microprocessor. Because the plaintext is the same in all cases, the microprocessor may store it; we do not need to represent the plaintext on the memory strands. Now, given a plaintext-ciphertext pair, the algorithm is performed in three steps:

1. *Input step: Initialize* the memory strands to form memory complexes representing all 2^{56} keys.
2. *Encryption step:* On each memory complex *compute the ciphertext* corresponding to the encryption of the plaintext under that complex's key.
3. *Output step: Select* the memory complex whose ciphertext matches the given ciphertext, and *read* the corresponding key.

The bulk of the work is performed during the second step, where DES data encryption occurs, so we outline it below. Our interest is in demonstrating how DES can be implemented on a molecular computer, and for these purposes the exact details of DES are unnecessary. For details, see [**4**] We will instead focus on the essential operations required in DES, and how these operations are combined to effect the full algorithm.

DES is a 16-round cipher. In each round a new 32 bit intermediate result is produced. These are designated $R_1, \ldots, R_{16}$. We store R_{15} and R_{16} in locations B_{57} through B_{120} (adjacent to the key), while $R_1, \ldots, R_{14}$ are stored in locations B_{121} through B_{568}. Essentially, R_{15} and R_{16}, taken together, form the desired ciphertext. (We encode the ciphertext adjacent to the key for implementation reasons explained in Subsection 3.4.) The left 32-bits and right 32-bits of the plaintext are referred to as R_{-1} and R_0, and are known to the controlling microprocessor.

Bits B_{569} through B_{578} are used as a *workspace* and are written and erased during the course of the computation. Hence, unlike the other bits which are used in a "write once" fashion, these bits may be cleared; for implementation reasons, we always clear the entire workspace at once.

Essentially, R_i is obtained from R_{i-1} and R_{i-2} by the following computation:

$$R_i = R_{i-2} \oplus S(E(R_{i-1}) \oplus K_i)$$

where $\oplus$ denotes exclusive or (x-or), K_i denotes a round dependent selection of 48 bits from the key, E denotes the *expand* function which takes the 32 bits of R_{i-1} and repeats or permutes them to yield 48 bits, and S denotes the *S-function* which takes a 48-bit input and maps it to a 32-bit output. The function E, the function S and the selection K_i are *hard-coded*, like the plaintext, into the microprocessor.

In fact, the S-function can be separated into eight independent 6-bit to 4-bit functions known as *S-boxes*. Hence, each R_i may be computed in eight independent operations each of which produces a 4 bit *chunk* of the result. A given chunk is a function of 16 *input bits*: 6 bits of R_{i-1}, 6 bits of K_i and 4 bits of R_{i-2}. We describe the computation of a chunk below:

1. 6 bits of R_{i-1} and 6 bits of K_i are x-ored to produce a 6-bit result which is then stored in the workspace locations $B_{569}, \ldots, B_{574}$.
2. One of the S-box functions is applied to bits $B_{569}, \ldots, B_{574}$ and the 4-bit result is stored in the workspace locations $B_{575}, \ldots, B_{578}$.
3. Bits $B_{575}, \ldots, B_{578}$ are x-ored with 4 bits of R_{i-2} to produce the desired chunk of R_i which is then stored in the appropriate four blocks of the intermediate result bits $B_{57}, \ldots, B_{568}$.
4. If the chunk being computed is not the last chunk of R_{16}, then the entire workspace, bits $B_{569}, \ldots, B_{578}$, is *cleared* in preparation for future use.

The positions on each memory complex of the 16 input bits required to compute a given chunk depend only on the chunk number (1, ... ,8) and the round number (1, ... ,16), though the 0/1 value of those bits will vary from memory complex to memory complex. The controlling microprocessor knows which positions contain these bits (they are hard-coded) and knows the x-or or S-box which it needs to apply.

We see, then, that encrypting a plaintext with DES comes down to a process of either (1) selecting 2 bits, producing their x-or, and writing the result in a new bit, or (2) selecting 6 bits, applying an S-box, and writing the resulting 4 bits.

3. Implementation

We now turn to implementing the algorithm on a stickers machine. Such a machine, as described in [**5**], may be thought of as a "parallel robotic workstation". It consists of a *rack* of tubes (*data tubes*, *sticker tubes*, and *operator tubes*), some *robotics* (arms, pumps, heater/coolers, connectors, etc.) and a microprocessor that controls the robotics. It is assumed in [**5**] that the components of the robotics and a set of three data or operator tubes may be arranged to perform any of the following four operations: *separate*, *combine*, *set* and *clear*.

We assume that the robotics are capable of an extended set of operations:

1. *Parallel separate.* The robotics can *separate* the DNA from each of 32 data tubes into two more data tubes by using 32 separation operator tubes at once.
2. *Parallel combine.* The robotics can *combine* the DNA from 64 different data tubes into one data tube at once. We assume that the *blank operator tube* used for a *combine* in [**5**] is really just a connector which is part of the robotics.

3. *Parallel set.* The robotics can, using one sticker tube with stickers S_k, *set* the bit B_k on the complexes in 64 different data tubes at once. We assume the *sticker operator tube* used for *set* in [**5**] is just a filter that can be built into the robotics.
4. *Clear.* The robotics can *clear* the workspace bits on all complexes in one data tube. We assume that the stickers on the workspace are removed simultaneously. Hence the workspace blocks may be implemented using so called *weak regions* [**5**]. Again, we assume the *sticker operator tube* used for *clear* in [**5**] is just a filter that can be built into the robotics.

Hence, we perform the above four operations using just *data tubes* that may hold DNA memory complexes, *sticker tubes* that are (for the purpose of the computation) an inexhaustible source of a particular sticker S_k, and *separation operator tubes* that hold probes for a particular block B_k.

In the following subsections we describe, where applicable, the implementation of the molecular algorithm using these operations and, for the purposes of estimating the *time* and *space* required by a stickers machine we keep track of the following three *resource quantities*:

1. Total *steps.* We define the number of *steps* as the number of *parallel separations*, *parallel combines*, *parallel sets* or *clears* that any given complex experiences *after* it has been initialized. Hence, we count the operation of the robotics on a large number of tubes in parallel as a single step and ignore the (perhaps serial) process of moving data and operator tubes.
2. Total *rack tubes.* We define the number of *rack tubes* to be the total number of data tubes, sticker tubes, and separation operator tubes used during the computation. All of the tubes are reusable, so we only need copies of a tube if a particular kind of tube must be used more than once in a parallel operation. We note that we never need duplicates of sticker tubes—our robotics are incapable of using more than one sticker tube at once. We will however need duplicate separation operator tubes and many data tubes since we often wish to separate complexes in several different data tubes on the same bit B_i at once.
3. Maximum number of *active tubes* per operation. We define the number of *active tubes*, for any time during the computation, as the number of tubes which the robotics have removed from the rack and are currently processing. Note that the maximum number of active tubes defines the width of the parallelism used by our algorithm and hence it must match the parallelism built into our robotics.

We note that these quantities are applicable for only for the *computation of the ciphertexts* (step 2 of the molecular algorithm) and the *selection* of the given ciphertext (the first part of step 3 of the molecular algorithm). These parts of the molecular algorithm require only the operations given above and hence are performed using the stickers machine. The *initialization* of the memory complexes and final *reading* of the key, however, use some operations that are not included above (*i.e.* dividing the contents of a tube into two tubes, PCR amplification, ligation, etc.) These special operations are used at most once during the implementation of the molecular algorithm and we assume that they could be performed by the human operator of the stickers machine in a reasonably short amount of time (a few hours) and in a small space (a few tubes). Hence they are ignored in the final discussion of the resources required by a stickers machine (Subsection 3.5).

3.1. Initialization of the memory strands. First we must create the initial tube encoding keys. Our desire is to have each of the 2^{56} memory strands store a different key. This might be accomplished, for example, in the following way:

1. Divide the memory strands into two tubes, A and B.
2. Add an excess of S_1 through S_{56} to tube A and allow them to saturate the first 56 blocks on each strand.
3. Use the complement of B_0 as a probe to *separate* the memory complexes in tube A from the excess stickers.
4. Add tube B to tube A.
5. Heat and cool tube A to randomly reanneal the stickers.

The memory complexes produced by this process appear to be reasonably modeled with a Poisson distribution; it is expected that approximately 63% of keys will be represented and that, on average, there is one of each key. Hence, if no errors are committed during the computation, we have a reasonable chance of recovering the key for a ciphertext of interest. Of course, the chance of succeeding can be increased by starting with more memory strands. To insure that approximately 95% of the keys are represented and that on average three copies of a key are present, we could simply use three times as much DNA. These issues are discussed in more detail in the Section 4.

3.2. Implementing the fundamental operations. As discussed in Section 2, the DES encryption algorithm is a composition of just two simple functions:

- x-ors which are 2-bit input to 1-bit output maps,
- S-boxes which are 6-bit input to 4-bit output maps.

The computation of the 2-bit to 1-bit x-or function is illustrated in Figure 1.

We now describe the computation of the x-or function $B_k = B_i \oplus B_j$ in greater detail, give the overhead required, and generalize this to an n-bit to m-bit function:

A. *Parallel separate* the sample two times to yield four data tubes, one for each possible value B_iB_j. This is accomplished by first using one separation operator tube specific for B_i and then, in parallel, using *two* separation operator tubes specific for B_j. (In Figure 1 the bits being considered are shaded gray.) In **[5]** a single separation is modeled as an operation involving three active tubes at once: a source data tube, a separation operator tube, and one of two destination data tubes picked up and filled by the robotics in sequence. Hence, during each single separation, three tubes are active at once and three data tubes are used. During the second *parallel separation* above, then, six data tubes are used and six tubes are active.

For an n-bit to m-bit function this generalizes to:

Parallel separate the sample n times to yield 2^n data tubes, one for each value of the n-bit input. This requires 2^{i-1} separation operator tubes for the ith *parallel separation* (a total of $2^n - 1$). Hence for the nth parallel separation $3 \times 2^{n-1}$ data tubes are required and $3 \times 2^{n-1}$ tubes are active.

B. *Parallel set* B_k to 1 with an S_k sticker for all tubes for which this is applicable. For an x-or this is only applicable when $B_iB_j = 01$ or 10 but for a general 2-1 function this may require the addition of a sticker to any subset

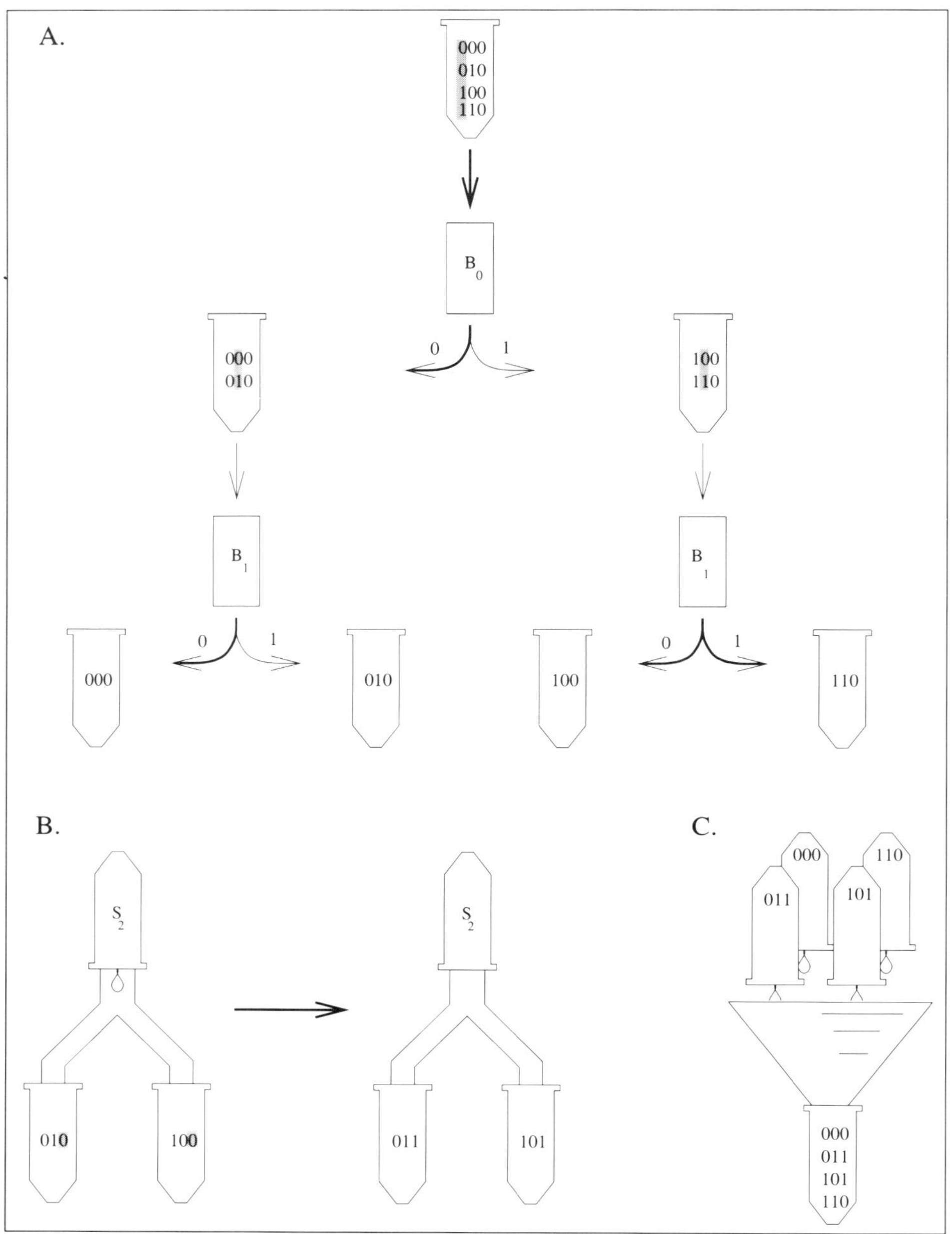

FIGURE 1. Computation of the 2-bit to 1-bit x-or function: $B_2 = B_0 \oplus B_1$.

of the four data tubes in parallel. This requires one sticker tube and four data tubes, a total of five active tubes.

For an n-bit to m-bit function this generalizes to:

Parallel set a (possibly different) subset of the 2^n data tubes m times using a total of m sticker tubes. This requires $2^n + 1$ active tubes. Note that the subset of data tubes to which the *parallel set* is applied is determined solely by the algorithm stored in the microprocessor.

C. *Parallel combine* the contents of all four data tubes into one data tube. This requires five data tubes and hence five active tubes.

For an n-bit to m-bit function this generalizes to:

Parallel combine the contents of all 2^n data tubes into one data tube. This requires $2^n + 1$ data tubes and $2^n + 1$ active tubes.

At the end of our x-or operation, all of our DNA has been returned to a single tube.

In general, an n-bit to m-bit function requires $n+m+1$ steps, $2^n - 1$ separation operator tubes (specific to various bits), m sticker tubes, a maximum of $3 \times 2^{n-1}$ data tubes, and a maximum of $3 \times 2^{n-1}$ active tubes.

At this point we can see that one of the resource quantities of interest, the maximum number of *active tubes* has already been specified. An 6-bit to 4-bit S-box is the biggest n-bit to m-bit function we implement in DES so we never use more than 96 active tubes. (The only other operation used in the algorithm, *clear*, uses but one active tube.)

The number of *rack tubes is* the sum of the data, sticker and separation tubes used but we may not simply sum up these numbers for an S-box and regard the result as its contribution to the computation. Data tubes are interchangeable so we know that the maximum number used in an S-box—96—will be the data tube contribution to the total *rack tube* count. However, sticker tubes and separator tubes have an identity, they are associated with a particular block of the memory complexes. For the calculation of *rack tubes* (and total steps as well) we need to consider more details of the molecular algorithm.

Thus, in the next subsection we consider how the x-ors and S-boxes used to compute DES are composed, count the total number of steps, and discuss ways to plan our use of separation operator tubes and sticker tubes to minimize the total number of rack tubes.

3.3. Computing the ciphertexts. We recall that the basic unit of computation in our molecular algorithm is a 4 bit *chunk* that is computed using a composition of x-ors and S-boxes which takes 16 bits of input. For any given chunk the positions of these 16 input bits are the same for *every* memory complex. Thus the positions of the input bits are *hard-coded* into the microprocessor and we ignore their exact values in what follows.

From our consideration of n-bit to m-bit functions, we know that 4 steps are required to compute an x-or and 11 steps are required to compute an S-box. From this it follows that to compute a chunk requires $6 \times 4 + 11 + 4 \times 4 = 51$ steps. After a chunk is computed, if the workspace is to be used again, it must be *cleared.* To complete the computation requires computing $16 * 8 = 128$ chunks, and *clearing* the workspace 127 times; hence the total number of steps required is 6655.

During the computation of an x-or, one separation operator tube is required to *separate* on the first bit, and two separation operator tubes are required to *separate* (in parallel) on the second bit. To economize on separation operator tubes when performing x-ors we choose to *separate* first on a bit for which there are many instances of that "kind" of bit on the memory complex, and second on a bit for which there are only a few instances of that "kind" of bit. Hence, for each x-or involving a bit of R_{i-2} (of which there are 448 possibilities from $R_1, \ldots, R_{14}$) and a bit from the workspace $B_{575}, \ldots, B_{578}$ (of which there are only 4), we *separate* on the R_{i-2} bit first and the workspace bit second. Likewise, for each x-or involving a bit of R_{i-1} (of which there are 480 possibilities from $R_1, \ldots, R_{15}$) and K_i (of which there are 56 possibilities), we *separate* on the R_{i-1} bit first and the key bit second. It follows that for each of the bits from $R_1, \ldots, R_{15}$ one separation operator tube specific to that bit is needed and, for each of the bits of K and each of the bits $B_{575}, \ldots, B_{578}$ two separation operator tubes specific to that bit are needed. Hence these bits require a total of $480 + 2 \times 56 + 2 \times 4 = 600$ separation operator tubes.

The implementation of the S-boxes demonstrates another way that we may economize on separation operator tubes: frequently used subcomputations do not require that we use up a new bit (and consequently another separation operator tube) every time they are run. Instead their input and output may be stored on a rewritable region of the memory complex—hence our placement of the input to the S-boxes in the first six workspace positions and the output in the last four workspace positions. Because of this, even though there are eight different S-boxes (one for each chunk) they all may use the same separation operator tubes and sticker tubes (though, under control of the microprocessor, the stickers are applied differently for each S-box). The S-boxes, then, only require an additional 63 separation operator tubes to *separate* the DNA into all possible 6-bit strings $B_{569}, \ldots, B_{574}$.

Thus, to complete the ciphertext generation stage of the algorithm, we require 663 separation operator tubes. In all, $512 + 10 = 522$ sticker tubes are required to hold the stickers used to write the intermediate results and workspace bits. (Recall that our robotics only use one sticker tube at a time.) The greatest number of data tubes used during the computation is 96, during the final parallel *separation* step of an S-box computation. In total $663 + 512 + 96 = 1271$ rack tubes are required to compute the ciphertexts.

3.4. Selecting the given ciphertext and reading the correct key. Once the ciphertexts have been computed, the desired key can be *selected* by searching for the complex which has the given ciphertext encoded next to its key. This requires 64 *separation* steps. Upon isolating the desired complex, it is necessary to *read* its key. Reading could be attempted using single molecule detection and a binary tree decoding as described in [**5**]. However, it is not clear that such an approach can be satisfactorily carried out in lab. Below we describe two additional approaches, each of which entails some modification of the methods described in the previous sections:

- In the first approach, the memory strands are 5′ biotinylated. Once the ciphertexts have been computed, as described in the previous sections, the memory complexes are transformed *en masse* into single-stranded form. One way to accomplish this is as follows:
  1. For each sticker $S_0, \ldots, S_{120}$ create a new *0-sticker*, S_i', which shares the 3′ and 5′ 8-mers with S_i but which differs from it in the middle 4-mer.
  2. Add an excess of the S_i' to the final solution under conditions which favor annealing despite the mismatches.
  3. Add ligase. Each memory complex now has a heteroduplex region in its key-ciphertext section where the non-memory strand consists of a sequence of (regular) stickers and 0-stickers. Note that our original decision to place the ciphertext next to the key on the memory

strands is no longer a mystery. Our placement minimizes the number of ligations which must be successful for a strand which encodes both the ciphertext and key to form.

4. *Separate* out the biotinylated memory strands using a streptavidin-coated solid support and retain the new strands as the *library* of solutions.

This process essentially converts Sticker memory complexes into *Lipton style* memory strands [**3**] for which each block has one of two unique sequences, one for 0 and another for 1. A ciphertext of interest may be *selected* (by applying the usual 64 *separation* steps), PCR amplified (using S_0' and the complement of the last ciphertext bit as primers), and *read* (with standard DNA sequencing).

Notice that, once created, the library of solutions can be replicated by PCR and hence multiple copies can be made. Each such copy is essentially a codebook consisting of (key, ciphertext) pairs. This codebook has approximately $2^{56} \times$ (56 key bits + 64 ciphertext bits) $= 2^{63}$ bits of information (the equivalent of approximately one billion 1 gigabyte CDs) but occupies a dry volume of approximately 1/7 of a teaspoon. As Boneh and Lipton have noted [**1**] such a codebook could be widely distributed and used to speed up subsequent attacks on DES.

- In the second approach to *reading*, instead of converting to Lipton style memory strands after the computation of ciphertexts, a Lipton-Sticker hybrid model for encoding the initial memory complexes is used.
  1. Make single-stranded DNAs representing all 2^{56} keys using the encoding of Lipton. Additionally, guarantee that each key strand begins and ends with the same short PCR primer.
  2. Ligate, to each of the Lipton key strands, identical 522 bit Sticker memory strands.
  3. Proceed to compute DES as usual.
  4. Perform 64 *separations* on the result to obtain the complex carrying the desired ciphertext.
  5. Perform PCR using the primers which bound the Lipton style key sequence. The key sequence will be amplified exponentially and may be sequenced.

We note that all of the techniques for *selecting* a desired ciphertext discussed here and in [**5**] would require 64 separation operations so we approximate that this process requires an additional 64 steps. Further, we note that to effect all 64 separations an additional 32 separation operator tubes specific to the bits of R_{16} (in addition to the 32 separation operator tubes already counted for R_{15}) are required.

3.5. Discussion. In summary, to find the key for a DES-encoded plaintext-ciphertext pair, we first create memory complexes representing all keys, then we compute the ciphertext corresponding to each key (6655 steps), and finally we select and read the key of interest (64 steps). This requires a total of 6719 steps, each of which is one of the operations described above.

The actual running time for the algorithm depends on how fast the operations can be performed. If we assume, as we might if a graduate student had to perform each operation, that each operation requires 1 day, then the computation will require 18 years. If each operation requires 1 hour ([**2**] assumes 2.4 hours) then the computation will require approximately 9 months. If each operation can be completed in 1 minute, perhaps using a robotic stickers machine, then the computation will take 5 days. Finally, if the effective duration of a step can be reduced to 1 second, perhaps by running the algorithm in a continuous flow *parallel refinery* [**5**], then the effort will require 2 hours.

The size of the rack is dictated by the amount of DNA used. When the 2^{56} memory complexes have half of their sticker positions occupied, as we expect will be the case at the end of the computation, they weigh approximately .7 g and, in solution at 5 g/liter, would occupy approximately 140 ml. Hence, the volume of the 1303 *rack tubes* (1271 for ciphertext computation, an additional 32 for ciphertext selection) need be no more than 140 ml each. It follows that the rack tubes occupy, at most, 182 L and can, for example, be arrayed in a rack 1 m (approximately 39 inches) long and wide and 18 cm (approximately 7 inches) deep. Since the robotics must be able to operate on 96 *active tubes* in parallel we approximate the volume required by the robotics, give or take some arms and pumps, as 13 L—1/14th the volume of the rack. The microprocessor is likely to be quite small. Thus, it is reasonable to assume the entire machine would fit on a desktop.

It is worth pointing out that, for much of the computation, the robotics are not processing at full capacity (96 tubes) and many of the tubes are sitting idle. Hence it may be possible to increase the parallelism significantly by pipelining the computation.

4. Analysis of Errors

We say that an error has occurred whenever a memory complex is transformed in an unintended way or ends up in an unintended place. Hence, there are many kinds of errors that can occur during the operation of our DNA computer: strands can break, stickers can fall off one memory complex and reanneal to another, complexes can be "lost" on the walls of a tube, complexes can end up in the wrong tube following a *separation*, etc.

For each operation, we define its error rate to be the fraction of molecules that commit an error during that operation. Some operations are more prone to errors than others. To simplify our analysis, we define E to be the error rate of the worst operation and assume that all of the operations have error rate E. Note that $1 - E$ corresponds to what chemists call yield. Hence an error rate of 10^{-4} corresponds to a step yield of 99.99%.

Given an input ciphertext-plaintext pair, it is possible that several different keys map the ciphertext into the plaintext (though for DES, it seems unlikely that the number of such keys would be large). All such keys will be called *winning keys*. Under ideal conditions, after the codebook is created and the separation on the input ciphertext performed, we are left with a *final tube* with the following properties:

1. For each winning key there is at least one complex encoding it.
2. All complexes that are there encode winning keys.

In reality this can fail for either of two reasons. First, complexes encoding winning keys may be missing, either because they were not created during initialization or because they encountered an error during the computation. Second, there may be *distractors*: complexes which do not encode winning keys, but due to errors end up in the final tube anyway. In subsection 4.1 we analyze the probability that a winning key has a complex encoding it in the final tube. In subsection 4.2 we calculate the expected number of distractors in the final tube.

We make the following assumptions:

1. After the initialization step, each complex encodes a 56-bit key chosen at random (*i.e.* chosen from the space of all 56-bit keys with equal probability).
2. DES with the input plaintext maps each of the keys to a random ciphertext.
3. A complex which encounters an error during the computation produces a random ciphertext (*i.e.* unrelated to the ciphertext normally associated with that complex's key).

4.1. Probability that a winning key has a complex encoding it in the final tube. In the computation above, we began with 2^{56} memory strands. It will be convenient to carry out the analysis in greater generality. We now assume that we begin with $2^{56}X$ memory strands, where X is a positive rational. Informally, X is the factor by which we multiply our original amount of DNA.

Let K_w be a winning key. Following initialization, the number of memory complexes which encode K_w is given by a binomially distributed random variable $(n, p) = (2^{56}X, \frac{1}{2^{56}})$.

The probability that a memory complex makes it correctly through all 6655 steps of the computation and 64 steps of the selection process (in total 6719 steps) is given by:

$$S = (1 - E)^{6719}$$

Hence, after the computation, the number of memory complexes in the final tube which encode K_w is given by a binomially distributed random variable $(n, p) = (2^{56}X, S\frac{1}{2^{56}})$. Because $2^{56}X$ is very large and $S\frac{1}{2^{56}}$ is very small, this distribution may be approximated by a Poisson distribution with Poisson parameter $\lambda = np = SX$. From this it follows that the expected number of complexes encoding K_w in the final tube is SX and that the probability that a complex encoding K_w is in the final tube is:

$$(1 - e^{-SX})$$

In particular when $X = 1/S$, the expected number of complexes encoding K_w in the final tube is one and the probability that a complex encoding K_w is in the final tube is 63%. We will refer to 63% as a *reasonable chance.*

4.2. Number of distractors in the final tube. For a memory complex M, let $H(M)$ denote the Hamming distance of the ciphertext encoded on M from the input ciphertext. For M to enter the final tube, $H(M)$ errors must occur during the final 64 separation steps. By our assumptions, after the computation of the codebook, each memory complex (whether it has encountered an error or not) encodes a ciphertext which is a random 64 bit string. It follows that the Hamming distances associated with memory complexes will be a binomially distributed random variable $(n, p) = (64, 0.5)$. Hence, the probability that $H(M) = L$ is:

$$\binom{64}{L}\left(\frac{1}{2}\right)^{64-L}\left(\frac{1}{2}\right)^{L} = \binom{64}{L}\frac{1}{2^{64}}$$

The probability of complex M with Hamming distance $H(M) = L$ making it through the 64 step selection process is given by the probability that it correctly negotiates $64 - L$ separations for which it matches the input ciphertext times the probability that it commits an error at the L separations for which it mismatches the correct ciphertext:

$$(1 - E)^{64-L} E^L$$

There are $2^{56}X$ complexes in the codebook so the expected number of distractors is given by:

$$\sum_{L=0}^{64} \frac{2^{56}X}{2^{64}} \begin{pmatrix} 64 \\ L \end{pmatrix} (1 - E)^{64-L} E^L = \frac{X}{256}$$

Perhaps surprisingly, the expected number of distractor molecules is independent of the error rate. Note, in particular, that for X=1 the expected number of distractors is less than one.

4.3. Feasibility. Combining the results from 3.2 and 3.3 gives Table 1 which shows for various error rates, the amount of DNA that must be pushed through the DES computation to insure that there is a *reasonable chance* (63%) of getting a winning key in the final tube. Table 1 also records the expected number of distractors which will be present in the final tube.

This indicates that for an error rate of 10^{-4}, a little more that 1 gram of DNA is needed and the final tube will usually have no distractors. If an error rate of 10^{-3} is achievable, then less than a kilogram of DNA is needed and only a small number of distractors need be dealt with before finding the correct answer. However, if an error rate of 10^{-2} is the best that is attainable, then huge amounts of DNA are needed (approximately 23 Earth masses) to have a *reasonable chance* that a winning key ends in the final tube, but even then it will have to be distinguished from a colossal number of distractors - clearly an unacceptable situation. In such cases, where only very high error rates are possible, techniques like those described in [**5**] (for example, a *refinery algorithm*) may be used to reduce the amount of DNA required. Section 5.8 of [**5**] gives a brief analysis of the application of a refinery algorithm to DES.

5. Conclusions

We wish to emphasize that our description of an attack on DES is, at this point, entirely theoretical and whether it can be carried out in the lab remains to be seen. Huge challenges remain. For example, as yet, we have been unable to perform *separations* (or any of the sticker operations) in the lab with error rates

achievable error rate E	0	10^{-4}	10^{-3}	10^{-2}
X	1	2	830	2.1×10^{29}
grams of DNA required	.7	1.4	580	1.5×10^{29}
distractors	.004	.008	3.2	8.3×10^{26}

TABLE 1. Quantity of DNA required and number of distractors as a function of error rate.

approaching 10^{-4}. Nonetheless, the analysis presented in this paper demonstrates (at least in principle) two things:

- "Real problems" can be solved with small machines which do not require huge amounts of DNA and use little or no enzymes.
- Error rates similar to those normally demanded of electronic computers are not required.

If the attack on DES described here can be carried out in the lab, then some other cryptosystems might also be vulnerable to this approach. Indeed, the small size of the machine we describe suggests that systems like the 64-bit key FEAL cryptographic system of Shimizu-Miyaguchi [**6**] might be susceptible to such an attack.

Finally, there are several messages for cryptography in these findings. First, it seems appropriate to reconsider one of the "axioms" of cryptography: Improvements in computational power always favor the cryptographer over the cryptanalyst. This is almost certainly untrue. The analysis presented here suggests the possibility of computers with super-parallelism that can help the cryptanalyst immensely, yet provide no help for the cryptographer. Even if DNA computers prove infeasible, it is possible that new machines machines capable of super-parallelism may make cryptosystems like DES insecure.

The DNA computer is an example of a super-parallel machine with very slow processors (complexes of DNA). The potential vulnerability of DES arises for two reasons. First, the key space is insufficiently large. Second, the DES algorithm is "too short." The fact that only 6655 steps are needed to do an encryption allows the slow DNA processors to finish their encryptions in (at least in theory) a reasonable amount of time. Hence the much valued throughput speed of DES and similar systems, may carry with it a potential vulnerability to super-parallel machines with slow processors.

References

[1] Boneh, D., Lipton, R., Batching DNA Computations. *Princeton CS Tech-Report CS-TR-489-95,* (1995).

[2] Boneh, D., Dunworth, C., Lipton, R., Breaking DES Using a Molecular Computer. In R. J. Lipton and E. B. Baum, editors, *DNA Computers: Proceeding of a DIMACS Workshop,* DIMACS Series in Discrete Mathematics and Theoretical Computer Science, American Mathematical Sociey, 37-65, 1996.

[3] Lipton, R., Using DNA to solve NP-Complete Problems. *Science*, **268** (1995) 542–545.

[4] National Bureau of Standards. Data Encryption Standard. *FIPS pub. 46,* U.S. Department of Commerce, (1977).

[5] Roweis, S., Winfree, E., Burgoyne, R., Chelyapov, N.V., Goodman, M.F., Rothemund, P.W.K., Adleman, L.M., A Sticker Based Model for DNA Computation. In L. Landweber and E. Baum, editors, *DNA Based Computers: Proceedings of the Second DIMACS Workshop*, DIMACS: Series in Discrete Mathematics and Theoretical Computer Science. American Mathematical Society. Providence, RI, 1-29, (1998).

[6] Shimizu, A., Miyaguchi, S., Fast Data Encipherment Algorithm FEAL, *In Lecture Notes in Computer Science,* **304** Springer-Verlag, New York, (1988), 267-278.

[7] Wiener, M., Efficient DES Key Search. *Tech Report-244,* School of Computer Science, Carleton University (1994).

LABORATORY FOR MOLECULAR SCIENCE, UNIVERSITY OF SOUTHERN CALIFORNIA

AND

(L. Adleman and P. Rothemund) Department of Computer Science, University of Southern California, Los Angeles, CA 90089

E-mail address: rothemun@pollux.usc.edu, phone: 626-584-1807 FAX: 213-740-8491

(S. Roweis and E. Winfree) Computation and Neural Systems Option, California Institute of Technology, Pasadena, CA 91125

DIMACS Series in Discrete Mathematics
and Theoretical Computer Science
Volume **44**, 1999

Massively Parallel DNA Computation: Expansion of Symbolic Determinants

Thomas H. Leete, Matthew D. Schwartz, Robert M. Williams, David H. Wood, Jerome S. Salem, and Harvey Rubin[†]

Abstract. A new type of algorithm is introduced for constructing DNA molecules which encode answers to mathematical problems. Examples include problems from the NP-Complete class, which are widely considered to be harder than those in the problem classes previously addressed. In particular, algorithms are presented that generate expansions of symbolic determinants given their patterns of zero entries. This is well-known to be exponentially more difficult than evaluating determinants whose entries are merely numerical. Prior approaches to DNA computation were impractical for large problems because they required processing vast quantities of DNA with steps associated with large error propagation. Our new approach to the production of the solution and reading the answer is based on reliable and automatable PCR steps and can solve large problems by processing up to 10^{15} or more distinct strands of DNA in parallel. The DNA algorithms described here should be applicable to a wide variety of problems that are intractable using conventional computers.

1. Introduction

Considerable attention has been recently focused on the design of a DNA based computer. Adleman first showed how a DNA computer might be utilized to solve the directed Hamiltonian circuit (DHC) problem for seven nodes [**1**]. This was followed by Lipton's demonstration of a hypothetical procedure using a DNA algorithm that could be used to solve the 3SAT problem, a versatile problem in NP-Complete with a natural equivalence to many other NP-Complete problems [**10**]. However, DNA had not been used to produce solutions to problems that are not solvable using standard computer architecture. In order to realize the full potential of the parallel processing technology of a DNA computer, the structure of the problem must fit the structure and limitations of DNA biochemistry.

Many problems in symbolic computation have natural expressions in terms of determinants, e.g., linear systems of equations, eigenvalues and eigenvectors, Jacobians, Hessians and conformal mapping. Solutions to such problems are critical

---

R. M. Williams and D. H. Wood were partially supported by the Office of Naval Research.

† To whom correspondence should be addressed.

components in a wide variety of applications including the design of robotic movements, flight pattern controls, animation effects, electrical circuit design, economic forecasting, optimization problems, biological simulations and surface simulation calculations.

In this paper we present a procedure for solving the problem of expanding symbolic determinants given their zero pattern (ESD0). This problem is harder than the problems in NP-Complete, which can be solved in 2^n steps, whereas ESD0 for an $n \times n$ matrix has to output up to n terms. The usual techniques for solving numerical determinants are much less effective if the elements are symbols. For example, Gaussian elimination for ESD0 would require $O((n!)^2)$ steps [**7**].

We describe here a solution to ESD0 using an algorithm that involves highly parallel operations on DNA. The answer to the expansion of the determinant is read directly from the set of DNA molecules generated by the algorithm. The system is easily automatable and robust.

2. Computing a determinant with DNA

With the technology and algorithms presented below, we generate only the components of the solution to ESD0. The results of the process that generates the expansion of the determinant are stored as fragments of double stranded DNA (oligomers). One highly advantageous property of the process is that we manipulate and store only the non-zero elements of the expanded determinant. This feature, coupled to the parallel processing nature of molecular hybridization, allows the DNA coprocessor to use fewer time steps than standard architecture computers [**12**]. The solution to ESD0 has three parts: encoding the data, expanding the determinant, and reading the solution.

3. Algorithms for encoding the data

Each element in the matrix is represented as a single stranded sequence of 60 nucleotides, composed of a 20 nucleotide 5' annealing sequence, a unique middle sequence of 20 nucleotides encoding an element value, and a 3' annealing sequence of 20 nucleotides. It is important to note that each element in the symbolic matrix is input defined by the user and is stored in an element value table. Each element will only anneal with elements from the preceding row at the upstream end and elements from the following row at the downstream end. Elements with a value equal to zero are represented by the absence of DNA. Element oligos are designed to avoid random cross-hybridization and hairpin formation. Also, the 5' and 3' hybridization regions have similar melting temperatures, Tm. Initializing the computer involves synthesizing the relevant oligos using widely available, inexpensive instruments and setting up $n \times n$ wells, with each well corresponding to one element of the matrix.

4. Algorithms for expanding the determinant

The oligomers that appear in the expansion of the determinant (hereafter referred to as OS, oligomer solutions) have one element from each row and one element from each column. The general form of each term in the OS is shown in Figure 1 general form

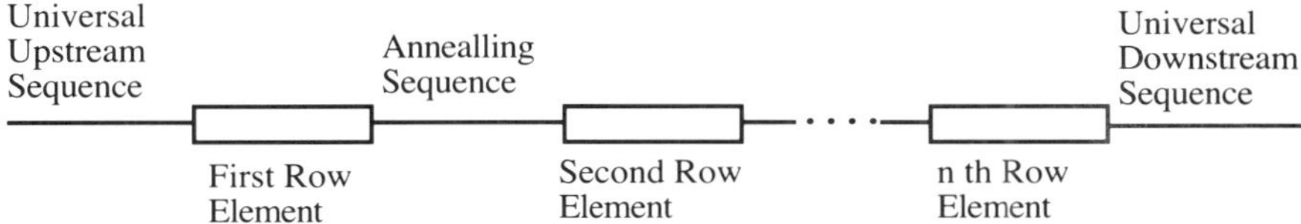

FIGURE 1. General form of each term in the oligomer solution.

4.1. Algorithm A_0. Algorithm A_0 expands the determinant by mixing all the wells (elements) together and then separating and reading the OS. Consider an $n \times n$ matrix with no zeros. Given the annealing properties of the elements, if all of the elements are mixed, the resulting DNA products would contain exactly one element from each row in the correct order from top to bottom. This generates n^n products, only $n!$ of which are OS. The non-OS oligomers must then be eliminated and they outnumber the OS oligomers by the ratio of $n^n/n!$. The biophysical processes that are available to carry out the elimination operation make this algorithm very inefficient and error prone and are therefore impractical [**1, 10**]. On a deeper level, this algorithm does not exploit the mathematical structure inherent in the problem. By contrast, we describe here two algorithms, A_1 and A_2, in which the oligomer elements are combined in processes that result in an ensemble of DNA oligomers which is a molecular representation of expansion of the determinant by minors. These new approaches generate only OS.

4.2. Algorithm A_1. Algorithm A_1 generates the OS from the top to the bottom row of the matrix; terms in the solution are built by adding one element at a time by sequential cycles of hybridization and amplification. Terms are generated in parallel, significantly reducing the amount of time necessary to complete the process. As an example of A_1, we present the steps in the expansion of the 5×5 determinant:

$$\begin{vmatrix} a_{11} & a_{12} & a_{13} & a_{14} & a_{15} \\ a_{21} & a_{22} & a_{23} & a_{24} & a_{25} \\ a_{31} & a_{32} & a_{33} & a_{34} & a_{35} \\ a_{41} & a_{42} & a_{43} & a_{44} & a_{45} \\ a_{51} & a_{52} & a_{53} & a_{54} & a_{55} \end{vmatrix}$$

4.2.1. *Anneal elements from each row with the row below it to generate two element DNA oligomers.* For example, consider the first and second rows. Each appropriate complementary second row element (a'_{22}, a'_{23}, a'_{24}, a'_{25}) is added to the first well of the first row (a_{11}) to generate a mix of hybrids ($a_{11}a_{22}$, $a_{11}a_{23}$, $a_{11}a_{24}$, $a_{11}a_{25}$), referred to as the "a_{11}" well. Hybrids are made double stranded by filling in with Vent DNA polymerase (NEB, MA). The same process is carried out for each first row element to generate 5 wells ($a_{11}, a_{12}, a_{13}, a_{14}, a_{15}$), each containing a mix of oligomers two elements in length. This step is done for every pair of adjacent rows in the matrix in parallel, generating wells representing rows $1+2$, $2+3$, $3+4$ and $4+5$.

4.2.2. *Combine adjacent 2-element DNA oligomers to form 3-element oligomers.* In the next and all following steps, oligomers of the same length from adjacent rows are added to each other, denatured, and allowed to anneal. Thus the "a_{11}" well

$$\begin{vmatrix} a_{11} & a_{12} & a_{13} & a_{14} \\ a_{21} & a_{22} & a_{23} & a_{24} \\ a_{31} & a_{32} & a_{33} & a_{34} \\ a_{41} & a_{42} & a_{43} & a_{44} \end{vmatrix}$$

TABLE 1. The program is written out in detail for the above 4 × 4 determinant.

$a_{11}a_{22}a_{33}a_{44}$	$a_{11}a_{22}a_{34}a_{43}$	$a_{11}a_{23}a_{32}a_{44}$	$a_{11}a_{23}a_{34}a_{42}$	$a_{11}a_{24}a_{32}a_{43}$	$a_{11}a_{24}a_{33}a_{42}$
$a_{12}a_{21}a_{33}a_{44}$	$a_{12}a_{21}a_{34}a_{43}$	$a_{12}a_{23}a_{31}a_{44}$	$a_{12}a_{23}a_{34}a_{41}$	$a_{12}a_{24}a_{31}a_{43}$	$a_{12}a_{24}a_{33}a_{41}$
$a_{13}a_{21}a_{32}a_{44}$	$a_{13}a_{21}a_{34}a_{42}$	$a_{13}a_{22}a_{31}a_{44}$	$a_{13}a_{22}a_{34}a_{41}$	$a_{13}a_{24}a_{31}a_{42}$	$a_{13}a_{24}a_{32}a_{41}$
$a_{14}a_{21}a_{32}a_{43}$	$a_{14}a_{21}a_{33}a_{42}$	$a_{14}a_{22}a_{31}a_{43}$	$a_{14}a_{22}a_{33}a_{41}$	$a_{14}a_{23}a_{31}a_{42}$	$a_{14}a_{23}a_{32}a_{41}$

TABLE 2. The terms (OS) of the determinant:

from the first step (containing $a_{11}a_{22}$, $a_{11}a_{23}$, $a_{11}a_{24}$, $a_{11}a_{25}$) has the products of the "a_{22}", "a_{23}, "a_{24}", and "a_{25}" wells added to it (containing $a_{22}a_{31}, a_{22}a_{33}$, $a_{22}a_{34}$, $a_{22}a_{35}$; $a_{23}a_{31}$, $a_{23}a_{32}$, $a_{23}a_{34}$, $a_{23}a_{35}$; $a_{24}a_{31}$, $a_{24}a_{32}$, $a_{24}a_{33}$, $a_{24}a_{35}$; and $a_{25}a_{31}$, $a_{25}a_{32}$, $a_{25}a_{33}$, $a_{25}a_{34}$, respectively). Hybridization occurs between polymers that share the same element; thus, $a_{11}a_{22}$ hybridizes to $a_{22}a_{31}$, $a_{22}a_{33}$, $a_{22}a_{34}$, and $a_{22}a_{35}$. Hybrids are again made double stranded by filling in with Vent polymerase.

4.2.3. *Selection of correct 3-element oligomers.* At this stage, products that contain inappropriate elements in the third position will be generated, such as $a_{11}a_{22}a_{31}$ which has two elements from the first column. To select for oligomer solutions, and also to amplify the amount of material available for subsequent steps, a PCR reaction is carried out on the filled in 3-element oligomers, where only the 3' end primers for OS terms are added. Thus in the "a_{11}" well we amplify with an "a_{11}" specific 5' end primer and a mix of "a_{32}," "a_{33}," "a_{34}," and "a_{35}" specific 3' end primers. This will ensure that we only amplify appropriate oligomers, while inappropriate oligomers will be diluted to an undetectable level [**8**]. As above, this process is carried out in parallel for every pair of rows to generate successive sets of three element oligomers.

Steps (*ii*) and (*iii*) are carried out repeatedly to generate full length terms for any size $n \times n$ matrix. The final product of this process (called the "solution row") consists of a row of n wells, each of which contains a mix of $(n-1)!$ OS and all the OS in a well have the same first element. We illustrate these principles of Algorithm A_1 by considering the 4×4 determinant in Table 1. Its terms are given in Table 2. For a 4×4 determinant, Algorithm A_1 requires 3 passes through the steps $(i) - (iii)$. The products of these steps are fully written out in Tables 3-5.

4.3. Algorithm A_2. The second algorithm (A_2) is similar to the first, but in this case oligomers are built starting from both the first row and the last row at the same time. Thus in addition to making 2 element oligomers that contain one second row element and all possible third row elements, we also make 2 element oligomers that contain one third row element and all possible second row elements. The algorithm uses the hybridization step instead of PCR to select for appropriate products. If we take the products of the "a_{11}" well generated by the first method ($a_{11}a_{22}$, $a_{11}a_{23}$, $a_{11}a_{24}$, and $a_{11}a_{25}$) and mix them with the products of a third row element with all possible second row elements (such as the "a_{32}" well containing $a_{21}a_{32}$, $a_{23}a_{32}$, $a_{24}a_{32}$, and $a_{25}a_{32}$), the resulting 3 element oligomer products will

		Products (Each line represents an individual well)
1 + 2:	a)	$a_{11}a_{21}$ $a_{11}a_{22}$ $a_{11}a_{23}$ $a_{11}a_{24}$
	b)	$a_{12}a_{21}$ $a_{12}a_{22}$ $a_{12}a_{23}$ $a_{12}a_{24}$
	c)	$a_{13}a_{21}$ $a_{13}a_{22}$ $a_{13}a_{23}$ $a_{13}a_{24}$
	d)	$a_{14}a_{21}$ $a_{14}a_{22}$ $a_{14}a_{23}$ $a_{14}a_{24}$
2 + 3:	e)	$a_{21}a_{31}$ $a_{21}a_{32}$ $a_{21}a_{33}$ $a_{21}a_{34}$
	f)	$a_{22}a_{31}$ $a_{22}a_{32}$ $a_{22}a_{33}$ $a_{22}a_{34}$
	g)	$a_{23}a_{31}$ $a_{23}a_{32}$ $a_{23}a_{33}$ $a_{23}a_{34}$
	h)	$a_{24}a_{31}$ $a_{24}a_{32}$ $a_{24}a_{33}$ $a_{24}a_{34}$
3 + 4:	i)	$a_{31}a_{41}$ $a_{31}a_{42}$ $a_{31}a_{43}$ $a_{31}a_{44}$
	j)	$a_{32}a_{41}$ $a_{32}a_{42}$ $a_{32}a_{43}$ $a_{32}a_{44}$
	k)	$a_{33}a_{41}$ $a_{33}a_{42}$ $a_{33}a_{43}$ $a_{33}a_{44}$
	l)	$a_{34}a_{41}$ $a_{34}a_{42}$ $a_{34}a_{43}$ $a_{34}a_{44}$

Step 1(i): PCR selection method (Algorithm A_1). Anneal rows 1 and 2, 2 and 3, 3 and 4, and fill in.

		Primers	Products
1 + 2	a)	$a_{11}, a_{22}, a_{23}, a_{24}$:	$a_{11}a_{22}$ $a_{11}a_{23}$ $a_{11}a_{24}$
	b)	$a_{12}, a_{21}, a_{23}, a_{24}$:	$a_{12}a_{21}$ $a_{12}a_{23}$ $a_{12}a_{24}$
	c)	$a_{13}, a_{21}, a_{22}, a_{24}$:	$a_{13}a_{21}$ $a_{13}a_{22}$ $a_{13}a_{24}$
	d)	$a_{14}, a_{21}, a_{22}, a_{23}$:	$a_{14}a_{21}$ $a_{14}a_{22}$ $a_{14}a_{23}$
2 + 3	e)	$a_{21}, a_{32}, a_{33}, a_{34}$:	$a_{21}a_{32}$ $a_{21}a_{33}$ $a_{21}a_{34}$
	f)	$a_{22}, a_{31}, a_{33}, a_{34}$:	$a_{22}a_{31}$ $a_{22}a_{33}$ $a_{22}a_{34}$
	g)	$a_{23}, a_{31}, a_{32}, a_{34}$:	$a_{23}a_{31}$ $a_{23}a_{32}$ $a_{23}a_{34}$
	h)	$a_{24}, a_{31}, a_{32}, a_{33}$:	$a_{24}a_{31}$ $a_{24}a_{32}$ $a_{24}a_{33}$
3 + 4	i)	$a_{31}, a_{42}, a_{43}, a_{44}$:	$a_{31}a_{42}$ $a_{31}a_{43}$ $a_{31}a_{44}$
	j)	$a_{32}, a_{41}, a_{43}, a_{44}$:	$a_{32}a_{41}$ $a_{32}a_{43}$ $a_{32}a_{44}$
	k)	$a_{33}, a_{41}, a_{42}, a_{44}$:	$a_{33}a_{41}$ $a_{33}a_{42}$ $a_{33}a_{44}$
	l)	$a_{34}, a_{41}, a_{42}, a_{43}$:	$a_{34}a_{41}$ $a_{34}a_{42}$ $a_{34}a_{43}$

Step 1(ii): Amplify

Step 1(iii): Dilute and denature

TABLE 3. The products of Algorithm A_1 during the first iteration

contain only appropriate oligomers. Since the products in the "a_{11}" well contain no "a_{21}" elements and the products in the "a_{32}" well contain no "a_{22}" elements, the only hybrids formed are $a_{11}a_{23}a_{32}$, $a_{11}a_{24}a_{32}$, and $a_{11}a_{25}a_{32}$, all of which are precursors of correct terms. We continue this process of annealing and filling in until the OS are formed. This method does not create any partial or full length oligomers that contain incorrect (non-OS) terms; however, it is more complicated than A_1 in the organization of wells and piping of samples. In both A_1 and A_2, the procedure generates excess unhybridized DNA in each step which must be removed so that they will not interfere with subsequent steps. A_1 does this by amplification of the product oligomers and dilution of the resulting PCR product. PCR could be used to select desired products from A_2, with the concomitant benefit that it would also increase the amount of material available for subsequent steps. The reaction products for A_2 are written out in Tables 6- 7 for each step of the expansion of the 4×4 determinant in Table 1.

One of the primary assumptions in these algorithms is that PCR will amplify all oligomers in a given well, so that none of the terms of the determinant are lost. This will hold if 1) the kinetics of the PCR reaction are the same for all oligomers, and 2) the concentration of each oligomer in the well is kept at a level that is detectable by PCR. The first criterion is satisfied because a) all of the correct oligomers are the same size in each step, and b) all amplification primers are designed with similar AT:GC content, so that the association constant of each primer is roughly equal. The second criterion is important to ensure that every oligomer is amplified in each step. Published reports have shown detection of as little as 1 aM DNA by PCR [**2**] which is 10^{-21} moles or about 600 molecules in a $100\mu L$ reaction. We can easily amplify as much as $500ng$ of DNA in a $100\mu L$ reaction, which is approximately 1

		Products:				
$(1+2)+(2+3)$	a)	$a_{11}a_{22}a_{31}$	$a_{11}a_{22}a_{33}$	$a_{11}a_{22}a_{34}$	$a_{11}a_{23}a_{31}$	$a_{11}a_{23}a_{32}$
		$a_{11}a_{23}a_{34}$	$a_{11}a_{24}a_{31}$	$a_{11}a_{24}a_{32}$	$a_{11}a_{24}a_{33}$	
	b)	$a_{12}a_{21}a_{32}$	$a_{12}a_{21}a_{33}$	$a_{12}a_{21}a_{34}$	$a_{12}a_{23}a_{31}$	$a_{12}a_{23}a_{32}$
		$a_{12}a_{23}a_{34}$	$a_{12}a_{24}a_{31}$	$a_{12}a_{24}a_{32}$	$a_{12}a_{24}a_{33}$	
	c)	$a_{13}a_{21}a_{32}$	$a_{13}a_{21}a_{33}$	$a_{13}a_{21}a_{34}$	$a_{13}a_{22}a_{31}$	$a_{13}a_{22}a_{33}$
		$a_{13}a_{22}a_{34}$	$a_{13}a_{24}a_{31}$	$a_{13}a_{24}a_{32}$	$a_{13}a_{24}a_{33}$	
	d)	$a_{14}a_{21}a_{32}$	$a_{14}a_{21}a_{33}$	$a_{14}a_{21}a_{34}$	$a_{14}a_{22}a_{31}$	$a_{14}a_{22}a_{33}$
		$a_{14}a_{22}a_{34}$	$a_{14}a_{23}a_{31}$	$a_{14}a_{23}a_{32}$	$a_{14}a_{23}a_{34}$	
$(2+3)+(3+4)$	e)	$a_{21}a_{32}a_{41}$	$a_{21}a_{32}a_{43}$	$a_{21}a_{32}a_{44}$	$a_{21}a_{33}a_{41}$	$a_{21}a_{33}a_{42}$
		$a_{21}a_{33}a_{44}$	$a_{21}a_{34}a_{41}$	$a_{21}a_{34}a_{42}$	$a_{21}a_{34}a_{43}$	
	f)	$a_{22}a_{31}a_{42}$	$a_{22}a_{31}a_{43}$	$a_{22}a_{31}a_{44}$	$a_{22}a_{33}a_{41}$	$a_{22}a_{33}a_{42}$
		$a_{22}a_{33}a_{44}$	$a_{22}a_{34}a_{41}$	$a_{22}a_{34}a_{42}$	$a_{22}a_{34}a_{43}$	
	g)	$a_{23}a_{31}a_{42}$	$a_{23}a_{31}a_{43}$	$a_{23}a_{31}a_{44}$	$a_{23}a_{32}a_{41}$	$a_{23}a_{32}a_{43}$
		$a_{23}a_{32}a_{44}$	$a_{23}a_{34}a_{41}$	$a_{23}a_{34}a_{42}$	$a_{23}a_{34}a_{43}$	
	h)	$a_{24}a_{31}a_{42}$	$a_{24}a_{31}a_{43}$	$a_{24}a_{31}a_{44}$	$a_{24}a_{32}a_{41}$	$a_{24}a_{32}a_{43}$
		$a_{24}a_{32}a_{44}$	$a_{24}a_{33}a_{41}$	$a_{24}a_{33}a_{42}$	$a_{24}a_{33}a_{44}$	

Step 2(i): Anneal $(1+2)$ to $(2+3)$, $(2+3)$ to $(3+4)$, and fill in.

	Primers	Products		
$(1+2)+(2+3)$	a) a_{11}, a_{32}, a_{33}, a_{34} :	$a_{11}a_{22}a_{33}$	$a_{11}a_{22}a_{34}$	$a_{11}a_{23}a_{32}$
		$a_{11}a_{23}a_{34}$	$a_{11}a_{24}a_{32}$	$a_{11}a_{24}a_{33}$
	b) a_{12}, a_{31}, a_{33}, a_{34} :	$a_{12}a_{21}a_{33}$	$a_{12}a_{21}a_{34}$	$a_{12}a_{23}a_{31}$
		$a_{12}a_{23}a_{34}$	$a_{12}a_{24}a_{31}$	$a_{12}a_{24}a_{33}$
	c) a_{13}, a_{31}, a_{32}, a_{34} :	$a_{13}a_{21}a_{32}$	$a_{13}a_{21}l$	$a_{13}a_{22}a_{31}$
		$a_{13}a_{22}l$	$a_{13}a_{24}a_{31}$	$a_{13}a_{24}a_{32}$
	d) a_{14}, a_{31}, a_{32}, a_{33} :	$a_{14}a_{21}a_{32}$	$a_{14}a_{21}a_{33}$	$a_{14}a_{22}a_{31}$
		$a_{14}a_{22}a_{33}$	$a_{14}a_{23}a_{31}$	$a_{14}a_{23}a_{32}$
$(2+3)+(3+4)$	e) a_{21}, a_{42}, a_{43}, a_{44} :	$a_{21}a_{32}a_{43}$	$a_{21}a_{32}a_{44}$	$a_{21}a_{33}a_{42}$
		$a_{21}a_{33}a_{44}$	$a_{21}la_{42}$	$a_{21}la_{43}$
	f) a_{22}, a_{41}, a_{43}, a_{44} :	$a_{22}a_{31}a_{43}$	$a_{22}a_{31}a_{44}$	$a_{22}a_{33}a_{41}$
		$a_{22}a_{33}a_{44}$	$a_{22}la_{41}$	$a_{22}la_{43}$
	g) g, a_{41}, a_{42}, a_{44} :	$a_{23}a_{31}a_{42}$	$a_{23}a_{31}a_{44}$	$a_{23}a_{32}a_{41}$
		$a_{23}a_{32}a_{44}$	$a_{23}la_{41}$	$a_{23}la_{42}$
	h) a_{24}, a_{41}, a_{42}, a_{43} :	$a_{24}a_{31}a_{42}$	$a_{24}a_{31}a_{43}$	$a_{24}a_{32}a_{41}$
		$a_{24}a_{32}a_{43}$	$a_{24}a_{33}a_{41}$	$a_{24}a_{33}a_{42}$

Step 2(ii): Amplify

Step 2(iii): Dilute and denature

TABLE 4. The products of Algorithm A_1 during the second iteration

For $(1+2+3)+(2+3+4)$, the Products are:

a)	$a_{11}a_{22}a_{33}a_{41}$	$a_{11}a_{22}a_{33}a_{44}$	$a_{11}a_{22}a_{34}a_{41}$	$a_{11}a_{22}a_{34}a_{43}$	$a_{11}a_{23}a_{32}a_{41}$	$a_{11}a_{23}a_{32}a_{44}$
	$a_{11}a_{23}a_{34}a_{41}$	$a_{11}a_{23}a_{34}a_{42}$	$a_{11}a_{24}a_{32}a_{41}$	$a_{11}a_{24}a_{32}a_{43}$	$a_{11}a_{24}a_{33}a_{41}$	$a_{11}a_{24}a_{33}a_{42}$
b)	$a_{12}a_{21}a_{33}a_{42}$	$a_{12}a_{21}a_{33}a_{44}$	$a_{12}a_{21}a_{34}a_{42}$	$a_{12}a_{21}a_{34}a_{43}$	$a_{12}a_{23}a_{31}a_{42}$	$a_{12}a_{23}a_{31}a_{44}$
	$a_{12}a_{23}a_{34}a_{41}$	$a_{12}a_{23}a_{34}a_{42}$	$a_{12}a_{24}a_{31}a_{42}$	$a_{12}a_{24}a_{31}a_{43}$	$a_{12}a_{24}a_{33}a_{41}$	$a_{12}a_{24}a_{33}a_{42}$
c)	$a_{13}a_{21}a_{32}a_{43}$	$a_{13}a_{21}a_{32}a_{44}$	$a_{13}a_{21}a_{34}a_{42}$	$a_{13}a_{21}a_{34}a_{43}$	$a_{13}a_{22}a_{31}a_{43}$	$a_{13}a_{22}a_{31}a_{44}$
	$a_{13}a_{22}a_{34}a_{41}$	$a_{13}a_{22}a_{34}a_{43}$	$a_{13}a_{24}a_{31}a_{42}$	$a_{13}a_{24}a_{31}a_{43}$	$a_{13}a_{24}a_{32}a_{41}$	$a_{13}a_{24}a_{32}a_{43}$
d)	$a_{14}a_{21}a_{32}a_{43}$	$a_{14}a_{21}a_{32}a_{44}$	$a_{14}a_{21}a_{33}a_{42}$	$a_{14}a_{21}a_{33}a_{44}$	$a_{14}a_{22}a_{31}a_{43}$	$a_{14}a_{22}a_{31}a_{44}$
	$a_{14}a_{22}a_{33}a_{41}$	$a_{14}a_{22}a_{33}a_{44}$	$a_{14}a_{23}a_{31}a_{42}$	$a_{14}a_{23}a_{31}a_{44}$	$a_{14}a_{23}a_{32}a_{41}$	$a_{14}a_{23}a_{32}a_{44}$

Step 3(i): Anneal $(1+2+3)$ to $(2+3+4)$, and fill in

For $(1+2+3)+(2+3+4)$ Primers	Products		
a) a_{11}, a_{42}, a_{43}, a_{44} :	$a_{11}a_{22}a_{33}a_{44}$	$a_{11}a_{22}a_{34}a_{43}$	$a_{11}a_{23}a_{32}a_{44}$
	$a_{11}a_{23}a_{34}a_{42}$	$a_{11}a_{24}a_{32}a_{43}$	$a_{11}a_{24}a_{33}a_{42}$
b) a_{12}, a_{41}, a_{43}, a_{44} :	$a_{12}a_{21}a_{33}a_{44}$	$a_{12}a_{21}a_{34}a_{43}$	$a_{12}a_{23}a_{31}a_{44}$
	$a_{12}a_{23}a_{34}a_{41}$	$a_{12}a_{24}a_{31}a_{43}$	$a_{12}a_{24}a_{33}a_{41}$
c) a_{13}, a_{41}, a_{42}, a_{44} :	$a_{13}a_{21}a_{32}a_{44}$	$a_{13}a_{21}a_{34}a_{42}$	$a_{13}a_{22}a_{31}a_{44}$
	$a_{13}a_{22}a_{34}a_{41}$	$a_{13}a_{24}a_{31}a_{42}$	$a_{13}a_{24}a_{32}a_{41}$
d) a_{14}, a_{41}, a_{42}, a_{43} :	$a_{14}a_{21}a_{32}a_{43}$	$a_{14}a_{21}a_{33}a_{42}$	$a_{14}a_{22}a_{31}a_{43}$
	$a_{14}a_{22}a_{33}a_{41}$	$a_{14}a_{23}a_{31}a_{42}$	$a_{14}a_{23}a_{32}a_{41}$

Step 3(ii): Amplify

TABLE 5. The products of Algorithm A_1 during the third and final iteration. The four final wells (a_{11}, a_{12}, a_{13}, a_{14}) form the Solution Row.

pmole of template for the average length OS. Thus, we can amplify 10^9 different OS (at a starting concentration of 10 aM each) in a single well. This can easily contain a 12×12 matrix with no zero elements, or much larger sparse matrices. Many of the PCR limits discussed here have not been rigorously tested, so the number of oligomers that can be amplified in a single well may be significantly higher.

4.4. Reading the answer. The process of accessing and reading the memory is based on the facts that 1) the results are linear arrays of element values and annealing regions of known sequence and 2) no information is lost in the process of assembling the OS, *i.e.*, information concerning the position and value of the elements of the determinant is maintained in the solutions.

Translating the OS into computer readable output involves subjecting each well of the solution row to a series of nested PCR reactions. A positive readout (*i.e.*, detection of a PCR product) indicates the presence of the particular element and causes promotion to the next round of nested PCR. A negative readout (the absence of a PCR product) indicates a zero term and causes that branch of the nested PCR to be pruned. The amplification signal is detected using a standard multisample UV spectrophotometer to measure ethidium bromide/double stranded DNA interaction. The algorithm for reading the memory is especially efficient when

1 down: $1+2$

$(a_{11}+a_{22}, a_{23}, a_{24}; a_{12}+a_{21}, a_{23}, a_{24}; a_{13}+a_{21}, a_{22}, a_{24}; a_{14}+a_{21}, a_{22}, a_{23})$

2 down: $2+3$

$(a_{21}+a_{32}, a_{33}, a_{34}; a_{22}+a_{31}, a_{33}, a_{34}; a_{23}+a_{31}, a_{32}, a_{34}; a_{24}+a_{31}, a_{32}, a_{33})$

3 up: $3+2$

$(a_{31}+a_{22}, a_{23}, a_{24}; a_{32}+a_{21}, a_{23}, a_{24}; a_{33}+a_{21}, a_{22}, a_{24}; a_{34}+a_{21}, a_{22}, a_{23})$

4 up: $4+3$

$(a_{41}+a_{32}, a_{33}, a_{34}; a_{42}+a_{31}, a_{33}, a_{34}; a_{43}+a_{31}, a_{32}, a_{34}; a_{44}+a_{31}, a_{32}, a_{33})$

Step 1(i): Anneal the above rows: (n wells/annealing)

1 down:				
a_{11} down:	$a_{11}a_{22}, a_{11}a_{23}, a_{11}a_{24}$	a_{12} down:	$c_{12}ba_{21}, a_{12}a_{23}, a_{12}a_{24}$
a_{13} down:	$a_{13}a_{21}, a_{13}a_{22}, a_{13}a_{24}$	a_{14} down:	$c_{14}a_{21}, a_{14}a_{22}, a_{14}a_{23}$
2 down:			
a_{21} down:	$a_{21}a_{32}, a_{21}a_{33}, a_{21}a_{34}$	a_{22} down:	$a_{22}a_{31}, a_{22}a_{33}, a_{22}a_{34}$
a_{23} down:	$a_{23}a_{31}, a_{23}a_{32}, a_{23}a_{34}$	a_{24} down:	$a_{24}a_{31}, a_{24}a_{32}, a_{24}a_{33}$
3 up:			
a_{31} up:	$a_{22}a_{31}, a_{23}a_{31}, a_{24}a_{31}$	a_{32} up:	$a_{21}a_{32}, a_{23}a_{32}, a_{24}a_{32}$
a_{33} up:	$a_{21}a_{33}, a_{22}a_{33}, a_{24}a_{33}$	a_{34} up:	$a_{21}a_{34}, a_{22}a_{34}, a_{23}a_{34}$
4 up:			
a_{41} up:	$a_{32}a_{41}, a_{33}a_{41}, a_{34}a_{41}$	a_{42} up:	$a_{31}a_{42}, a_{33}a_{42}, a_{34}a_{42}$
a_{43} up:	$a_{31}a_{43}, a_{32}a_{43}, a_{34}a_{43}$	a_{44} up:	$a_{31}a_{44}, a_{32}a_{44}, a_{33}a_{44}$

Step 1(ii): Fill in (amplify - optional)

Step 1(iii): Dilute and denature.

TABLE 6. The products of the first stage of Algorithm A_2: the Hybridization Selection Method. We note that down refers to a polymer built from the first row down and up refers to a polymer built from the last row up.

the determinant is large and has zero elements. When the matrix is sparse there may be elements which are not zero, but which do not appear in any OS.

An advantage in reading the solution using DNA computation is that information about the presence or absence of terms in a solution is generated rapidly and early in the procedure. This information cannot be captured with the same facility with algorithms used in conventional computers.

The algorithm for reading the results of the solution is illustrated by the 5×5 case as follows: Recall that the first well (the a_{11} solution well) of the Solution Row contains all terms of the form $a_{11} \ldots a_{52}$, $a_{11} \ldots a_{53}$, $a_{11} \ldots a_{54}$ and $a_{11} \ldots a_{55}$; the second well in the Solution Row contains all entries of the form $a_{12} \ldots a_{51}$, $a_{12} \ldots a_{53}$, $a_{12} \ldots a_{54}$ and $a_{12} \ldots a_{55}$; etc. To identify the individual terms in each well we divide the well into subsets using PCR with nested primers

1. For the a_{11} solution well, the first step is to amplify with each possible combination of primers specific to second row and last row elements. There are only four possible second row elements (a_{22}, a_{23}, a_{24}, a_{25}) with which to

1 down + 3 up:	
a_{11} down + a_{32} up, a_{33} up, a_{34} up:	$a_{11}a_{22}a_{33}$ $a_{11}a_{22}a_{34}$ $a_{11}a_{23}a_{32}$
	$a_{11}a_{23}a_{34}$ $a_{11}a_{24}a_{32}$ $a_{11}a_{24}a_{33}$
a_{12} down + a_{31} up, a_{33} up, a_{34} up:	$a_{12}a_{21}a_{33}$ $a_{12}a_{21}a_{34}$ $a_{12}a_{23}a_{31}$
	$a_{12}a_{23}a_{34}$ $a_{12}a_{24}a_{31}$ $a_{12}a_{24}a_{33}$
a_{13} down + a_{31} up, a_{32} up, a_{34} up:	$a_{13}a_{21}a_{32}$ $a_{13}a_{21}a_{34}$ $a_{13}a_{22}a_{31}$
	$a_{13}a_{22}a_{34}$ $a_{13}a_{24}a_{31}$ $a_{13}a_{24}a_{32}$
a_{14} down + a_{31} up, a_{32} up, a_{33} up:	$a_{14}a_{21}a_{32}$ $a_{14}a_{21}a_{33}$ $a_{14}a_{22}a_{31}$
	$a_{14}a_{22}a_{33}$ $a_{14}a_{23}a_{31}$ $a_{14}a_{23}a_{32}$

4 up + 2 down:	
a_{41} up + a_{22} down, a_{23} down, a_{24} down:	$a_{23}a_{32}a_{41}$ $a_{24}a_{32}a_{41}$ $a_{22}a_{33}a_{41}$
	$a_{24}a_{33}a_{41}$ $a_{22}a_{34}a_{41}$ $a_{23}a_{34}a_{41}$
a_{42} up + a_{22} down, a_{23} down, a_{24} down:	$a_{23}a_{31}a_{42}$ $a_{24}a_{31}a_{42}$ $a_{21}a_{33}a_{42}$
	$a_{24}a_{33}a_{42}$ $a_{21}a_{34}a_{42}$ $a_{23}a_{34}a_{42}$
a_{43} up + a_{22} down, a_{23} down, a_{24} down:	$a_{22}a_{31}a_{43}$ $a_{24}a_{31}a_{43}$ $a_{21}a_{32}a_{43}$
	$a_{24}a_{32}a_{43}$ $a_{21}a_{34}a_{43}$ $a_{22}a_{34}a_{43}$
a_{44} up + a_{22} down, a_{23} down, a_{24} down:	$a_{22}a_{31}a_{44}$ $a_{23}a_{31}a_{44}$ $a_{21}a_{32}a_{44}$
	$a_{23}a_{32}a_{44}$ $a_{21}a_{33}a_{44}$ $a_{22}a_{33}a_{44}$

Step 2(i)-(ii): Mix 1 down + 3 up, 2 down + 4 up, Anneal, fill in and amplify:

Step 2(iii): Dilute and denature

TABLE 7. The products of Algorithm A_2 during the second iteration.

1 down + 4 up:	
a_{11} down + a_{42} up, a_{43} up, a_{44} up:	$a_{11}a_{22}a_{33}a_{44}$ $a_{11}a_{22}a_{34}a_{43}$ $a_{11}a_{23}a_{32}a_{44}$
	$a_{11}a_{23}a_{34}a_{42}$ $a_{11}a_{24}a_{32}a_{43}$ $a_{11}a_{24}a_{33}a_{42}$
a_{12} down + a_{41} up, a_{43} up, a_{44} up:	$a_{12}a_{21}a_{33}a_{44}$ $a_{12}a_{21}a_{34}a_{43}$ $a_{12}a_{23}a_{31}a_{44}$
	$a_{12}a_{23}a_{34}a_{41}$ $a_{12}a_{24}a_{31}a_{43}$ $a_{12}a_{24}a_{33}a_{41}$
a_{13} down + a_{41} up, a_{42} up, a_{44} up:	$a_{13}a_{21}a_{32}a_{44}$ $a_{13}a_{21}a_{34}a_{42}$ $a_{13}a_{22}a_{31}a_{44}$
	$a_{13}a_{22}a_{34}a_{41}$ $a_{13}a_{24}a_{31}a_{42}$ $a_{13}a_{24}a_{32}a_{41}$
a_{14} down + a_{41} up, a_{42} up, a_{43} up:	$a_{14}a_{21}a_{32}a_{43}$ $a_{14}a_{21}a_{33}a_{42}$ $a_{14}a_{22}a_{31}a_{43}$
	$a_{14}a_{22}a_{33}a_{41}$ $a_{14}a_{23}a_{31}a_{42}$ $a_{14}a_{23}a_{32}a_{41}$

TABLE 8. The products of A_2 during steps *(i)-(ii)* of the third iteration. We mix 1 down with 4 up, anneal, fill in, and amplify. The four final wells (a_{11}, a_{12}, a_{13}, a_{14}) form the the "solution row."

amplify. Similarly, for each second row element there remain three possible fifth (last) row elements (a_{52}, a_{53}, a_{54}, or a_{55}), making the total number of PCR reactions for this step (including all of the other solution row wells) equal to $5 \cdot 4 \cdot 3$, or $n(n-1)(n-2)$. Thus after one step we have 60 wells. The OS in each well have the same first, second and last terms, and the remaining terms are yet to be identified.

2. To complete the identification of the OS, we subdivide and amplify with each remaining combination of third and fourth row primers. Thus, if DNA from the $a_{11}a_{22}\ldots a_{53}$ well amplifies with primers a_{35} and a_{44}, the original a_{11} solution well must have contained an oligomer with the identity $a_{11}a_{22}a_{35}a_{44}a_{53}$. In this step there are two possible third row primers for each well, and only one possible fourth row primer for each third row primer. Thus the number of PCR reactions required for this step is: $5\cdot 4\cdot 3\cdot 2\cdot 1$, or $n(n-1)(n-2)(n-3)(n-4)$. The total number of PCR reactions necessary to identify all OS is: $(5)(4)(3)+(5)(4)(3)(2)(1)=180$.

For a general $n\times n$ matrix, this becomes:

$$(n)(n-1)(n-2)+n(n-1)(n-2)(n-3)(n-4)+\ldots$$

$$=\frac{n!}{(n-3)!}+\frac{n!}{(n-5)!}+\cdots=\sum_{i=1}^{[(n-1)/2]}\frac{n!}{(n-1-2i)}$$

$$\approx n!sinh(1)$$

The number of PCR reactions required to read the solution of ESD0 is a function of the number and positions of the zeros in the determinant. For example, if the a_{44} element had the value 0, the only products we would get from the $a_{11}a_{22}..a_{53}$ well would be from the a_{34} and a_{45} primer pair, indicating that $a_{11}a_{22}a_{34}a_{45}a_{53}$ was the only term containing a_{11}, a_{22}, and a_{53}. When a PCR reaction generates no products, there are no OS containing that combination of elements, and no further analysis is necessary. The computational advantage of this approach over a standard computer is the reduction of the number of PCR reactions (calculations) necessary to solve the ESD0 problem.

Readout Example: Consider the example with the 10×10 matrix in Table 9 with 2 elements in each row having a non-zero value. Having worked out the solution to this determinant in advance we now give the answer to whether or not a PCR product is generated. In the DNA computer, the result of the ethidium bromide based detection system provides the answer.

Step 1: Amplify a_{14} and a_{15} wells with all possible second and tenth row primers. The results are worked out in Table 10. We note that the pair of

$$\begin{pmatrix}
0 & 0 & 0 & a_{14} & a_{15} & 0 & 0 & 0 & 0 & 0\\
0 & 0 & 0 & 0 & a_{25} & a_{26} & 0 & 0 & 0 & 0\\
a_{31} & 0 & 0 & a_{34} & 0 & 0 & 0 & 0 & 0 & 0\\
0 & a_{42} & a_{43} & 0 & 0 & 0 & 0 & 0 & 0 & 0\\
0 & 0 & a_{53} & 0 & 0 & 0 & 0 & a_{58} & 0 & 0\\
0 & 0 & 0 & a_{64} & 0 & 0 & a_{67} & 0 & 0 & 0\\
0 & 0 & 0 & 0 & 0 & 0 & 0 & a_{78} & a_{79} & 0\\
a_{81} & 0 & 0 & 0 & 0 & a_{86} & 0 & 0 & 0 & 0\\
0 & a_{92} & 0 & 0 & 0 & 0 & a_{97} & 0 & 0 & 0\\
0 & 0 & 0 & 0 & 0 & 0 & 0 & 0 & a_{10,9} & a_{10,10}
\end{pmatrix}$$

TABLE 9. A 10×10 matrix

DNA	primers	PCR product?	Result
a_{14}	$a_{25} + a_{10,9}$	no	
	$a_{25} + a_{10,10}$	yes	$a_{14}a_{25} \dots a_{10,10}$
	$a_{26} + a_{10,9}$	no	
	$a_{26} + a_{10,10}$	no	
a_{15}	$a_{26} + a_{10,9}$	no	
	$a_{26} + a_{10,10}$	yes	$a_{15}a_{26} \dots a_{10,10}$

TABLE 10. Amplification of a_{14} and a_{15} wells with all possible second and tenth row primers

DNA	primers	PCR product?	Result
$a_{14}a_{25}a_{10,10}$	$a_{3,1} + a_{9,2}$	yes	$a_{14}a_{25}a_{31} \dots a_{92}a_{10,10}$
	$a_{31} + a_{97}$	no	
$a_{15}a_{26}a_{10,10}$	$a_{31} + a_{92}$	no	
	$a_{31} + a_{97}$	no	
	$a_{34} + a_{92}$	yes	$a_{15}a_{26}a_{34} \dots a_{92}a_{10,10}$
	$a_{34} + a_{97}$	no	

TABLE 11. Amplification of products of Table 10 with all possible third and ninth row primers

primers a_{15} and a_{25} are omitted because they are in the same column and can not appear in the solution, an easily programmed and automated step.

Step 2: Amplify products of first step with all possible third and ninth row primers. The results are given in Table 11.

At this point, this particular problem can be finished by inspection or by handing off the remainder of the calculation to a standard architecture computer. For each product there is only one possible permutation of the remaining values –

$$a_{14}a_{25}a_{31}a_{43}a_{58}a_{67}a_{79}a_{86}a_{92}a_{10,10}$$

and

$$a_{15}a_{26}a_{34}a_{43}a_{58}a_{67}a_{79}a_{81}a_{92}a_{10,10}.$$

In each step there were PCR reactions that gave no product, even though the primers used corresponded to non-zero elements. In these cases it was quickly learned that there were no permutations that included the elements tested without reaching a zero element. In practice the hand-off point will be different for different matrices. In this sense the DNA computer functions as a coprocessor.

5. A special case for small n.

In the process of generating the OS to a full $n \times n$ determinant, we generate all possible solutions. In certain situations, *i.e.*, for small n (*e.g.* $n < 15$), once this information is stored as a solution of double stranded DNA, a specific zero pattern can be read into the computer and the terms containing zero elements can be removed from the solution. The remainder of the solution can then be read as described above. Terms that contain zero elements can be removed from the solution by designing a unique restriction site into each element value region. The biological step corresponding to inputting the zero pattern is adding the appropriate restriction enzyme(s) to the solution mix, eliminating particular oligomers from participating in the amplification processes that comprise the reading function. This method obviates separation steps based on DNA affinity chromatography and gel electrophoresis proposed in other DNA algorithms. Alternatively, inputting the zero pattern can correspond to not entering a particular PCR primer during the read mode of the computation.

The advantage to this approach is that OS are built only once instead of building the custom OS for each zero pattern. However, even if a reading method more sensitive than PCR (one that could detect single molecules) were available, this procedure would reach a limit when n is approximately 22. At this size, the volume of solution required to contain the OS is prohibitively large (see Figure 2).

6. The case where n is large.

The maximum OS generated by algorithms A_1 and A_2 is limited by the maximum length of DNA amplifiable by PCR. Long PCR [**4**] enables direct amplification of up to approximately 10 kb which corresponds to matrices of $n \approx 250$. For matrices that generate OS longer than this, the solutions may be built in modular units and annealed in a fashion similar to A_2. To read these large matrices, an algorithm consisting of overlapping nested PCR can be used.

7. Building the Coprocessor

We have designed the algorithms described here to adhere to the following constraints, 1) each step can be automated to run on standard robotic piping systems, 2) the solution is complete and accurate, 3) the approach is robust and 4) it is adaptable to new technologies of DNA detection. In fact, PCR reactions like those described in the section above have been carried out in the course of other work in our lab and many others.

The slowest processes in expanding or reading the determinant are the PCR reactions. A custom robot [**9**] can perform 6144 simultaneous PCR reactions, which, we call a single PCR run. In expanding the determinant, there are $n-1$ steps, each one consisting of at most $n(n-1)$ PCR reactions. The number of PCR reactions at any step for matrices up to $n = 78$ will be less than 6144 PCR reactions, so each step will require only a single PCR run. Therefore, the time required to generate OS will be $(n-1)p$, where p is approximately 2 hours, the time required for a single PCR run.

Read time depends on both the size of the matrix and the number of OS. For matrices with many zeros, there will be few terms to amplify in each step and the rate-limiting factor is the number of steps. The number of steps required to read the determinant is $n/2$, so that the total read time is $(n/2)p$. However, the number of PCR reactions for the final step can be as large as $n!$. In this case the read time for each step is $Tp/(6144R)$, where T is the total number of PCR reactions in each step and R is the number of robots used. For a matrix with 3.6×10^6 terms , the final step will involve 3.6×10^6 PCR reactions, or roughly 591 PCR runs (50 machine-days) on a single robot. By increasing PCR throughput, either by running more than one robot or designing a robot with larger capacity, the

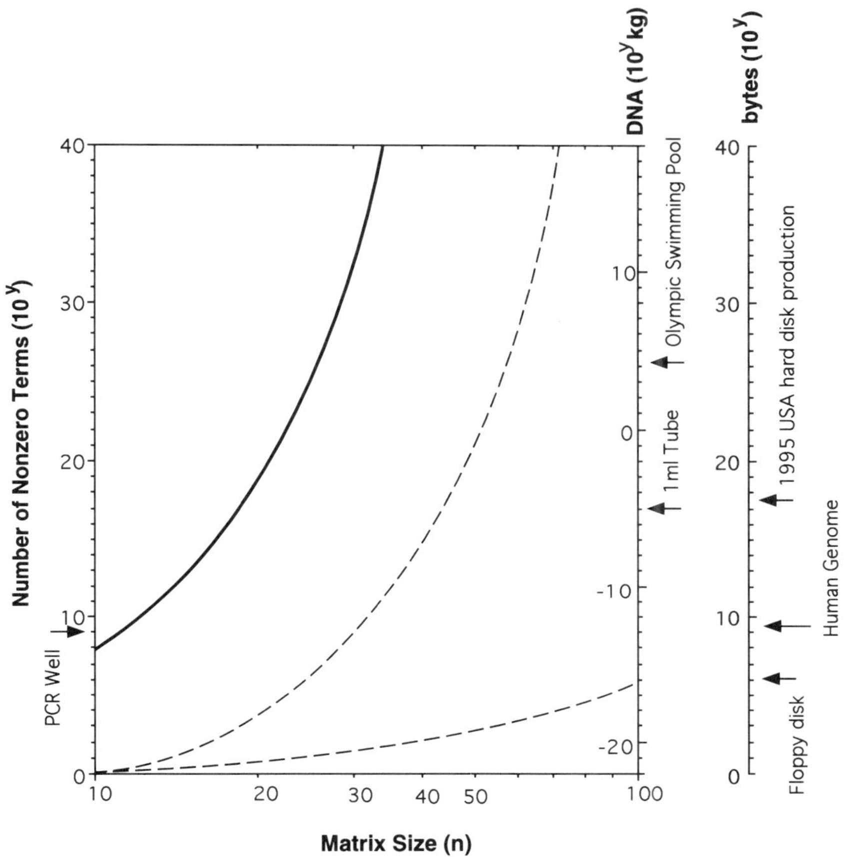

FIGURE 2. Applicability of DNA computation for solving matrices of various sizes. The number of non-zero terms in the determinant of an $n \times n$ matrix are shown for both a matrix with no zeros (solid line) and the upper and lower bounds for a matrix with 90% zeros (dashed lines). The vertical axis on the right shows the space complexity of the number of non-zero terms. Storage space required is shown in both kilograms of DNA and in bytes. Markers on the DNA axis correspond to a test tube and a swimming pool of DNA at 10 mg/ml, where each term is represented by a single molecule. The PCR well marker represents the maximum number of non-zero terms that can be amplified in a single PCR well.

number of PCR runs required can be reduced. For example, with 10 machines in the current configuration, the total time for the last step of a full 10×10 matrix is reduced to 60 runs, or roughly 5 machine-days. A very large capacity PCR robot [**9**] has been constructed which can perform over $145,000$ PCR reactions at once, and could complete the final step in the expansion of a 10×10 determinant with no zeros in 25 runs or approximately 2-3 days.

The architecture of DNA has certain properties that we have not yet exploited in solving ESD0. For example, eventually we may use certain techniques developed for single photon counting [**5**] and fluorescence correlation spectroscopy [**11**] to separate and identify very large numbers of molecules on very fast time scales.

Ultimately, we imagine that an efficient computer could combine bio-molecular and conventional architecture, instead of working entirely in one domain. For example, in processing large matrices there will be a point at which it may be more efficient to institute a handoff from the DNA computer to a conventional computer to complete the solution to the problem. The key is to provide the conventional computer information it cannot obtain efficiently.

8. Conclusion

We have shown that DNA computation of expanded symbolic determinants given their zero pattern is both practical and useful. The limits of DNA computation technology lie within the bounds of our ability to capture the mathematical structures of certain problems in the molecular structures of the biomolecules and in creating algorithms that manipulate these relationships.

References

[1] L. M. Adleman, Molecular Computation of Solutions to Combinatorial Problems. *Science* 266: 1021-1024 (1994).

[2] M. Becker-André, and K. Hahlbrock, *Nuc. Acid. Res.* 17:9437 (1989).

[3] R.A. Brualdi and H.J. Ryser, *Combinatorial Matrix Theory.* Encyclopedia of Mathematics and Its Applications, **39** Cambridge Univ. Press. Cambridge. (1991).

[4] S. Cheng, C. Fockler, W. M. Barnes, R. Higuchi, Effective Amplification of Long Targets from Cloned Inserts and Human Genomic DNA. *Proc. Natl. Acad. Sci.* 91: 5695-5699 (1994).

[5] M. Eigen and R. Rigler, Sorting Single Molecules – Application to Diagnostics and Evolutionary Biotechnology. *Proc. Natl. Acad. Sci.* 91: 5740-5747 (1994).

[6] M.R. Garey and D.S. Johnson, *Computers and Intractability: A Guide to the Theory of NP-Completeness.* W.H.Freeman, San Francisco. (1979).

[7] W.M. Gentleman and S.C. Johnson, *ACM Transactions of Mathematical Software*, 2:232 (1976).

[8] G. Gilliland, S. Perrin, K. Blanchard, H. F. Bunn, Analysis of Cytokine Messenger-RNA and DNA-Detection and Quantitation by Competitive Polymerase Chain-Reaction. Proc. Natl. Acad. Sci. 87:2725-2729 (1990).

[9] T. J. Hudson L. D. Stein, S. S. Gerety, et. al., An STS-Based Map of the Human Genome. *Science* 270:1945-1954 (1995).

[10] R.J. Lipton, DNA Solution of Hard Computational Problems, *Science* 268: 542-545 (1995).

[11] A. C. Pease, D. Solas, E.J. Sullivan, M.T. Cronin, C.P. Holmes, S.P.A Fodor, Light Generated Oligonucleotide Arrays For Rapid DNA-Sequence, *Proc. Natl. Acad. Sci.* 91:5022-5026 (1994).

[12] Z. Zlatev, *Computational Methods for General Sparse Matrices.* Mathematics and Its Applications. **65** Kluwer Academic Publishers, Dordrecht The Netherlands. (1991).

(T. Leete, D. Schwartz, and J. Salem) Department of Medicine, University of Pennsylvania, 536 Johnson Pavilion, Philadelphia, PA 19104

(R. Williams) Naval Air Systems Team, Warminster, PA 18974

(D. Wood) Department of Computer and Information Sciences, University of Delaware, Newark, DE 19716

(H. Rubin) Department of Medicine, University of Pennsylvania, 536 Johnson Pavilion, Philadelphia, PA 19104

E-mail address: rubinh@mail.med.upenn.edu, (215)-662-6475, Fax:(215)-662-7842

DIMACS Series in Discrete Mathematics
and Theoretical Computer Science
Volume **44**, 1999

Universal DNA Computing Models Based on the Splicing Operation

Gheorghe Păun

Abstract. We briefly present several types of language generating mechanisms based on the *splicing operation* (a model of the recombinant behavior of DNA sequences under the influence of restriction enzymes and ligases). All these models characterize the recursively enumerable languages, hence are equal in power to the Turing machines. On the basis of the constructions in the proofs of this assertion, one can obtain *universal* (hence programmable) computing devices.

1. Introduction

This paper is mathematically (automata and language theory) oriented, it deals with *computing models* rather than with *computers*. Therefore, we do not recall here the abstraction process leading from the actual DNA recombination to the formal operation of *splicing*. The reader is referred to [**16, 18**] for details. We only mention that the *site* where a restriction enzyme can cut a DNA sequence is encoded as a context (pair of words) (u, v). When two enzymes produce DNA fragments with matching ends we encode this as a quadruple $(u_1, u_2; u_3, u_4)$: (u_1, u_2) defines the place where the first enzyme cuts, (u_3, u_4) defines the place where the second enzyme cuts, and the fact that the two pairs are related means that the fragments match and a ligase can recombine them, producing new sequences. A quadruple $(u_1, u_2; u_3, u_4)$ is named, following [**16**], a *splicing rule* (in the rest of the paper, we will write the splicing rules as strings, in the style of [**24**]).

We consider here the splicing operation at an abstract level, working with *strings* (not with double stranded helicoidal DNA sequences), consisting of *abstract symbols* (not necessarily the four nucleotides A, C, G, T), whereas the *length of sites* is not bounded in advance. The first two assumptions are not essential (from a practical point of view): due to the precise Watson-Crick complementarity we can work with single stranded sequences and our abstract symbols can be codified using the four DNA letters. In constructions below, the length of sites is kept as small as possible. However, other features of our models seem quite liberal, at least according to the present day possibilities of handling the DNA. On the positive side, because the splicing is an abstract operation, it might be implemented on other chemical structures, not necessarily on DNA.

We present several generative devices based on the splicing operation, all of them characterizing the family of recursively enumerable (RE) languages, hence equal in power to Turing machines. When introducing each device we shall also

Research supported by the Academy of Finland, project 11281.

informally present the underlying ideas of the construction. Suggestions from the regulated rewriting area (checking the presence/absence of certain symbols in the processed strings) and from grammar systems theory (distributed architectures) are followed, as well as multisets (sets with multiplicities associated to their elements).

This variety of architectures is the major contribution this paper has to offer "DNA computer scientists," who, hopefully, can find here suggestions for designing and implementing universal DNA computers.

A key feature of our models is the fact that (at the theoretical level, of course) they work "automatically," without any external control. This is in a definite contrast with most of the DNA simulations of Turing machines already presented in the field literature, where the models work ... in the presence of their authors, who control the sequence of the involved operations. See a typical example in [**32**].

2. The splicing operation

Consider an alphabet V and two special symbols, $\#, \$$, not in V. A *splicing rule* (over V) is a string of the form $r = u_1\#u_2\$u_3\#u_4$, where $u_i \in V^*, 1 \leq i \leq 4$ (V^* is the free monoid generated by V under the operation of concatenation). For such a rule r and strings $x, y, z \in V^*$ we write

$$(x, y) \vdash_r z \quad \text{iff} \quad x = x_1u_1u_2x_2,\ y = y_1u_3u_4y_2,\ z = x_1u_1u_4y_2,$$
$$\text{for some } x_1, x_2, y_1, y_2 \in V^*.$$

An *H scheme* is a pair $\sigma = (V, R)$, where V is an alphabet and

$$R \subseteq V^*\#V^*\$V^*\#V^*$$

is a set of splicing rules.

For a given H scheme $\sigma = (V, R)$ and a language $L \subseteq V^*$, we define

$$\sigma(L) = \{z \in V^* \mid (x, y) \vdash_r z, \text{ for some } x, y \in L, r \in R\},$$
$$\sigma^0(L) = L,\ \sigma^{i+1}(L) = \sigma^i(L) \cup \sigma(\sigma^i(L)), i \geq 0,$$
$$\sigma^*(L) = \bigcup_{i \geq 0} \sigma^i(L).$$

Thus, $\sigma^*(L)$ is the closure of L under the splicing operation, that is the set of strings which can be obtained by starting from the strings of L and applying an arbitrary number of splicing operations.

We denote by *FIN, REG, CF, RE* the families of finite, regular (recognized by finite automata), context-free, and recursively enumerable languages (we use here as a characterizing class of devices the Chomsky type-0 grammars, known to be equivalent to Turing machines). Moreover, for two families of languages, F_1, F_2, we denote by $H(F_1, F_2)$ the set of all languages $\sigma^*(L)$, for $L \in F_1$ and $\sigma = (V, R), R \in F_2$ (because R is a set of strings, hence a language, the definition makes sense).

Lemma 1. ([**8, 30**]) $H(REG, FIN) \subseteq REG$.

Lemma 2. ([**24**]) *Let F be a family of languages closed under intersection with regular sets and restricted morphisms. For every $L \subseteq V^*, L \notin F$, and $c, d \notin V$, we have $L' \notin H(F, RE)$, for $L' = (dc)^*L(dc)^* \cup c(dc)^*L(dc)^*d$.*

Corollary. $REG - H(FIN, RE) \neq \emptyset$.

3. Extended H systems; the Basic Universality Theorem

The results above show that the power of splicing schemes has serious limitations: they cannot directly provide computationally complete mechanisms. This goal can be achieved in the following way.

An *extended H system* is a quadruple $\gamma = (V, T, A, R)$, where V is an alphabet, $T \subseteq V$ (terminal alphabet), $A \subseteq V^*$ (axioms), and $R \subseteq V^*\#V^*\$V^*\#V^*$, where $\#, \$$ are special symbols not in V; $\sigma = (V, R)$ is the *underlying H scheme* of γ.

The *language generated* by γ is defined by $L(\gamma) = \sigma^*(A) \cap T^*$. For two families of languages, F_1, F_2, we denote by $EH(F_1, F_2)$ the family of languages $L(\gamma)$ generated by extended H systems $\gamma = (V, T, A, R)$, with $A \in F_1, R \in F_2$.

Note that the intersection with T^* in the definition of $L(\gamma)$ can be practically implemented: *extract* from the tube all strings containing at least a symbol not in T (the procedure is used/described in many places, e.g., in [**2**]).

Lemma 3. $EH(FIN, FIN) = REG$.

The inclusion $\subseteq$ follows from Lemma 1 and the closure of REG under intersection; the opposite inclusion is proved in [**28**], where the extended H systems were introduced.

Lemma 4. (*The Basic Universality Theorem*; [**25**]) $RE \subseteq EH(FIN, REG)$.

Although the result is not "applicable" (it is not realistic to deal with either infinite or even regular sets of splicing rules), it is very useful. On the basis of the construction in its proof, suitably modified, a series of devices were proposed that have finitely many rules, with the application regulated by certain additional mechanisms. We will discuss some such variants in the subsequent sections; to this aim, we present here the core of the proof of the Basic Universality Theorem.

Consider a type-0 grammar $G = (N, T, S, P)$ and construct the extended H system $\gamma = (V, T, A, R)$, where

$$\begin{aligned} V &= N \cup T \cup \{X, X', B, Y, Z\} \cup \{Y_\alpha \mid \alpha \in N \cup T \cup \{B\}\}, \\ A &= \{XBSY,\ ZY, XZ\} \cup \{ZvY \mid u \to v \in P\} \\ &\cup\ \{ZY_\alpha,\ X'\alpha Z \mid \alpha \in N \cup T \cup \{B\}\}, \end{aligned}$$

and R contains the following groups of rules:

Simulating rules:

1. $Xw\#uY\$Z\#vY$, for $u \to v \in P, w \in (N \cup T \cup \{B\})^*$;

Rotating rules:

2. $Xw\#\alpha Y\$Z\#Y_\alpha$, for $\alpha \in N \cup T \cup \{B\}, w \in (N \cup T \cup \{B\})^*$,
3. $X'\alpha\#Z\$X\#wY_\alpha$, for $\alpha \in N \cup T \cup \{B\}, w \in (N \cup T \cup \{B\})^*$,
4. $X'w\#Y_\alpha\$Z\#Y$, for $\alpha \in N \cup T \cup \{B\}, w \in (N \cup T \cup \{B\})^*$,
5. $X\#Z\$X'\#wY$, for $w \in (N \cup T \cup \{B\})^*$;

Terminating rules:

6. $\#ZY\$XB\#wY$, for $w \in T^*$,
7. $\#Y\$XZ\#$.

The idea behind this construction is the following one. We start from $XBSY$. By the simulation rules we simulate the rewriting rules of P on the rightmost position of the current string: $XwuY$ leads to $XwvY$, for $u \to v \in P$, by the operation $(Xw|uY,\ Z|vY) \vdash_1 XwvY$ (we have indicated by a vertical bar the place of the splicing). In order to make possible the simulation of rules in P on any position, not necessarily on a suffix, the current string is *rotated*, using the rules in the second group: a symbol α from the right-hand end is removed by a rule of type 2 and this information is stored in the subscript α of Y_α $((Xw|\alpha Y, Z|Y_\alpha) \vdash_2 XwY_\alpha)$. Then α is introduced in the left-hand end, together with X', by a rule of type 3 $((X'\alpha|Z, X|wY_\alpha) \vdash_3 X'\alpha wY_\alpha)$; in the presence of X', a rule of type 4 can replace Y_α by Y $((X'\alpha w|Y_\alpha, Z|Y) \vdash_4 X'\alpha wY)$, and finally a rule of type 5 returns X' to X $((X|Z, X'\alpha wY) \vdash_5 X\alpha wY)$. We have started with $Xw\alpha Y$ and we have obtained $X\alpha wY$, hence the symbol α has been moved from the right-hand end to the left-hand end of the string. The markers X, Y are reproduced. The process can be iterated, hence we can obtain in this way any circular permutation of the string.

The symbol B is used to indicate at every step where the *actual left-hand end* of the string is placed. It is always present, circulates from the right-hand end to the left-hand end as any other symbol, cannot be rewritten by rules in P, and can be removed only in the presence of X, by a rule of type 6. This means both that the string is in the correct permutation and that it is terminal. Rule 7 simply removes the marker Y; note that all the other rules need the presence of Y, hence the work of γ is blocked if rule 7 is used prematurely.

From these explanations, the reader should be able to check that $L(G) = L(\gamma)$; further details can be found in [**25, 18**].

4. Some finite and computationally complete models

The language R of splicing rules in the previous proof is infinite, but of a very particular type. In fact, the part w of rule strings carries no information. Thus, following suggestions from regulated rewriting area (*random context grammars*, [**10**]), we can consider H systems with the rules used only in the presence (absence) of certain associated symbols.

An *extended H system with permitting context conditions* is a quadruple $\gamma = (V, T, A, R)$, where V, T, A are as above and R is a set of triples of the form $p = (r; C_1, C_2)$, where $r = u_1\#u_2\$u_3\#u_4$ is a splicing rule over V and $C_1, C_2 \subseteq V$.

For $x, y \in V^*$ and $p \in R$ as above, we write $(x, y) \vdash_p z$ if and only if $(x, y) \vdash_r z$, each symbol in C_1 appears in x and each symbol in C_2 appears in y. The language generated by γ is defined in the natural way. We consider here only extended H systems with permitting context conditions having finite sets of axioms and of splicing rules; the family of languages generated by such systems is denoted by $EH(FIN, pFIN)$.

From the above mentioned proof of the Basic Universality Theorem and from the Turing-Church thesis, the reader can naturally expect the following result:

Theorem 1. $RE = EH(FIN, pFIN)$.

In fact, only triples $(r; C_1, C_2)$ with one of the sets C_1, C_2 empty and the other one consisting of only one symbol can be used. For the sake of completeness, we "translate" the construction in the proof of the Basic Universality Theorem in the new framework, specifying only the splicing rules and their context conditions (the axioms are the same as above):

Simulating rules:

$$1.\ (\#uY\$Z\#vY;\ \{X\},\emptyset),\ \text{for } u \to v \in P;$$

Rotating rules:

$$\begin{array}{lll} 2. & (\#\alpha Y\$Z\#Y_\alpha;\ \{X\},\emptyset), & \text{for } \alpha \in N \cup T \cup \{B\}, \\ 3. & (X'\alpha\#Z\$X\#;\ \emptyset,\{Y_\alpha\}), & \text{for } \alpha \in N \cup T \cup \{B\}, \\ 4. & (\#Y_\alpha\$Z\#Y;\ \{X'\},\emptyset), & \text{for } \alpha \in N \cup T \cup \{B\}, \\ 5. & (X\#Z\$X'\#;\ \emptyset,\{Y\}); & \end{array}$$

Terminating rules:

$$\begin{array}{ll} 6. & (\#ZY\$XB\#;\ \emptyset,\{Y\}), \\ 7. & (\#Y\$XZ\#;\ \emptyset,\emptyset). \end{array}$$

Because, as one can see, the set of the checked symbols is well specified,

$$\begin{aligned} &Q_1 = \{X, X'\} \text{ in the first string} \\ &Q_2 = \{Y\} \cup \{Y_\alpha \mid \alpha \in N \cup T \cup \{B\}\} \text{ in the second string,} \end{aligned}$$

we can also check the *non-appearance* of symbols in $Q_1 - C_1, Q_2 - C_2$, respectively, and the result is the same. That is, we can consider *extended H systems with forbidding context conditions*, as quadruples $\gamma = (V, T, A, R)$ with V, T, A as above and R a set of triples of the form $p = (r; D_1, D_2)$, $r = u_1\#u_2\$u_3\#u_4, D_1, D_2 \subseteq V$. For $x, y, z \in V^*$ and such a triple $p \in R$ we write $(x, y) \vdash_p z$ if $(x, y) \vdash_r z$; no element of D_1 appears in x and no element of D_2 appears in y. The language generated by γ is defined as above. We denote by $EH(FIN, fFIN)$ the family of languages generated by extended H systems with forbidding context conditions and finite sets of axioms and of splicing rules.

From the previous discussion we get

Theorem 2. $RE = EH(FIN, fFIN)$.

This time, however, we cannot obtain the same "normal form" as that given by the construction above: the forbidding contexts D_1, D_2 in each triple $(r; D_1, D_2)$ still have the property that one of D_1, D_2 is empty, but the other one cannot be taken as a singleton. Indeed, when $C_i = \emptyset$, then we can also take $D_i = \emptyset, i = 1, 2$, because no condition is to be checked, but C_i a singleton leads to a non-singleton $Q_i - C_i$.

In short, starting from a finite set of axioms and iteratively applying splicing rules from a finite set, we can characterize the family of recursively enumerable languages providing that each splicing operation is performed only in certain conditions: either (1) certain symbols ("favourizing items," "catalysts," "promotors") are present in the spliced strings, or (2) certain symbols ("inhibitors") are not present in the spliced strings (these symbols are associated with the splicing rules).

A way to implement the permitting context condition, in the form needed above, could be the following one. Suppose we have a single stranded DNA sequence, w, to which we want to apply the splicing rule r only providing that w starts with a specified symbol X. The rule r acts on the first term of the splicing by an enzyme E_1. The enzymes act on double stranded sequences only. Associate with r a primer which can be attached to w only if X is present in w (hence the primer is complementary to X). Thus, by polymerization we get a double stranded sequence only when X is present in w and only in such a case the enzyme E_1 will work. If, after splicing, we want to return to single stranded sequences, we can melt the result.

The construction in the proof of the Basic Universality Theorem can be reformulated also in terms of *H systems with targets*, in the sense of [**28**].

An *extended H system with local targets* is a construct $\gamma = (V, T, A, R)$, where V, T, A are as in a usual extended H system and R is a set of pairs $p : (r, Q_p)$, where r is a splicing rule over V and Q_p is a regular language over V. For $x, y, w \in V^*$ we write $(x, y) \vdash_p w$ iff $(x, y) \vdash_r w$ and $w \in Q_p$ (the result of the splicing with respect to r belongs to the associated language Q_p – this can be seen as a model of the Lamarckian evolution, see also [**32**]).

If for all p the target languages Q_p are equal to a given Q, then we say that γ has a global target, and we write it in the form $\gamma = (V, T, A, P, Q)$.

We denote by $EH(FIN, ltFIN), EH(FIN, gtFIN)$ the families of languages generated by systems with local, rspectively, global targets, with finite sets of axioms and of rules. We get

Theorem 3. $RE = EH(FIN, ltFIN) = EH(FIN, gtFIN)$.

It is enough to prove the relation $RE \subseteq EH(FIN, gtFIN)$ and such a proof can be obtained by a modification of the proofs above. In fact, only the *rotating* phase needs a modification; it can be realized by using the following rules (the notations are as above; all strings containing the symbol Z are axioms):

$$\begin{array}{lll} \textit{Rotating}: & 2. & \#\alpha Y\$Z\#Y_\alpha, & Q_{2,\alpha} = XU^*Y_\alpha, \\ & 3. & X'\alpha\#Z\$X\#, & Q_{3,\alpha} = X'U^*Y_\alpha, \\ & 4. & \#Y_\alpha\$Z\#Y', & Q_{4,\alpha} = X'U^*Y', \text{ for all } \alpha \in U, \\ & 5. & X''\#Z\$X'\#, & Q_5 = X''U^*Y', \\ & 6. & \#Y'\$Z\#Y, & Q_6 = X''U^*Y, \\ & 7. & X\#Z\$X''\#, & Q_7 = XU^*Y. \end{array}$$

The targets are written locally, but their union can be used as a global target.

Another possible variant is to use fitness mappings, as in genetic algorithm area (see [**15, 19**], etc) and to allow only the splicing of strings with "high enough" fitness. In our case, we can define a boolean fitness mapping, depending on the presence/absence in the strings of the checking symbols considered above. We do not enter here into details.

From the previous constructions, one see that all rotating and finishing rules have a *radius* bounded by two: for $r = u_1\#u_2\$u_3\#u_4$ we say that the radius of r is the maximum of $|u_1|, |u_2|, |u_3|, |u_4|$. Starting the constructions from a type-0 grammar G with rules in the Kuroda normal form, that is with rules of the forms $X \to u, |u| \leq 2$, and $XY \to UZ, X, Y, U, Z$ nonterminals, we get simulating rules of radius at most 3. In fact, as shown in [**27**], rules of radius two suffice. Thus, from this point of view, the constructions above are realistic: the enzymes have short recognition patterns, usually around six symbols.

Because the DNA is present in both a linear and a circular form, the splicing operation has been considered also for circular strings; see, e.g., [**17, 31, 33**]. Then in [**34**] the core idea of the proof of the Basic Universality Theorem has been used to prove that extended H systems for circular strings are computationally universal. We refer to [**34**] for details and we mention here only the fact that three types of splicing rules are used, for *simulating*, *rotating*, and *cutting* the circular words. A very interesting result of [**34**] (also present in [**12**], where Post normal systems are used as a starting characterization of RE) is the fact that characterizations of certain families intermediate between REG and RE are obtained (for instance, CF), by restricting the form of the used splicing rules.

5. The existence of "programmable" EH systems

From Theorems 1, 2, 3, via the equivalence of type-0 grammars with Turing machines, we obtain the fact that extended H systems of the types (FIN, REG), $(FIN, pFIN)$, and $(FIN, fFIN)$ have the same power as Turing machines. The result can be strengthened. Indeed, examining the previous constructions, one sees that *if we start from a universal type-0 grammar, then we get an extended H system which is universal*, too, in the following sense:

A universal type-0 grammar, for an alphabet T, is a construct

$$G_u = (N_u, T, -, P_u),$$

where N_u is a fixed nonterminal alphabet and P_u is a fixed set of rewriting rules. Given an arbitrary type-0 grammar $G = (N, T, S, P)$, it is possible to construct a string $w(G)$, called *the code* of G, such that $L(G_u(w(G))) = L(G)$, where

$$G_u(w(G)) = (N_u, T, w(G), P_u)$$

(the derivations in this grammar start from the axiom string $w(G)$). That is, yG_u can simulate G, providing the code of G is introduced in G_u as a starting string. Universal type-0 grammars can be obtained starting from universal Turing machines, or can be directly constructed as in [**3**].

It is easy to see that, starting from a universal type-0 grammar $G_u = (N_u, T, -, P_u)$, the previous constructions provide extended H systems γ_u with all components fixed (they depend on N_u, T and P_u) and without the axiom $XBSY$. Taking an arbitrary type-0 grammar $G = (N, T, S, P)$ and its code $w(G)$, and adding the axiom $XBw(G)Y$ to γ_u, we get a system $\gamma_u(w(G))$ such that $L(\gamma_u(w(G))) = L(G)$. Therefore, we obtain

Theorem 4. *For each alphabet T, there are extended H systems $\gamma_u = (V_u, T, A_u, R_u)$, of the types (FIN, REG) and $(FIN, \alpha FIN)$, $\alpha \in \{p, f, lt, gt\}$, which are universal with respect to the type-0 grammars with the terminal alphabet T.*

Note that the "program" necessary to simulate a given grammar G on such a "computer" consists of one string only, $XBw(G)Y$, added as an axiom to γ_u.

In exactly the same way we can obtain "programmable universal computers" from the proofs in the next sections, but we shall not repeat this fact any more. From a "practical" point of view, however, this is the main result of these investigations.

6. Communicating distributed H systems

We consider now some variants of H systems with an architecture suggested by the grammar systems area, see [**4**]. The basic idea of a grammar system is to put several grammars to work together, according to a specified protocol, for generating a common language. For our purposes here – the elimination of the infiniteness of the set of splicing rules in the Basic Universality Theorem – the *parallel communicating grammar systems* [**29**] architecture turned out to be useful

From now on, we work with the "symmetric" splicing operation defined by

$$\begin{aligned}(x,y)\vdash_r\ (w,z)\quad \text{iff}\quad & x = x_1u_1u_2x_2,\ y = y_1u_3u_4y_2,\\ & w = x_1u_1u_4y_2,\ z = y_1u_3u_2x_2,\\ & \text{for some } x_1,x_2,y_1,y_2 \in V^*,\end{aligned}$$

and $r = u_1\#u_2\$u_3\#u_4$. Then, for a splicing scheme $\sigma = (V,R)$ and a language $L \subseteq V^*$, we put

$$\begin{aligned}\sigma(L) = \{w \in V^* \mid & (x,y)\vdash_r (w,z) \text{ or } (x,y)\vdash_r (z,w),\\ & \text{for some } x,y \in L \text{ and } r \in R\}.\end{aligned}$$

The iterated splicing, σ^*, is defined as before. (All the results obtained in the more general case considered in the previous sections remain true also for this new version of the relation $\vdash_r$.)

A *communicating distributed* (CD) *H system* (of degree $n, n \geq 1$) is a construct $\Gamma = (V,(A_1,R_1,V_1),\ldots,(A_n,R_n,V_n))$, where V is an alphabet, $A_i \subseteq V^*$, $R_i \subseteq V^*\#V^*\$V^*\#V^*$, and $V_i \subseteq V$, for each $1 \leq i \leq n$.

Each triple (A_i,R_i,V_i) is called a *component* of the system, or a *test tube*; A_i is the set of axioms of the tube i, R_i is the set of splicing rules of the tube i, V_i is the *selector* of the tube i. We denote

$$W = V^* - \bigcup_{i=1}^{n} V_i^*.$$

The pair $\sigma_i = (V,R_i)$ is the underlying H scheme associated with the component i of the system. An n-tuple $(L_1,\ldots,L_n), L_i \subseteq V^*, 1 \leq i \leq n$, is called a *configuration* of the system; L_i is also called the *contents* of the ith tube.

For two configurations $(L_1,\ldots,L_n),(L'_1,\ldots,L'_n)$, we define

$$\begin{aligned}(L_1,\ldots,L_n) \Longrightarrow (L'_1,\ldots,L'_n)\quad \text{iff}\quad & \text{for each } i, 1 \leq i \leq n,\\ & L'_i = \bigcup_{j=1}^{n}(\sigma_j^*(L_j)\cap V_i^*)\cup(\sigma_i^*(L_i)\cap W).\end{aligned}$$

In words, the contents of each tube are spliced according to the associated set of rules (we pass from L_i to $\sigma_i^*(L_i), 1 \leq i \leq n$), and the result is redistributed among the n components according to the selectors $V_1,\ldots,V_n$; the part which cannot be redistributed (does not belong to some $V_k^*, 1 \leq k \leq n$) remains in the tube. Because we have imposed no restrictions on the alphabets V_i, for example, we did not suppose that they are pairwise disjoint. When a string in $\sigma_j^*(L_j)$ belongs to several languages V_i^*, then copies of this string will be distributed to all tubes i with this property. (In some sense, we involve here all the operations *merge, separate, amplify*, used as basic operations in the "programming language" described in [**20**, **1**].)

The *language generated* by Γ is

$$L(\Gamma) = \{w \in V^* \mid w \in L_1 \text{ for } (A_1,\ldots,A_n) \Longrightarrow^* (L_1,\ldots,L_n), t \geq 0\},$$

where $\Longrightarrow^*$ is the reflexive and transitive closure of the relation $\Longrightarrow$.

We denote by CDH_n the family of languages $L(\Gamma)$, for $\Gamma = (V, (A_1,R_1,V_1), \ldots, (A_m,R_m,V_m))$, with $m \leq n$, A_i and R_i being finite sets for each $i, 1 \leq i \leq m$. When n is not specified, we write CDH_*.

Theorem 5. $CDH_1 \subset REG \subset CDH_2 \subseteq CDH_3 \subseteq \ldots \subseteq CDH_* = RE$.

The relation $RE \subseteq CDH_*$ is the core result and the construction in its proof is also important for obtaining universal CD H systems:

Take a type-0 grammar $G = (N, T, S, P)$, take a new symbol B, assume $N \cup T = \{\alpha_1, \ldots, \alpha_n\}$, write for uniformity $B = \alpha_{n+1}$, and construct the CD H system

$$\begin{aligned}\Gamma = {} & (V, (A_1, R_1, V_1), (A_2, R_2, V_2), (A_{3,1}, R_{3,1}, V_{3,1}), \ldots, (A_{3,n+1}, R_{3,n+1}, V_{3,n+1}), \\ & (A_4, R_4, V_4), (A_5, R_5, V_5), (A_6, R_6, V_6), (A_7, R_7, V_7)),\end{aligned}$$

with

$$\begin{aligned}
V &= N \cup T \cup \{B, X, X', Y, Z, Z'\} \cup \{Y_i \mid 1 \leq i \leq n+1\}, \\
A_1 &= \emptyset, R_1 = \emptyset, V_1 = T, \\
A_2 &= \{XBSY, Z'Z\} \cup \{ZvY \mid u \to v \in P'\} \cup \{ZY_i \mid 1 \leq i \leq n+1\}, \\
R_2 &= \{\#uY\$Z\#vY \mid u \to v \in P'\} \cup \{\#\alpha_i Y\$Z\#Y_i \mid 1 \leq i \leq n+1\} \\
&\cup \{Z'\#Z\$XB\#\}, \\
V_2 &= T \cup \{B, X, Y\}, \\
A_{3,i} &= \{X'\alpha_i Z\}, R_{3,i} = \{X'\alpha_i\#\$X\#\}, \\
V_{3,i} &= T \cup \{B, X, Y_i\}, \\
&\qquad \text{for } 1 \leq i \leq n+1, \\
A_4 &= \{ZY\}, R_4 = \{\#Y_i\$Z\#Y \mid 1 \leq i \leq n+1\}, \\
V_4 &= T \cup \{B, X'\} \cup \{Y_i \mid 1 \leq i \leq n+1\}, \\
A_5 &= \{XZ\}, R_5 = \{X\#Z\$X'\#\}, V_5 = T \cup \{B, X', Y\}, \\
A_6 &= \{ZZ\}, R_6 = \{\#Y\$ZZ\#\}, V_6 = T \cup \{Y, Z'\}, \\
A_7 &= \{ZZ\}, R_7 = \{\#ZZ\$Z'\#\}, V_7 = T \cup \{Z'\}.
\end{aligned}$$

The reader may check that $L(\Gamma) = L(G)$, or can find details in [**6**]. The idea is the same as in the proof of the Basic Universality Theorem. In short, the first component selects the terminal strings, the second one simulates rules in P' on the right-hand end of the current string, and also starts "rewinding" the string: the rightmost symbol of the string, α_i, is removed, the marker Y_i memorizes that, then tube $(3.i)$ introduces α_i in the left-hand end of the string. Iterating this procedure, the string of G is circularly permuted, making possible the simulation of any rule in P' on the right-hand end of a string produced in Γ. The actual beginning of the string is marked with B, and only when this symbol is in the leftmost position, can P_1 start finishing the derivation; the last two tubes remove the auxiliary symbols Y, Z'. If the string is terminal, it can be communicated to the first tube and accepted.

An attractive feature of the previous model is the fact that we can modify it in such a way to have only one splicing rule in each tube (each enzyme works in well specified conditions, hence it is not realistic to assume that several enzymes can work together in an efficient way). Of course, the main difficulty here is the communication of the strings from a tube to another one.

7. Synchronized distributed H systems

The H systems in Sections 4 and 6 are somewhat hybrid, they mix splicing with language-theoretic operations, such as checking permitting or forbidding contexts.

In search for a *homogeneous* model, the *synchronized distributed* H systems were proposed in [**26**].

A *synchronized distributed* (SD) *H system* (of degree $n, n \geq 1$), is a construct $\Gamma = (V, T, C_1, \ldots, C_n)$, where V is an alphabet, $T \subseteq V$, and $C_i = (w_i, A_i, I_i, E_i)$, $1 \leq i \leq n$, where $w_i \in V^*$, $A_i \subseteq V^*$, I_i, $E_i \subseteq V^*\#V^*\$V^*\#V^*$, for $\#$, $\$$ symbols not in V.

The meaning of these elements is as follows: V is the *total alphabet*, T is the *terminal alphabet*, and $C_i, 1 \leq i \leq n$, are the *components* of the system (we also call them *test tubes*); for each component, w_i is the *active axiom*, A_i is the set of *passive axioms*, I_i is the set of *internal splicing rules*, and E_i is the set of *external splicing rules*.

The *contents* of a tube $C_i, 1 \leq i \leq n$, are described by a pair (x_i, M_i), where $x_i \in V^*$ is the *active* string and $M_i \subseteq V^*$ is the set of *passive* strings. An n-tuple $\pi = [(x_1, M_1), \ldots, (x_n, M_n)]$ is called a *configuration* of the system. For $1 \leq i \leq n$ and a given configuration π as above, we define

$$\mu(x_i, \pi) = \begin{cases} \textit{external}, & \text{if there are } r \in E_i \text{ and } x_j, j \neq i, \\ & \text{such that } (x_i, x_j) \vdash_r (u, v), u, v \in V^*, \\ \textit{internal}, & \text{otherwise.} \end{cases}$$

Then, for two configurations $\pi = [(x_1, M_1), \ldots, (x_n, M_n)]$ and $\pi' = [(x'_1, M'_1), \ldots, (x'_n, M'_n)]$, we write $\pi \Longrightarrow_{ext} \pi'$ if the following conditions hold:

1. there is $i, 1 \leq i \leq n$, such that $\mu(x_i, \pi) = \textit{external}$,
2. for each $i, 1 \leq i \leq n$, with $\mu(x_i, \pi) = \textit{external}$, we have $(x_i, x_j) \vdash_r (x'_i, z)$, for some $j, 1 \leq j \leq n, j \neq i$, and $r \in E_i$; moreover, $M'_i = M_i \cup \{z\}$,
3. for each $i, 1 \leq i \leq n$, with $\mu(x_i, \pi) = \textit{internal}$, we have $(x'_i, M'_i) = (x_i, M_i)$.

For two configurations π and π' as above, we write $\pi \Longrightarrow_{int} \pi'$ if the following conditions hold:

1. for all $i, 1 \leq i \leq n$, we have $\mu(x_i, \pi) = \textit{internal}$,
2. for each $i, 1 \leq i \leq n$, either $(x_i, z) \vdash_r (x'_i, z')$, for some $z \in M_i, z' \in V^*, r \in I_i$, and $M'_i = M_i \cup \{z'\}$, or
3. no rule $r \in I_i$ can be applied to (x_i, z), for any $z \in M_i$, and then $(x'_i, M'_i) = (x_i, M_i)$.

The relation $\Longrightarrow_{ext}$ defines an external splicing, $\Longrightarrow_{int}$ defines an internal splicing. Note that in both cases all the splicing operations are performed in parallel and the components not able to use a splicing rule do not change their contents. We stress the fact that the external splicing has priority over the internal one and that all operations have as the first term an active string; the first string obtained by splicing becomes the new active string of the corresponding component, the second string becomes an element of the set of passive strings of that component.

We write $\Longrightarrow$ for both $\Longrightarrow_{ext}$ and $\Longrightarrow_{int}$, and $\Longrightarrow^*$ for the reflexive and transitive closure of $\Longrightarrow$. The *language generated* by an SD H system Γ is defined by

$$L(\Gamma) = \{w \in T^* \mid [(w_1, A_1), \ldots, (w_n, A_n)] \Longrightarrow^* [(x_1, M_1), \ldots, (x_n, M_n)], \text{ for } w = x_1, x_i \in V^*, 2 \leq i \leq n, \text{ and } M_i \subseteq V^*, 1 \leq i \leq n\}.$$

We denote by SDH_n the family of languages generated by SD H systems with at most n components, $n \geq 1$, all of them having *finite sets of axioms and finite sets* *of splicing rules.* When no restriction is imposed on the number of components, we write SDH_*.

From the proof of the inclusion $REG \subseteq EH(FIN, FIN)$ in [**28**], one can see that $REG \subseteq SDH_1$; therefore, $REG = SDH_1 = EH(FIN, FIN)$. The reader can check that SDH_2 contains non-regular languages. It is not known whether or not the inclusion $SDH_2 \subseteq SDH_3$ is proper, but, anyway, the hierarchy collapses at this level:

Theorem 6. $RE = SDH_3$.

The proof of the inclusion $RE \subseteq SDH_3$, [**26**], is no more based on the same simulate-and-rotate idea as the proof of the Basic Universality Theorem, but on the idea of using certain strings in a specified number of copies, which are "consumed" by splicing. Here, the active string of each component is of this type, because it is unique at each step.

Consider a language $L \in RE$, generated by a type-0 Chomsky grammar $G = (N, T, S, P)$. We construct the SD H system $\Gamma = (V, T, C_1, C_2, C_3)$, with

$$\begin{aligned}
V &= N \cup T \cup \{B, B', C, D, E, F, H, J, U, X, X', Y, Z\},\\
w_1 &= SXXE,\\
A_1 &= \{HvYY \mid u \to v \in P\} \cup \{HZZ\alpha,\ H\alpha UU \mid \alpha \in N \cup T\} \cup \{JXX, F\},\\
I_1 &= \{\#uXX\$H\#vYY \mid u \to v \in P\} \cup \{\#\alpha XX\$H\#ZZ\alpha \mid \alpha \in N \cup T\}\\
&\cup \{\#XX\alpha\$H\#\alpha UU \mid \alpha \in N \cup T\}\\
&\cup \{\#YY\$J\#XX,\ \#ZZ\$J\#XX,\ \#UU\$J\#XX,\ \#XXE\$F\#\},\\
E_1 &= \{Z\alpha\#\$XX\# \mid \alpha \in N \cup T\} \cup \{UU\#\$XX\alpha\# \mid \alpha \in N \cup T\}\\
&\cup \{YY\#\$XX\#,\ XX\#\$BZ\#,\ XX\#\$BU\#,\ XX\#\$BY\#\},\\
w_2 &= CXX', A_2 = \emptyset, I_2 = \emptyset,\\
E_2 &= \{X\#X'\$X\#X,\ C\#X\$\#B,\ C\#B\$D\#X,\\
&\quad B\#B'\$Z\#Z,\ B\#B'\$Y\#Y,\ B\#B'\$U\#U\},\\
w_3 &= DXX', A_3 = \{BB', DXX'\},\\
I_3 &= \{\#DXX'\$\#BB',\ \#BB'\$\#DXX'\}, E_3 = \emptyset.
\end{aligned}$$

The intuition behind this construction is the following one. To a sentential form w of G corresponds in Γ a string of the form w_1XXw_2E, where $w_1w_2 = w$; XX indicates the "working place" (the place of the read-write head in the style of Turing machines); a rule of G can be simulated only in the presence of XX: if $w_1 = w_1'u$, for some $u \to v \in P$, and $w_1'uw_2 \Longrightarrow w_1'vw_2$ in G, then a derivation of the form $w_1'uXXw_2E \Longrightarrow w_1'vXXw_2E$ will be realized in Γ. The symbol E is a marker for the right-hand end of the string. Only in its presence can we remove XX (by the last rule in I_1). After removing XX, no further splicing is possible in C_1, hence the string is lost if it contains nonterminal symbols. In order to simulate rules in P on any position of the current string, we have to freely move XX to the left or to the right. All these operations – simulation of a rule in P, moving XX one step to the right, moving XX one step to the left – are performed in a similar way: First, we copy in the tube C_2 the suffix of the active string of C_1 starting with the second occurrence of X. Then, on the first component we start performing the desired operation. This means changing XX for YY (simulation of a rule), ZZ (moving one step to the left), or UU (moving one step to the right). This is done by an internal splicing. Then again an external splicing is performed, bringing together the active strings of the first and the second components. The initial string is reconstructed, modulo two changes: XX is replaced by YY, ZZ, UU and the desired operation is performed. At the same time, the second tube executes an external splicing which "cleans up" its active string. This is done with the help of the third tube, which has only this role: providing an appropriate string for splicings in the second tube. Again the second tube makes a copy of the second "half" of the active string of the first tube, starting now with the second occurrence of the double symbol Y, Z, U, respectively. By an internal splicing, the first component reintroduces the substring XX; then it asks for the end of the string from the second tube, reconstructing the string. The whole procedure can be iterated, because we are again in the same situation as at the beginning. At every step, the passive string sets of C_1 and C_2 are increased; that of C_3 remains constantly equal to $\{BB', DXX'\}$. However, C_2 never performs an internal splicing, whereas the newly introduced passive strings of C_1 never participate in a splicing, due to the blocking symbols H and J occurring in the internal splicing rules of C_1.

From these explanations, it should be clear that $L(\Gamma) = L(G)$. Further details can be found in [**26**].

8. Grammar systems communicating by splicing

In the previous section we have looked for *homogeneous* H systems. We discuss now a quite opposite approach, started in [**9**], namely of using splicing instead of communication in a parallel communicating (PC) grammar system (we obtain in this way a hybrid system, including both rewriting and splicing operations). As considered in [**29**] (see also [**4**]), the work of a PC grammar system consists of *synchronized componentwise rewriting steps* and *communication steps*, when the components can send their strings from one to another, on request. (Recently, also PC grammar systems with *communication by command* have been introduced, [**7**], as models of the WAVE communication paradigm in parallel computing, with the communication protocol similar to that used in communicating distributed H systems discussed in Section 6.) The following variant has been considered in [**9**].

A *splicing grammar system* is a construct

$$\Gamma = (N, T, (S_1, P_1), (S_2, P_2), \dots, (S_n, P_n), M),$$

where

(*i*) N, T are disjoint alphabets and $P_i, 1 \leq i \leq n$, are finite sets of production rules over $N \cup T$,

(*ii*) M is a finite subset of $(N \cup T)^*\#(N \cup T)^*\$(N \cup T)^*\#(N \cup T)^*$, with $\#, \$$ two distinct symbols which are not in $N \cup T$.

The sets P_i are called the *components* of Γ.

For two n-tuples (we call them *configurations*) $x = (x_1, x_2, \dots, x_n), y = (y_1, y_2, \dots, y_n)$, $x_i, y_i \in (N \cup T)^*, 1 \leq i \leq n$, we define $x \Longrightarrow y$ iff one of the following two conditions holds:

(i) for each $1 \leq i \leq n,\ x_i \Longrightarrow_{P_i} y_i$,

(ii) there exist $1 \leq i, j \leq n$ such that $x_i = x_i' u_1 u_2 x_i''$, $x_j = x_j' u_3 u_4 x_j''$, and $y_i = x_i' u_1 u_4 x_j''$, $y_j = x_j' u_3 u_2 x_i''$, for $u_1 \# u_2 \$ u_3 \# u_4 \in M$; for $k \neq i, j$, we have $y_k = x_k$.

In the above definition, (i) defines a rewriting step, whereas (ii) defines a splicing step, corresponding to a communication step in PC grammar systems. Note that no priority of any of these operations over the other one is assumed.

The language generated by the first component of Γ (the *master*) is the language of the system:

$$L(\Gamma) = \{x_1 \in T^* \mid (S_1, \ldots, S_n) \Longrightarrow^* (x_1, \ldots, x_n),\ x_j \in (N \cup T)^*, j \neq 1\},$$

where $\Longrightarrow^*$ is the reflexive and transitive closure of the relation $\Longrightarrow$.

We denote by $SGS_n(X)$ the family of languages $L(\Gamma)$ generated by splicing grammar systems of degree at most $n, n \geq 1$, with components of type X. Following [**9**], we consider here $X \in \{REG, CF\}$. When no restriction is imposed on the number of components, then we replace the subscript n with $*$.

In [**9**] the authors proved that splicing grammar systems with three context-free components characterize the recursively enumerable languages. The result is strenghtened in [**23**]: two components are enough.

Theorem 7. $RE = SGS_2(CF)$.

The proof in [**23**] makes use of the Geffert characterization of recursively enumerable languages [**14**]: each $L \in RE, L \subseteq T^*$, can be generated by a grammar $G = (\{S, A, B, C\}, T, S, P)$ with P containing only context-free rules of the form $S \rightarrow x, x \in (\{S, A, B, C\} \cup T)^*$ and the rule $ABC \rightarrow \lambda$. Starting from such a grammar, the following splicing grammar system is constructed in [**23**]:

$$\begin{aligned}
\Gamma &= (\{A, B, C, S, X, Y, Z, S_2\}, (S, P_1), (S_2, P_2), M),\\
P_1 &= (P - \{ABC \rightarrow \lambda\}) \cup \{A \rightarrow A\},\\
P_2 &= \{S_2 \rightarrow YXXABCZ,\ X \rightarrow XX, X \rightarrow X\},\\
M &= \{\#ABC\$YX\#XABC,\ \#XABC\$YXABC\#\}.
\end{aligned}$$

The reader can verify that $L(G) = L(\Gamma)$.

Note that in view of the definition of the relation $\Longrightarrow$ between configurations of a splicing grammar system, we implicitly use multiplicities of strings, because each component of the system has in every moment – both after a rewriting and after a splicing step – exactly one current string.

For the case of regular splicing grammar systems we have $REG = SGS_1(REG) \subset SGS_2(REG)$, because, as it is proved in [**23**], $SGS_2(REG)$ contains non-context-free languages.

9. Universality using multisets

The idea of using multiplicities in controlling the splicing operation is already used in [**11**], where the range of a Turing machine is encoded as the language $\sigma^*(L)$ of a splicing scheme iterated on a finite language L. The result is strenghtened and proved in a much simpler way in [**22**]; then [**13**] obtained a characterization of recursively enumerable languages on this basis. We recall the result in the form presented in [**18**].

Take an alphabet V. A multiset over V^* is a mapping $M : V^* \longrightarrow \mathbf{N} \cup \{\infty\}$; $M(x)$ is the number of copies of $x \in V^*$ in this multiset. The set $\{w \in V^* \mid M(w) > 0\}$ is called the *support* of M and it is denoted by $supp(M)$. For two multisets M_1, M_2 we define the *union* by $(M_1 \cup M_2)(x) = M_1(x) + M_2(x)$ and the *difference* by $(M_1 - M_2)(x) = \max\{0, M_1(x) - M_2(x)\}$.

Consider now an extended H system $\gamma = (V, T, A, R)$, with A a multiset over V^* and $R \subseteq V^*\#V^*\$V^*\#V^*$. For two multisets M_1, M_2 over V^* we define

$$\begin{array}{lll} M_1 \Longrightarrow M_2 & \text{iff} & \text{there are } x, y, z, w \in V^* \text{ such that} \\ & (i) & M_1(x) > 0, (M_1 - \{(x,1)\})(y) > 0, \\ & (ii) & x = x_1u_1u_2x_2, y = y_1u_3u_4y_2, z = x_1u_1u_4y_2, w = y_1u_3u_2x_2, \\ & & \text{for some } x_1, x_2, y_1, y_2 \in V^* \text{ and } u_1\#u_2\$u_3\#u_4 \in R, \\ & (iii) & M_2 = (((M_1 - \{(x,1)\}) - \{(y,1)\}) \cup \{(z,1)\}) \cup \{(w,1)\}. \end{array}$$

The reflexive and transitive closure of $\Longrightarrow$ is denoted by $\Longrightarrow^*$. The language generated by γ is defined by

$$L(\gamma) = \{w \in V^* \mid A \Longrightarrow_\sigma^* M \text{ such that } w \in supp(M)\}.$$

For two families of languages, F_1, F_2, we denote by $H(mF_1, F_2)$ the family of languages $L(\gamma)$, for $\gamma = (V, T, A, R)$, with $R \in F_2$, and A a multiset over V^* such that $supp(A) \in F_1$.

Theorem 8. $RE = EH(mFIN, FIN)$.

Indeed, consider a type-0 Chomsky grammar $G = (N, T, S, P)$, with the rules in P of the form $u \to v$ where $1 \le |u| \le 2$, $0 \le |v| \le 2$, and $u \ne v$ (for instance, we can take G in Kuroda normal form). Also assume that the rules in P are labelled in a one-to-one manner with elements of a set Lab; we write $r : u \to v$, for r being the label of $u \to v$. By U we denote the set $N \cup T$ and we construct the extended H system with multiplicities $\gamma = (V, T, A, R)$, where

$$V = N \cup T \cup \{X_1, X_2, Y, Z_1, Z_2\} \cup \{(r), [r] \mid r \in Lab\},$$

and the set R contains the following splicing rules:

1. $\delta_1\delta_2Yu\#\beta_1\beta_2\$(r)v\#[r]$, for $r : u \to v \in P$, $\beta_1, \beta_2 \in U \cup \{X_2\}$, $\delta_1, \delta_2 \in U \cup \{X_1\}$,
2. $Y\#u[r]\$(r)\#v\alpha$, for $r : u \to v \in P$, $\alpha \in U \cup \{X_2\}$,
3. $\delta_1\delta_2Y\alpha\#\beta_1\beta_2\$Z_1\alpha Y\#Z_2$, for $\alpha \in U$, $\beta_1, \beta_2 \in U \cup \{X_2\}$, $\delta_1, \delta_2 \in U \cup \{X_1\}$,
4. $\delta\#Y\alpha Z_2\$Z_1\#\alpha Y\beta$, for $\alpha \in U$, $\delta \in U \cup \{X_1\}$, $\beta \in U \cup \{X_2\}$,
5. $\delta\alpha Y\#\beta_1\beta_2\beta_3\$Z_1Y\alpha\#Z_2$, for $\alpha \in U$, $\beta_1 \in U$, $\beta_2, \beta_3 \in U \cup \{X_2\}$, $\delta \in U \cup \{X_1\}$,
6. $\delta\#\alpha YZ_2\$Z_1\#Y\alpha\beta$, for $\alpha \in U$, $\delta \in U \cup \{X_1\}$, $\beta \in U \cup \{X_2\}$,
7. $\#YY\$X_1^2Y\#w$, for $w \in \{X_2^2\} \cup T\{X_2^2\} \cup T^2\{X_2\} \cup T^3$,
8. $\#X_2^2\$Y^3\#$.

The multiset A contains the string

$$w_0 = X_1^2 Y S X_2^2,$$

with the multiplicity $A(w_0) = 1$, and the following strings with infinite multiplicity:

$$\begin{array}{lcll} w_r & = & (r)v\,[r]\,, & \text{for } r : u \to v \in P, \\ w_\alpha & = & Z_1 \alpha Y Z_2, & \text{for } \alpha \in U, \\ w'_\alpha & = & Z_1 Y \alpha Z_2, & \text{for } \alpha \in U, \\ w_t & = & YY. & \end{array}$$

The system works as follows. The rules in groups 1 and 2 simulate rules in P, but only in the presence of the symbol Y. The rules in groups 3 and 4 move the symbol Y to the right, the rules in groups 5 and 6 move the symbol Y to the left. The "starting string" in A is w_0. All the rules of types 1 – 6 involve a string derived from w_0 and containing such a symbol Y introduced by the string w_0; note that each such rule can only use one string in A which is different from w_0. At any splicing step we have two occurrences of X_1 at the beginning of a string and two occurrences of X_2 at the end of a string (maybe even the same string). The rules in groups 1, 3, and 5 separate strings of the form $X_1^2 z X_2^2$ into two strings $X_1^2 z_1$, $z_2 X_2^2$, each having multiplicity one; the rules in groups 2, 4, and 6 bring together these strings, leading to a string of the form $X_1^2 z' X_2^2$. The rules in groups 7 and 8 remove the auxiliary symbols X_1, X_2, Y. If the remaining string is in T^*, then it is an element of $L(G)$. The symbols $(r), [r]$ are associated with the labels in Lab, and Z_1 and Z_2 are associated with *moving* operations.

Using these explanations, the reader can verify that each derivation in G can be simulated in γ and conversely, hence $L(G) = L(\gamma)$.

Note that we always count the copies of one or two strings only, distinguishing between having *one* or *none* copies of them, and that having strings with infinite multiplicity does not mean *actual* infinity: the number of copies of such strings is *large enough* for our needs (when one more copy is needed, we can have it – e.g., amplifying by PCR techniques existing copies).

10. Concluding remarks

We have presented several "computers," some of them based on splicing rules controlled by permitting or by forbidding conditions, some based on distributed architectures, and one based on multisets. All of them are computationally complete (equal in power to the Turing machines) and can lead to universal devices, hence "programmable DNA computers." However, all these models are based on several assumptions. Whether or not these assumptions can be relaxed, and whether or not a further model can be devised, which combines the realistic features of the previous models, remains to be investigated. Anyway, the simple splicing operation, incorporated in the specific architectures discussed above, is enough to equal the power of Turing machines, which seems quite promising.

References

[1] L. M. Adleman, On constructing a molecular computer. in **[21]**

[2] D. Boneh, C. Dunworth, R. J. Lipton, J. Sgall, On the computational power of DNA. *Discrete Applied Math.*, 71:79-94, 1996.

[3] C. Calude, Gh. Păun, Global syntax and semantics for recursively enumerable languages. *Fundamenta Informaticae*, 4(2):245-254, 1981.

[4] E. Csuhaj-Varju, J. Dassow, J. Kelemen, Gh. Păun. *Grammar Systems. A Grammatical Approach to Distribution and Cooperation*, Gordon and Breach, London, 1994.

[5] E. Csuhaj-Varju, L. Freund, L. Kari, Gh. Păun, DNA computing based on splicing: universality results, *Proc. First Annual Pacific Symp. on Biocomputing*, Hawaii, Jan. 1996 (L. Hunter, T. E. Klein, eds.), World Sci. Publ, Singapore, 1996, 179-190.

[6] E. Csuhaj-Varju, L. Kari, Gh. Păun, Test tube distributed systems based on splicing. *Computers and AI*, 15, 2-3:211-232, 1996.

[7] E. Csuhaj-Varju, J. Kelemen, Gh. Păun, Grammar systems with WAVE-like communication. *Computers and AI*, 15(5):419-436, 1996.

[8] K. Culik II, T. Harju, Splicing semigroups of dominoes and DNA. *Discrete Appl. Math.*, 31:261-277, 1991.

[9] J. Dassow, V. Mitrana, Splicing grammar systems. *Computers and AI*, 15(2-3):109-122, 1996.

[10] J. Dassow, Gh. Păun, *Regulated Rewriting in Formal Language Theory*. Springer-Verlag, Berlin, Heidelberg, 1989.

[11] K. L. Denninghoff, R. W. Gatterdam, On the undecidability of splicing systems. *Intern. J. Computer Math.*, 27:133-145, 1989.

[12] C. Ferretti, S. Kobayashi, T. Yokomori, DNA splicing systems and Post systems, *Proc. First Annual Pacific Symp. on Biocomputing*, Hawaii, Jan. 1996 (L. Hunter, T. E. Klein, eds.), World Sci. Publ, Singapore, 1996, 288-299.

[13] R. Freund, L. Kari, Gh. Păun, DNA computing based on splicing: The existence of universal computers, *Technical Report 185-2/FR-2/95*, TU Wien, 1995, and *Theories of Computing Science*, to appear.

[14] V. Geffert, Normal forms for phrase-structure grammars. *RAIRO, Th. Inform. and Appl.*, 25:473-496, 1991.

[15] D. E. Goldberg, *Genetic Algorithms in Search, Optimization and Machine Learning*, Addison-Wesley, Reading, Mass., 1989.

[16] T. Head, Formal language theory and DNA: an analysis of the generative capacity of specific recombinant behaviors. *Bull. Math. Biology*, 49:737-759, 1987.

[17] T. Head, Splicing schemes and DNA, *Lindenmayer Systems: Impacts on Theoretical Computer Science and Developmental Biology* (G. Rozenberg, A. Salomaa, eds.), Springer-Verlag, Berlin, 1992, 371-383.

[18] T. Head, Gh. Păun, D. Pixton, Language theory and molecular genetics. Generative mechanisms suggested by DNA recombination, chapter 2 in vol. 2 of *Handbook of Formal Languages* (G. Rozenberg, A. Salomaa, eds.), Springer-Verlag, Berlin, Heidelberg, 1997, 295-360.

[19] J. H. Koza, J. P. Rice, *Genetic Algorithms. The Movie*, MIT Press, Cambridge, Mass., 1992.

[20] R. J. Lipton, Speeding up computations via molecular biology, in [**21**], 67-74.

[21] R. J. Lipton, E. B. Baum, editors, *DNA Based Computers: Proceedings of a DIMACS Workshop*, DIMACS Series in Disc. Math. and Theor. Comp. Sci., Amer. Math. Soc., 1996.

[22] Gh. Păun, On the power of the splicing operation, *Intern. J. Computer Math.*, 59 1995), 27-35.

[23] Gh. Păun, On the power of splicing grammar systems. *Ann. Univ. Buc., Matem.-Inform. Series*, 45(1):93-106, 1996.

[24] Gh. Păun, On the splicing operation. *Discrete Applied Math.*, 70:57-79, 1996.

[25] Gh. Păun, Regular extended H systems are computationally universal. *J. Aut., Languages, Combinatorics*, 1(1):27-36, 1996.

[26] Gh. Păun: Two-level distributed H systems. *Proc. of the Third Conf. on Developments in Language Theory, Proc. of Developments in Language Theory Conf., III*, Thessaloniki, 1997 (S. Bozapalidis, ed.), Aristotle Univ. of Thessaloniki, 1997, 309-327.

[27] Gh. Păun, Computing by splicing: How simple rules ? *Bulletin of EATCS*, 60:145-150, 1996.

[28] Gh. Păun, G. Rozenberg, A. Salomaa, Computing by splicing. *Theor. Computer Sci.*, 168(2):321-336, 1996.

[29] Gh. Păun, L. Sântean (now Kari), Parallel communicating grammar systems: the regular case. *Ann. Univ. Buc., Matem.-Inform. Series*, 38:55-63, 1989.

[30] D. Pixton, Regularity of splicing languages, *Discrete Appl. Math.*, 69 1996), 101-124.

[31] D. Pixton, Linear and circular splicing systems, *Proc. 1st Intern. Symp. on Intell. in Neural and Biological Systems*, IEEE, Herndon, 1995, 38-45.

[32] W. D. Smith, DNA computing in vitro and vivo, in [**21**], 121-186.

[33] R. Siromoney, K. B. Subramanian, V. Rajkumar Dare, Circular DNA and splicing systems, *Proc. of Parallel Image Analysis*, Berlin, *LNCS 654*, Springer-Verlag, 1992, 260-273.

[34] T. Yokomori, S. Kobayashi, C. Ferretti, On the power of circular splicing systems and DNA computability. *Report CSIM 95-01*, Univ. of Electro-Comm., Chofu, Tokyo, 1995, and *IEEE Intern. Conf. on Evolutionary Computing*, Indianapolis, 1997, 219-224.

Institute of Mathematics of the Romanian Academy, PO Box 1 – 764, 70700 București, Romania

E-mail address: gpaun@imar.ro

DIMACS Series in Discrete Mathematics
and Theoretical Computer Science
Volume **44**, 1999

Running Dynamic Programming Algorithms On a DNA computer

Eric B. Baum and Dan Boneh

ABSTRACT. In this paper we show that DNA computers are especially useful for running algorithms which are based on dynamic programming. This class of algorithms takes advantage of the large memory capacity of a DNA computer. We present algorithms for solving certain instances of the knapsack problem using a dynamic programming approach. Unlike other algorithms[**1, 12**] for DNA computers, which are brute force, dynamic programming is the same algorithm one would use to solve (smaller) problems on a conventional computer.

1. Introduction

In a recent seminal paper [**1**], Adleman introduced the idea of computing using DNA molecules. Adleman's techniques were then generalized by Lipton [**12**] who showed that formula-SAT can be solved on a DNA computer. These algorithms essentially use a brute force approach to solve hard combinatorial problems. This approach is interesting due to the massive parallelism available in DNA computers. However, the brute force approach is limited by the number of DNA molecules in solution. Breaking the DES cryptosystem [**3**] is the only "real world" application yet suggested for this approach.

In this paper we propose a new class of algorithms which can be implemented on a DNA computer, namely some algorithms based on dynamic programming. Graph reachability and knapsack are classical problems solvable in this way. These problems are solvable by conventional computers in polynomial time (or pseudo-polynomial time for knapsack), but only so long as they are small enough to fit in memory. DNA computers using dynamic programming could solve substantially larger instances because of their large memory capacity than either conventional computers or previous brute force algorithms on DNA computers.

Throughout the paper we assume an error free model. This means that all molecular biology experiments work perfectly without any errors. There are at least two hurdles to real world DNA computing: finding "killer applications" where they can substantially exceed the capabilities of standard computers, and overcoming errors and other real world departures from the ideal. As a working hypothesis we regard these two hurdles as independent. One may reasonably hope that errors that arise during the operation of a DNA computer can be dealt with. For example, fault tolerant techniques enable error-free computations on conventional machines in the presence of faulty processors. We note that several researchers [**5, 9**] have already suggested some techniques for making DNA computers resistant to errors.

2. Dynamic programming

The basic idea of the dynamic programming technique is the following: given a combinatorial problem, we iteratively solve harder and harder problems until the required problem is solved. Thus, to solve the original problem we first solve many easier sub-problems. Classical problems that can be solved using this approach include the shortest path in a graph problem, finding a minimal cost triangulation, the knapsack problem and many others [**6**].

The reason dynamic programming algorithms are suitable for DNA computers is that the sub-problems can be solved in parallel. Thus, a DNA computer might be useful for solving large instances of the problems mentioned above. Note that this application of a DNA computer is radically different than the one suggested by Adleman [**1**]. The algorithm we run on the DNA computer is not a brute force algorithm. It is exactly the same algorithm which one would use to solve the problem on a conventional machine. We hope that this application of DNA computers can be used to illustrate their advantages (and disadvantages) with respect to conventional machines.

Currently we only know how to use a DNA computer for dynamic programming algorithms of a very special form. The generic problem which we can solve is that of *graph reachability*: given a directed graph $G = (V, E)$ and two vertices S and T test if there is a path from S to T. The graph reachability problem can be solved using a simple dynamic programming algorithm known as breadth-first-search or BFS for short [**6**].

As was stated above, our objective is to solve the graph reachability problem for very large graphs. The graph will be described by a set of m functions $f_i : V \to V$ for $i = 1, \dots, m$. Given a vertex $v \in V$, the neighbors of v are $\{f_1(v), \dots, f_m(v)\}$. Thus, each vertex can have at most m neighbors. We denote by $C(f_i)$ some fixed circuit computing the function f_i. Using this notation, an instance of the graph reachability problem is $(S, T, m, C(f_1), \dots, C(f_m))$. Using this formulation it is not difficult to see that the graph reachability problem is PSPACE complete [**8**].

The BFS algorithm will solve the graph reachability problem in time $O(|E|)$ on a conventional machine. Let d be the diameter of the graph G, i.e. $d = \max\{\text{distance}(u, v) \mid u, v \in V\}$. We give an algorithm by which a DNA computer can solve the graph reachability problem in time $O(d)$. As we shall see, in many applications the diameter of the graph is small even though its size is too large to fit in a conventional computer. Such applications are ideal for a DNA computer.

3. Solving graph reachability using DNA

In this section we show how to run the BFS algorithm on a DNA computer. We first need to recall a basic result from [**4**].

THEOREM 3.1 ([**4**]). *Let $f : \{0,1\}^n \to \{0,1\}^m$ be a function computable by a circuit of size s. Let T be a solution of DNA strands representing*[1] *a set $S \subseteq \{0,1\}^n$. Then in s "biological-steps" it is possible to construct a solution of DNA strands representing all pairs $\{(x, f(x)) \mid x \in S\}$.*

Let $(S, T, m, C(f_1), \dots, C(f_m))$ be an instance of graph reachability. The algorithm for solving graph reachability using DNA can be described as follows:

1. Start with a solution containing strands representing the vertex S.
2. Iterate the following procedure d times where d is the diameter of the graph:
   (a) Let D be the current solution. Apply PCR to create m copies of the solution D. Let $D_1, \dots, D_m$ be m separate vats containing the replicas of D.
   (b) Use Theorem 3.1 to compute the function f_i in D_i. We can arrange things so that the strands in solution D_i now look like:
   $$3' \ - \ v \ - \ B_1 \ - \ f_i(v) \ - \ B_2 \ - \ 5'$$
   where B_1 is the restriction site for some enzyme and B_2 is some fixed DNA strand.
   (c) In each D_i apply the restriction enzyme which cuts the strand at B_1. Then extract from D_i all strands containing the string B_2. Denote the set of extracted strands by D'_i.
   (d) Pour all of $D'_1, \dots, D'_m$ into one vat denoted by D'. Note that D' now represents the set of all neighbors of vertices in D.
   (e) Test if the target vertex T appears in the solution D'. If it does stop. Otherwise set $D \leftarrow D'$ and go to step 2.

Note that in step 2d we pour the contents of m vats into one vat. Since we assume that the volume of a vat is constant (e.g. 1 liter) this step requires that we only pour one m'th fraction out of each of the D'_i. In what follows we analyze the number of DNA strands that need be used so that no information will be lost with high probability.

Unfortunately there is a technical problem which causes this algorithm to fail on certain instances of graph reachability. At every step of the algorithm above, the number of strands representing a vertex V in the graph is proportional to the numbers of paths from S to V. This is fine, as long as the number of paths from S to any vertex V is roughly the same for all vertices V. However, if some vertices have many more paths leading to them then those vertices will dominate the solution. Reachable vertices with only a few paths leading to them might be lost. Thus our algorithm can only work for graphs which are balanced in this sense.

DEFINITION 3.1. *Let G be a directed graph with a distinguished vertex S. For an integer $r > 0$ denote by $P_r(V)$ the number of paths from S to V of length exactly r. Let*

$$M(r) = \max_V \{P_r(V)\} \quad \text{and} \quad N(r) = \min_{V : P_r(V) > 0} \{P_r(V)\}$$

The graph G is called ϵ-balanced if for any $r > 0$ we have

$$N(r)/M(r) > \epsilon$$

[1] When we say that a solution of DNA strands represents a set S of binary strings we mean that for every $x \in S$ there is at least one strand in the solution representing x. The encoding of binary strings as DNA strands is explained in [**4**].

We now analyze the number of DNA molecules needed to make this algorithm work. Our final figure will depend on $1/\epsilon$ which shows that the algorithm will not work on badly balanced graphs.

Let d be the diameter of the graph G. Let W_r be the number of vertices V in G for which there is a directed path from S to V of length exactly r. Define W to be

$$W = \max\{W_r \quad \text{where} \quad 0 < r \leq d\}$$

One can think of W as the size of the widest level in the graph. Note that throughout the algorithm the number of different types of DNA molecules in the solution is bounded by W. However, the total number of DNA molecules is always fixed. This follows since at each step we duplicate each strand m times and then pour out all but an m'th fraction of the solution.

Let G be an ϵ balanced directed graph. We claim that if the initial solution used by the algorithm contains

$$P = \frac{W}{\epsilon}\frac{d}{m}\log^2 W$$

DNA molecules each representing the vertex S then the algorithm will find a path to the vertex T if it exists. This means that throughout the algorithm each vat will contain P DNA molecules. Let V be some vertex for which there is a path of length r from S to V. We show that if P strands are used then vertex V will be present in the solution generated at iteration r of the algorithm. This is sufficient to ensure that the algorithm will work since if a path from S to T exists then vertex T will be found after at most d steps, where d is the diameter of the graph G.

Consider the solution generated at iteration r of the algorithm for some $r > 0$. Recall that there are at most W different types of molecules in the solution. Let k be the number of copies of some molecule in the solution. At step 2d of the algorithm we throw away all but an m'th of the solution. After this step the expected number of copies of the molecule in the solution is k/m. Let $k' = k/m$. Due to random drift the actual number of copies of the molecule would deviate from k' by at most $\sqrt{k'/m}\log W$ with high probability. The number of copies of each molecule should be large enough to ensure that the random drift does not wipe out that molecule.

We now let k be the number of copies of the least frequent molecule in the vat generated at step r of the algorithm. The random drift in the number of copies is at most $\sqrt{rk/m}\log W$ which is bounded by $\sqrt{dk/m}\log W$. Taking k to be $d/m\log^2 W$ ensures that random drift does not kill the least frequent molecule.

Recall that the graph G is ϵ-balanced. This means that the most frequent molecule has at most k/ϵ copies. Since there are at most W different types of molecules in the solution, the total number of molecules is bounded by Wk/ϵ. Substituting the value for k above, we obtain the value P specified above. This proves that having P molecules in each vat generated by the algorithm will ensure that the path to T will be found in d steps if it exists.

4. Communication Issues

In a DNA computer, one has very many "processors", i.e. many DNA molecules. But these have essentially no communication whatever. We know how to compute a function based on the information encoded on each individual string, but in general not how to utilize information from two strings[2]. The balanced graph limitation arises because we have no way to compare strings, and thus to remove duplicate molecules encoding the same node. One can propose several possible methods for removing such duplicate molecules, but it is unclear whether they would be effective.

For example, here is an idea for how to remove duplicate molecules. Devise an encoding whereby if string S codes for a given node on the graph, so does string $\bar{S}$, its Watson-Crick complement. (This is not hard.) One might then arrange to produce the encoding S and $\bar{S}$ in roughly equal numbers, or alternatively to randomly to replace strings by their complements. This could easily be accomplished by including some marker sequence M on all strings created. Extracting on M would get you a tube full of sequences coded in the "positive". PCR and separation and extraction on $\bar{M}$ would get you a tube coded in the complement. You can mix these. Then one might hope that complementary strings might mate, and could then be removed somehow, for example by cutting them with a restriction enzyme acting only on double strings. Since the reaction of combination of S molecules with $\bar{S}$ molecules might be expected to proceed at a rate proportional to the square of their concentrations, one might hope to remove molecules with high concentration, i.e. multiple representation, without removing all the representatives of nodes with low representation. Unfortunately, a rate analysis presented in Appendix A seems to indicate that the time it takes for this reaction to balance the number of copies of each vertex is exponential in the size of the problem. Furthermore, such a procedure seems prone to various kinds of errors.

In general dynamic programming algorithms, one might update a node of a graph as some function of its neighbors. For example each node might store a number, and the number at a given node might be updated as the minimum of its neighbors. We do not know how to realize such general forms of dynamic programming because we do not now how to communicate numbers between DNA strands. We now remark that somewhat more general forms of dynamic programming than simple graph reachability can, in fact, be realized. Let n be some node of the graph represented by DNA string s Let s also encode some Boolean vector x. In general, the circuit evaluation procedure allows one to produce a DNA string s' representing a neighboring node n', and also encoding a Boolean vector $f(x')$ for f some circuit. (What we find hard to do is to encode a Boolean vector $g(x, y)$ where y is some vector stored on a different DNA molecule!)

For example, consider the problem of shortest path in a graph. This is ordinarily solved by a Breadth First Search procedure as discussed above, where however at each node one maintains the distance to the source. This can readily be done by adding 1 each time one produces a new DNA molecule. Adding 1 can of course be represented by a simple circuit. When one produces a string representing the goal node T, it will encode the distance traveled. The first time one produces such a string, its distance will be the shortest. Alternatively, if one has produced many, one can search them (using a binary search) to find the shortest distance encoded.

[2] Reif[**13**] and Beaver [**2**] proposed a solution to this problem. This solution, however, involves using enough DNA molecules to encode all pairs of virtual processors. If one might (optimistically) have 10^{18} DNA molecules, then, this proposal would restrict one to 10^9 virtual processors. But since DNA processors are at least 10^9 times slower than electronic ones, this is of little practical interest.

5. Solving the knapsack problem

The dynamic programming approach for solving the knapsack problem [**6**] can be easily stated as a graph reachability problem. The diameter of the resulting graph is linear in the size of the knapsack problem. In this section we analyze the types of knapsack problems which can be solved on a DNA computer.

The knapsack problem we are interested in is the following: given positive integers $b_1, \ldots, b_n$ and an integer B, does there exist a subset $S \subseteq \{1 \ldots n\}$ such that $\sum_{i \in S} b_i = B$. This problem is known to be NP-complete[**9**]. The knapsack problem comes up often in optimization (see [**7**] and references therein).

We note that Lagarias and Odlyzko [**11**] showed that random instances of knapsack in which the integers b_i are extremely large (larger than 2^{n^2}) can be solved in polynomial time on a conventional machine. Their approach is based on lattice basis reductions. We intend to show that DNA computers can solve certain instances of the knapsack problem even when the integers b_i are smaller than 2^n.

The knapsack problem can easily be posed as a graph problem as follows: Let $M = \sum_{i=1}^{n} b_i$. Consider an $M \times n$ grid where there is an edge from point (x, i) to $(y, i+1)$ if $y = x$ or $y = x + b_i$. The knapsack problem has a solution if and only if there is a path in the above graph from $(0, 1)$ to (B, n).

Note that the amount of memory required to run the BFS algorithm on the knapsack graph on a conventional machine is $O(M)$. Thus, to solve a knapsack problem with a hundred numbers each of size 10^{14} one needs a memory of size 10^{16}. This is far more than is currently attainable by a conventional computer. A standard computer disk might store 10^{10} bits. An optimistic upper bound on the amount of memory available with current technology is 10^{13} bits. On the other hand, a liter of water can easily contain 10^{16} DNA strands.

To solve the knapsack problem using DNA we intend to use the graph reachability algorithm of the previous section. Recall that for the algorithm to work the graph must be balanced. This means that we can only solve instances of knapsack for which the corresponding graph is ϵ-balanced where ϵ is not too small. The exact bound on ϵ depends on the actual physical apparatus used.

We now discuss how to add numbers using DNA. Let T be a solution of DNA strands representing a set of integers $\{\alpha_1, \ldots, \alpha_k\}$. We assume that the numbers are encoded using decimal digits. That is, the strand representing the number α_i simply encodes the decimal digits of α_i. Without loss of generality we assume that all of $\alpha_1, \ldots, \alpha_k$ contain m decimal digits (if this is not the case, one can always pad the numbers with zeros on the left).

We wish to transform the solution T to a solution T' representing all numbers $\{\alpha_1, \alpha_1 + X, \ldots, \alpha_k, \alpha_k + X\}$ for some integer X with m decimal digits. This can be done as follows:

1. Let d be the least significant digit of X. Separate the solution T into 10 solutions $T_0, T_1, \ldots, T_9$ such that T_i contains all numbers in T whose least significant digit is d.
2. In each T_i append to all strands in T_i the oligo representing the digit $i + d \bmod 10$.
3. Pour the solutions $T_0, \ldots, T_{9-d}$ into one vat T^0. Pour the remaining solutions into another vat T^1. The strands in T^0 represent those numbers which have not generated a carry. Similarly, the strands in T^1 represent those which have generated a carry.

4. Iterate this procedure appropriately for each of the m digits.
5. At the end of the process all strands in the solution look like:

$$3' - \alpha_i - S - \alpha_i + X - 5'$$

where S is some predetermined separator.
6. Use a restriction enzyme to cut all strands at the string S. This will produce the required solution containing all numbers

$$\{\alpha_1, \dots, \alpha_n, \alpha_1 + X, \dots, \alpha_n + X\}$$

Suppose we allow 10 extraction operations to be carried out simultaneously. Then the above procedure requires m steps where a step involves 10 extractions in parallel and an append.

To solve the knapsack problem using DNA apply the above procedure for each of the numbers $b_1, \dots, b_n$ starting with a solution containing only the number 0. If the final solution contains the number B then the answer to the given instance of the knapsack problem is positive. Otherwise the given instance has no solution. It is not difficult to modify the algorithm so that the set S such that $\sum_{i \in S} b_i = B$ is also output by the algorithm.

It is straightforward to generalize this approach to solve other versions of the Knapsack problem. For example, a frequent optimization problem is to find the largest number less than B which can be represented as a sum $\sum_{i \in S} b_i$. The procedure above produces a solution containing all numbers which can be represented as $\sum_{i \in S} b_i$ for any S. By extracting sequentially on each component value it is straightforward to separate this solution into two parts, those less than B and those no smaller than B. By extracting from the first of these sequentially on each component it is straightforward to derive a solution containing only the largest number in it. Sequencing this answers the optimization problem.

The number of steps the algorithm requires is nm. This means that the above example where we are given a hundred numbers each of size 10^{14} can be solved in 1400 "biological-steps". This figure is (hopefully) within the reach of a DNA computer.

6. Conclusions

In this paper we attempted to extend the set of problems to which DNA computers might apply. We noted that DNA computers might be useful for running dynamic programming algorithm since such algorithms require large amounts of memory for solving relatively simple tasks.

As an example we saw that the graph reachability problem can be solved efficiently on a DNA computer. A useful instance of the graph reachability problem is the knapsack problem. Our algorithm was able to solve large instances of the knapsack problem. We estimated that a conventional machine would not be able to solve such large problems using this type of algorithms. We hope that this example will be useful in demonstrating the *raison d'être* of DNA computers.

It is more difficult to implement those dynamic programming algorithms which require communicating information between nodes of a graph, because the DNA molecules in a DNA computer are essentially oblivious of one another, and have no way to communicate. This example may also thus illustrate a limitation of DNA computers. We did suggest some approaches to extending the power of our algorithms.

We note that since the knapsack problem is in NP, one could try to use the circuit satisfaction algorithm of [**4**] to solve the problem. However, the circuit satisfaction algorithm is a brute force approach. The algorithm will try all possible subsets of numbers $\{b_1, \ldots, b_n\}$ to test if one of them sums up to B. In our example, the input was made up of 100 numbers. This means that the algorithm would have had to test 2^{100} different subsets. Since a liter of water can contain at most 2^{70} DNA strands, the circuit-satisfaction algorithm would not be able to solve this instance of knapsack.

Our algorithm for solving knapsack is not a brute force approach. In fact, it is the exactly the same algorithm that one would use to solve the problem on a conventional machine. The fact that DNA has far more memory than a conventional machine enables a DNA computer, at least in principle, to solve larger instances of the problem.

Acknowledgments: We are grateful to F. Guzman for remarking that our techniques extend to shortest path and to S. Kurtz for the calculation in Appendix A.

Appendix A

One idea for removing molecules duplicating representation of a vertex is to set it up so that they can stick to each other (or two of them stick to a killer molecule), so that they can then be removed. This was discussed in §4. The hope was that since the sticking rate would be proportional to the square of the concentration, multiply represented molecules would rapidly be removed. Unfortunately, the time this reaction takes to balance the number of copies of each vertex is too large. The following analysis of reaction times was presented to us by S. Kurtz (private communication).

Let y be the concentration of a molecule. Then we have

$$dy/dt = -y^2$$

because the reaction rate is proportional to the square of the concentration, as molecules must collide to react. We have absorbed a rate constant into the time units. Solving this equation we find

$$y = 1/(t + c)$$

for c a constant. Let's say we have two molecule types. Let A represent the concentration of the first type and B the second. Then

$$A = 1/(t + 1/c_A) \qquad B = 1/(t + 1/c_B)$$

where c_A and c_B are respectively the initial concentrations of A and B. Assume $c_A >> c_B$. Now

$$A/B = (c_A c_B t + c_A)/(c_A c_B t + c_B)$$

We ask for the time at which A/B will be as small as α. Solving we find

$$t = (c_A - \alpha c_B)/(c_A c_B(\alpha - 1))$$

Since $c_A >> \alpha c_B$ we can approximate

$$t \sim 1/(c_B(\alpha - 1))$$

This is the amount of time the reaction takes to balance the concentrations of all vertices. Note that this time depends only on the concentration of the rarest molecules. Since the graph contains an exponential number of vertices the concentration of the rarest vertex must be one over exponential. Consequently, the time t is exponential in the size of the problem showing that the reaction time may be too big.

It is interesting to note that the above anlysis neglected to take the reaction rate constant into account. In reality we obtain a differential equation of the form

$$dy/dt = -\beta y^2$$

where β is some reaction rate constant. Above we normalized β to 1. The solution is now

$$y = 1/(\beta t + c)$$

giving t as $t \sim 1/(c_B(\alpha - 1)\beta)$. It is possible that β is sufficiently large so as to kill the exponential in the expression for t for *reasonable size problems.* For instance, suppose one tries to use our algorithm on a graph with 10^{18} nodes. Then, if time is measured in seconds and β is 10^{15} then the reaction time is reasonable after all. The value of β can only be determined through experimentation.

References

[1] L. Adleman, Molecular Computation of Solutions to Combinatorial Problems. *Science*, 266:1021-1024, 1994.

[2] D. Beaver, A Universal Molecular Computer. In R. J. Lipton and E. B. Baum, editors, *DNA Computers: Proceeding of a DIMACS Workshop,* DIMACS Series in Discrete Mathematics and Theoretical Computer Science, American Mathematical Sociey, 29-36, 1996.

[3] D. Boneh, C. Dunworth, R. Lipton, Breaking DES using a molecular computer. In R.J. Lipton and E. B. Baum (editors), *DNA Based Computers: Proceedings of a DIMACS Workshop*, DIMACS Series in Discrete Math. and Theoretical Computer Science, Amer. Math. Soc., (1996), 37-66.

[4] D. Boneh, C. Dunworth, R. Lipton, J. Sgall, On Computational Power of DNA. *Princeton CS Tech-Report CS-TR-499-95*, `ftp://ftp.cs.princeton.edu/reports/1995/499.ps.Z`

[5] D. Boneh, C. Dunworth, R. Lipton, J. Sgall, Making DNA Computers Error Resistant. In L. Landweber and E. Baum, editors, *DNA Based Computers: Proceedings of the Second DIMACS Workshop*, DIMACS Series in Discrete Math. and Theoretical Computer Science, Amer. Math. Soc., 163-170, 1998.

[6] T. Corman, C. Leiserson, R. Rivest, "Introduction to Algorithms", McGraw Hill, 1991.

[7] G. Dantzig, Discrete-variable extremum problems, *Operations Research* 5: 266–277.

[8] M. Garey, D. Johnson, "Computers and Intractability", Freeman, 1979.

[9] R. M. Karp, Reducibility among combinatorial problems, *Complexity of Computer Computations*, R. E. Miller and J. W. Thatcher (eds), Plenum Press, NY, pp 85–103, 1972.

[10] R. Karp, C. Kenyon, O. Waarts, Error-resilient DNA computations. *Proceedings of Symposium on Discrete Algorithms*, SODA 1996.

[11] J. Lagarias, A. Odlyzko, Solving low density subset sum problems. *J. Assoc. Comp. Mach.*, 32: 229–246, 1985.

[12] R. Lipton, Using DNA to solve NP-Complete Problems. *Science* 268:542-545 1995.

[13] J. Reif, "Parallel Molecular Computation", Proceedings of SPAA 95.

[14] W. D. Smith, DNA Computers in vitro and vivo, In R. J. Lipton and E. B. Baum, editors, *DNA Based Computers: Proceedings of a DIMACS Workshop*, DIMACS Series in Discrete Math. and Theoretical Computer Science, Amer. Math. Soc., 121-185, 1996.

NEC Research Institute, 4 Independence Way, Princeton, NJ 08540

E-mail address: `eric@research.nj.nec.com`

Department of Computer Science, Princeton University, 35 Olden Street, Princeton, NJ 08544

E-mail address: `dabo@cs.princeton.edu`

DIMACS Series in Discrete Mathematics
and Theoretical Computer Science
Volume **44**, 1999

A Molecular Computation of the Road Coloring Problem

Natasha Jonoska and Stephen A. Karl

ABSTRACT. Two algorithms for molecular computation of the road coloring problem are presented. We present in detail the laboratory techniques to implement these algorithms. In both of these algorithms a new operation of substring matching in the process of separating molecules is introduced. The laboratory techniques of its implementation are discussed.

1. Introduction

Recently, mathematicians and computer scientists have turned to molecular biological techniques as possible methods for solving NP-complete problems. A small instance of a well known NP-complete problem, "The Hamiltonian Path Problem" was solved in a test tube using standard methods of molecular biology [**1**]. In the same article it was asserted that for some problems it might be the case that molecular computers can outperform the electronic ones. In [**2**], Adleman proposes a model for constructing a practical molecular computer. Many authors have followed upon Adleman's ideas. In [**14**], Lipton shows how one might solve a large class of problems using molecular computation. Others have built on the ideas of massive parallel computation using molecular biology [**19, 16, 4**]. Nonetheless, except in two cases where Adleman's experiment was repeated, there have been no laboratory demonstrations of any of the new algorithms proposed.

In this paper we present an algorithm using the techniques of molecular biology for solving the Road Coloring Problem (RCP). It is not known if this problem is NP-complete, but for a given graph, there is no known polynomial time algorithm that solves it. Here, using the massive parallelism of DNA computation we present a polynomial time algorithm for solving the RCP for a given graph. Two main molecular procedures have been suggested as a possible way of separating molecules. The first one, used by Adleman [**1**] in his experiments, employs magnetic beads, and the second uses restriction endonucleases [**4**]. In this paper we extensively use the hybridization of DNA as a part of the separation process which introduces a new string operation of substring matching. In Section 2 we present two procedures that are involved in solving the RCP and a detailed description of the laboratory techniques is given in Section 3. We end our presentation with a

This work was supported, in part, by the University of South Florida Research and Creative Scholarship Grant Program under Grant No. 124933RO.

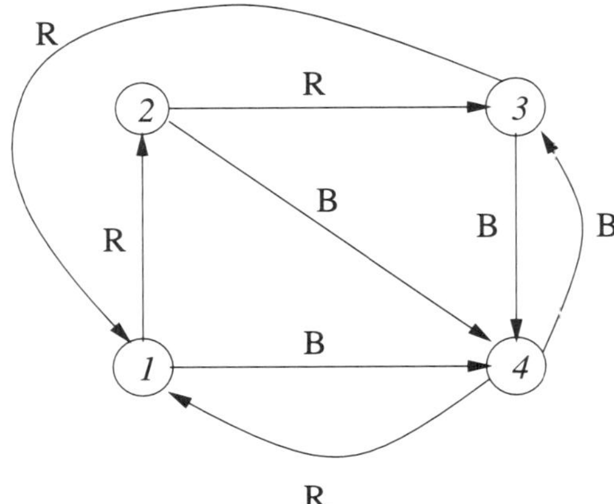

Figure 1.

formal description of the new operation of substring matching and a discussion on the physical limitations of the laboratory techniques in DNA computation (Section 4).

A detailed laboratory implementation of the algorithms for the graph presented on Figure 1 is planned and throughout the text we will refer to this example.

First, we describe the Road Coloring Problem. Let $G = (V, E)$ be a directed graph (digraph) with constant outdegree k, (k-out). Let $A = \{c_1, c_2, \ldots, c_k\}$ be a k-letter alphabet whose elements will be called colors. A *k-coloring* (or just *coloring*), of G is a labeling of the edges with the k colors from A. We say that a coloring of G is *deterministic* if the outgoing edges are colored distinctly. A vertex $v \in V$ is called *synchronizing* if there is a sequence of colors x_v such that every path in G with label x_v ends at v. In that case x_v is a synchronizing word for v. We say that G is *synchronizable* if there is a deterministic k-coloring of G such that every vertex of G is synchronizing.

A graph is called *aperiodic* if the *gcd* of the cycle length is 1. We say that G is *strongly connected* if for every two vertices v and w, there is a path from v to w and from w to v.

We note that in the case of deterministic coloring of a strongly connected graph if there is one synchronizing vertex, then every vertex is synchronizing. Figure 1 presents a strongly connected aperiodic 2-out digraph with a synchronizing coloring (the colors are R for Red and B for Blue). A synchronizing word is RBR.

The **Road Coloring Problem** asks if every strongly connected aperiodic k-out digraph is synchronizable. The name of the problem can be associated in the following way. If we interpret the vertices of the graph as cities, and the edges of the graph as one way roads, then a synchronizing coloring of the graph guarantees that a traveler can never be lost, i.e., by following the home town synchronizing sequence of colors, the traveler will arrive home regardless of the starting point.

The problem originally was posed in [**3**] where it appeared in a scope of a problem in symbolic dynamics. There, it was conjectured that the answer to the question is affirmative. Solutions have been provided for some specific classes of graphs [**15, 11, 13**], which support the conjecture, but the general case remains open.

Each colored graph can be viewed as a finite state automaton or encoder. Synchronizing sequences have been studied since the beginning of the automata theory and long before the RCP had appeared. Great interest for such automata especially has been in the coding theory where synchronizing encoders are error resistant (a synchronizing word returns the automaton in the same state as if no error had occurred).

2. Procedures

Algorithmically, there are two questions in the RCP which need to be addressed. First, if presented with a graph that is already colored, can it be determined whether the coloring is synchronizing? Second, if presented with an uncolored graph, is the graph synchronizable?

It has been conjectured by Černy [**9**] that if there is a synchronizing sequence of colors for a given coloring of a graph, then the shortest sequence must be at most $(n-1)^2$ long (where n is the number of vertices). Examples have indicated that this bound cannot be improved (see for ex. [**17**]), but to date, only a cubical bound has been proven. Providing that the conjecture is right, in a worst case for a two color graph, $2^{(n-1)^2}$ sequences must be evaluated in order to determine if the given coloring scheme is synchronizing. Currently, there is no known algorithm which can solve this problem in a polynomial time relative the number of vertices.

A molecular computational approach can avoid this exponential increase. We propose the following procedure to be implemented in a molecular computation (assume that $V = \{v_1, \dots, v_n\}$ is the set of vertices):

Procedure 1.

Step 1: Generate random paths through the graph.

Step 2: For each $i = 1, \dots, n$, keep the paths that start at vertex v_i and end at vertex v_n in set P_i.

Step 3: For $i = 1, \dots, n$ keep all paths in P_i with length $(n-1)^3$.

Step 4: For $i = 2, \dots, n$, keep the paths in P_i that start with v_i that have the same color sequence as paths in P_{i-1} starting with v_{i-1}.

Step 5: If any paths remain in P_n, (that start at v_n and end at v_n), then the answer is "yes" otherwise, the answer is "no."

Steps 2 and 3 require n and Step 4 requires $n-1$ operations. Therefore the above procedure requires only linear number of operations, i.e. the algorithm is $O(n)$.

In order to solve the second, and more general question above, all possible colorings of the graph need to be generated. For a 2-out digraph there are $(2)^n$ such colorings, where n is the number of vertices in the graph. For each coloring, in the worst case, one should check at least $2^{(n-1)^2}$ paths. In general, for k-out digraph, there are $(k!)^n$ colorings.

We suggest the following procedure for solving the second problem. We assume that the graph is 2-out and so there are only two colors (red and blue). For each vertex v_i an edge e_i is fixed starting at v_i. The other edge starting at v_i is denoted with f_i. The notation e_i^r is used to denote that e_i is colored "red" and notation e_i^b for "blue."

Procedure 2.

Step 1: Generate random paths through the graph with a random choice of red or blue colors for the edges.
Step 2: Keep all paths with length $(n-1)^3$.
Step 3: For $i = 1, \ldots, n$ do
keep in Q_1 only the paths that contain e_i^r but not e_i^b or f_i^r.
keep in Q_2 only the paths that contain e_i^b but not e_i^r or f_i^b.
join Q_1 and Q_2.
Step 4: For each $i = 1, \ldots, n$ keep the paths that start at vertex v_i and end at vertex v_n in set P_i.
Step 5: For $j = 2, \ldots, n$ do
Step 5.1: Keep in R the paths from $P_j, \ldots, P_n$ that start at $v_j, \ldots, v_n$ respectively, and have the same color sequence as paths in P_{j-1} that start with v_{j-1}.
Step 5.2: For $i = 1, \ldots, n$ do
From R
keep in Q_1 the paths from R that color match paths from P_{j-1} and contain e_i^r but not e_i^b or f_i^r
keep in Q_2 the paths from R that color match paths from P_{j-1} and contain e_i^b but not e_i^r or f_i^b
join Q_1 and Q_2 in R.
Step 6: If any paths remain in P_n, then the answer is "yes," otherwise, the answer is "no."

The above algorithm requires n number of operations in Steps 3 and 4, and $n(n-1)$ in Step 5, so the algorithm is $O(n^2)$.

3. Laboratory Techniques

Protocols for Procedure 1.

The laboratory techniques needed to implement either of the above Procedures are straight forward and commonly used in molecular biology (see [**5**] for specific protocols). Here, we outline in detail the biochemical techniques associated with each step in the protocols for the problem associated with the graph in Figure 1.

3.1. Coding of the vertices, colors and edges: The vertices are coded as 20 nucleotide long single-stranded DNA molecules (oligos). Each vertex (v_i) has a unique sequence. The sequences for the vertices were chosen to have minimal similarity to each other as well as to the sequences which specify the color of the edges. We follow the recommendation from [**10**] to set the bound for the Hamming distance between the molecules to be 6.

The colors are specified by a 20 nucleotide long sequence. Both colors were designed to be rich in guanine (G) and cytosine (C) to maximize the potential dsDNA stability of these regions [**8**]. The cross-color DNA similarity also was minimized. Each edge of the given graph (e_i) is represented by a specific, 60 nucleotide ssDNA molecule. As in Adleman [**1**], the first 10 nucleotides of an edge will be identical to the 10 nucleotides of the $3'$ end of the starting vertex oligo. The next 20 nucleotides are an edge and color specific coding sequence. The following

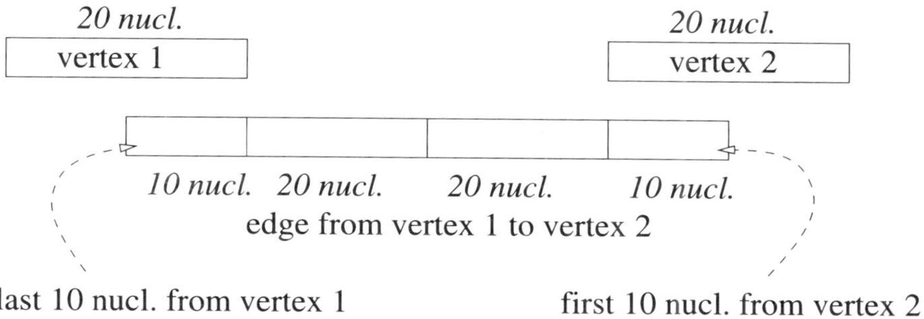

Figure 2.

20 nucleotides are the color coding sequence and the remaining 10 nucleotides of an edge correspond to the 10 nucleotides of the $5'$ end of the ending vertex. An additional five oligos are needed to specify the starting and ending points of the paths. Four alpha (α) oligos consist of an arbitrary 20 nucleotide sequence followed by ten nucleotides of the $5'$ end of each of the vertex sequences. An additional omega (ω) oligo consists of ten nucleotides from the $3'$ end of the common ending vertex (e.g., v_4) followed by an arbitrary 20 nucleotide sequence.

3.2. Assembling random paths: Oligonucleotides that are anti-parallel and complimentary to the vertices (but not α or ω) also are synthesized. These will function as linkers to the edges so that random paths can be generated in Step 1. Once generated, the edges of the paths will be permanently bonded together with DNA ligase.

3.3. Selecting potential paths: The following is a description of the protocols that will be implemented to solve the graph illustrated in Figure 1. After path assembly, Steps 2 and 3 of Procedure 1 are accomplished by a combination of vertex specific polymerase chain reaction (PCR) amplification and agarose gel electrophoresis. Four separate PCR amplification reaction are performed. Each PCR reaction has one of the vertex specific α oligos and the common vertex ω oligo. The resulting DNAs from these amplifications are subjected to agarose gel electrophoresis. DNA fragments $[60(4-1)^3 + 40 = 1,660]$ base pairs long are isolated from the gel and serve as templates for a second round of PCR amplifications. At this point there are 4 tubes each with potential solutions representing paths that start at each of the four vertices respectively, and all of them end at vertex 4. The molecules represent paths with a length of 27 edges plus α and ω (i.e., 1660 base pairs).

3.4. Isolating paths with identical color sequences: Step 4 requires that the order of colors be determined and is somewhat more complex. Pair-wise tests will be made between paths that start at each of the vertices and end at vertex 4. A small amount of DNA (referred to as test DNA) corresponding to the paths that begin at a vertex (e.g., v_2) is mixed with an excess of DNAs (referred to as driver DNA) representing paths that begin at another vertex (e.g., v_1). The dsDNAs are denatured by heating and then allowed to renature slowly by cooling. This predominantly will result in the single stranded driver DNA molecules reforming dsDNA with their original compliments. A much smaller number of partially dsDNA molecules will form between single stranded test DNA and single stranded driver DNA (we refer to these dsDNA molecules as heterologous molecules). Since driver DNA sequences are in great excess relative to test DNA sequences, there should be little if any fully complimentary double stranded test DNA molecules. Heterologous molecules that represent paths with shared color sequences will have all of the color nucleotides in dsDNA form. Any paths that do not have a common sequence of colors will have color segments that are not dsDNA. These partially dsDNA molecules are removed from the pool by hybridization with biotin labeled oligos, complimentary in sequence to the color sequence. Under appropriate conditions, the biotin labeled oligos will bind to the available single stranded but not double stranded color segments. The resulting complex is then reacted with antibodies to biotin which have been conjugated to streptavidin paramagnetic beads. After treating the mixture with a magnetic field, the unbound DNA molecules are recovered. The recovered DNA is subjected to vertex specific PCR using the α primer for the test DNA sequences (e.g., α-v_2) and the ω primer. This results in the isolation of paths that start at v_2 and have the same color sequences with the paths starting at v_1. This DNA is then used as the driver to color match paths starting at another vertex (e.g. v_3) and the procedure is repeated as above. In each step of this process we are accumulating the paths that have common color sequences with all of the driver DNAs previously used.

3.5. Determining color sequence of solutions: The final step is to recover any DNA strands that remain (i.e., paths through the graph with common color sequences from all vertices) and to determine the exact sequence of each of the paths. To do this, any remaining DNA molecules are ligated into a bacterial cloning vector and the cloned paths are used to transforming bacterial cells. Vector DNA from these clones are then sequenced by standard manual or automated sequencing protocols. Alternatively, the cloned paths can be subjected to PCR amplification using color specific and ω primers. Separate "red" and "blue" PCR reactions are performed. The color sequence of the paths can be deduced by size sorting of the resulting amplified DNA.

Protocols for Procedure 2.

The laboratory protocols to implement Procedure 2. are similar to those for Procedure 1. with the following modifications. First, edges which appear in all possible colorings of the graph are included in the initial ligation step. This creates the possibility that some paths contain *both* colorings for the same edge. To remove these illegitimate path codings, each edge was coded with an edge and color specific 20 nucleotide sequence (denoted with e_i^r or e_i^b...) After the second step, the molecules are amplified through PCR with α and ω primers. The DNA is then separated in two tubes Q_1 and Q_2 mixed with edge and color specific biotin labeled oligos (Q_1 with e_1^b and f_1^r and Q_2 with e_1^r and f_1^b). The molecules are denatured by heating and cooled. with a magnetic field (as in Protocol 3.4). The unbound molecules are recovered and are subjected to PCR amplification with α and ω primers. Then, Q_1 will contain paths which include only e_1^r and/or f_1^b and Q_2 will contain paths which include e_1^b and/or f_1^r. This step is repeated for each vertex $i = 1, \ldots, n$.

In the process of matching the colors, two distinct paths with the same color sequence may contain the same edge colored in different colors. In other words, if p and p' are paths with same color sequence, but p contains e_i^r and p' contains e_i^b or f_i^r, the heterologous molecules from Protocol 3.4. may contain partly dsDNAs with one strand encoding p and the other strand encoding p'. We will call these molecules inconsistent heterologous matches. The inconsistent heterologous molecules containing impossible edge combinations are removed by hybridization with biotin labeled oligos complimentary to the edge and color specific sequences in a manner similar to Procedure 1. In the following description we will again use the graph in Figure 1. as an example. For Step 5.1., an excess of driver DNA corresponding to the paths that start at v_1 is mixed with small amounts of test DNAs corresponding to paths that start at v_2, v_3, v_4. They are denatured and allowed to hybridize as in Protocol 3.4. The unmatched color sequences are removed as before. The isolated DNA is subjected to PCR amplification in three tubes with the α-v_2, α-v_3 and α-v_4 respectively and ω primers.

For Step 5.2, aliquots of these test DNAs are mixed in two separate tubes with an excess of the driver DNA and then heated and cooled. The renatured DNAs are again mixed with edge and color specific biotin labeled oligos. For example, one of the tubes will contain the e_1^b and f_1^r specific biotin labeled oligos and the other will contain e_1^r and f_1^b labeled oligos. This results in tubes Q_1 and Q_2 from Step 5.2. in Procedure 2. The magnetic bead separation is repeated for each vertex, i.e. edge e_i. This needs to be done while the molecules are hybridized. We assume that molecules exist in many copies and they have possibilities of forming different consistent and inconsistent heterologous matchings. Denaturing molecules by heating and allowing them to cool again could yield new inconsistent heterologous matching (eg. for edge e_1) which might be omitted in the second and the following steps of edge and color specific separations (separations using the edges e_2 to e_n). After magnetic bead separation (Step 5.2.), the unbound DNAs from each tube are pooled and reamplified in three tubes with the α-v_2, α-$v_3, \dots, \alpha$-v_n respectively and ω primers.

The entire above procedure is repeated using $v_2, \dots, v_{n-1}$ as driver DNA and $v_3, \dots, v_n$ as test DNA. If any molecules remain, they should encode synchronizing paths of synchronizing colorings of the graph.

4. Discussion

There have been two suggestions for a general model of molecular based computations ([**2**] or [**4**]). Both of these models are based on basically the same 4 operations: (1) *separate* or *remove*, (2) *merge* or *union*, (3) *amplify* or *copy* and (4) *detect* or *select*. The need to "match" the colors of the paths is particular to the RCP. We were unable to solve this problem by using only the four operations mentioned above. The structure of the DNA molecules is such that they store or encode information, but cannot transmit it within the molecule. In the procedures above, we perform another operation which is *matching* the molecules containing the same color sequence. We believe that this is possible by careful coding, especially of the colors. This leads to a new operation added to the list of four. A possible way of describing this operation is as follows: Let $S_1, S_2 \subseteq \Sigma^*$ where Σ is an alphabet (in particular, Σ can be taken to be $\{C, G, A, T\}$), $S_1 \cap S_2 = \emptyset$ and S_2 is finite. In the RCP for the graph in Figure 1., S_1 is the set of sequences encoding the vertices and S_2 has two strings encoding the two colors. Let $U = \{\, v_1c_1 \cdots v_nc_n \mid v_i \in S_1, c_i \in S_2\}$. Let $U_1, U_2 \subseteq U$ be two subsets of strings. Then $\mathbf{Match}(U_1, U_2, S_2) = \{w = v_1c_1 \cdots v_nc_n \in U_2 \mid \exists w' \in U_1, w' = v_1'c_1 \cdots v_n'c_n\,\}$.

In other words, **Match**(U_1, U_2, S_2) gives a subset of U_2 in which for every string w in this set there is a string w' in U_1 such that the c_i-subsequences in w and w' are equal.

From a molecular point of view, the operation **Match**(U_1, U_2, S_2) will be successful only if careful coding of the strings $c_i \in S_2$ is planned, i.e. sufficient length oligos (about 20bp) and rich in guanine (G) and cytosine (C) relative to the strings v_i. We also assume that the molecules that are possible outcomes of this operation exist in many copies and by taking small amount of DNA from U_2 we are not omitting solutions. This process also involves separation of the unmatched molecules (see Protocol 3.4.) which may be done in parallel for all $c_i \in S_2$ at once.

Separating molecules by using the magnetic beads and, in particular, repeating this process many times may yield errors which will not give us 100% efficiency. In Procedure 2., the repetitive use of **Match** involving separation and PCR amplification (up to 24 times for the graph in Figure 1.) increases the likelihood of error. In particular, in Step 5.2. in Procedure 2 we mix an excess of DNA molecules representing paths that start at one vertex (say v_1) with a small amount of test DNAs from some of the rest of the tubes (paths starting at v_2, v_3 and v_4). By increasing the number of test DNAs we are increasing the possibility that in the process of color matching, test DNAs will meet with each other and have consistent edge colors. These molecules will not be removed and will be recovered after Step 5.2. Thus generate errors. In this step we are, in essence, performing **Match** in parallel. At this moment we believe that for small graphs of 10 or so vertices (and in particular for the graph in Figure 1.) these molecules would be almost non-existent and the PCR amplification would not affect them. We have no measure for the accuracy of the parallelism of this step and we believe that after experimental laboratory implementation of the Procedure 2 we will have some estimate for the limits of this step.

An alternative way of separating molecules is presented in [**4**]. It is suggested that restriction endonucleases that recognize and cleave at specific sequences in the DNA may be used in removing unwanted molecules. Although, it was asserted that this process is more error-resistant, a successful laboratory implementation of this technique has not been reported. We believe that a similar technique of removing unwanted molecules might be used in the process of **Match**, but we have yet to pursue this line of research.

In both models, [**4**] and [**14**], it is assumed that the input has an exponential number of distinct molecules (as suggested by Lipton in [**14**]). Step 1., in both of our procedures, generates the input for our algorithms and we assume that this step generates an exponential number of molecules which encode the paths needed for a potential solution. As the number of vertices increases, the length and the number of the paths that contain a solution of the RCP increases with a cubic power relative to the number of vertices, i.e. if we have 11 vertices, each edge is encoded with 60bp oligo, the length of the molecule used in our procedures that encodes a solution to the RCP is $60 \cdot 10^3$bp. At this length, it is difficult, by gel electrophoresis, to separate molecules that are 60bp apart. The algorithms, however, can be performed even when the precise size separation of the DNA molecules is not possible. Unfortunately, PCR amplification is also difficult (if not impossible) for very long molecules [**6**]. At this time the length barrier is another of the limiting factors to the molecular computations.

The coding of the oligos needed for molecular computations is a critical step. The application of the Hamming distance bound for the oligo sequences as in [**10**] was not a sufficient constraint for our purposes. During the hybridization required to generate paths in the first step of the procedures, illegitimate partial molecule matching by sliding of the reverse complements of the vertex oligos can introduce errors in the construction of the paths. This situation was not considered by [**10**]. A good code satisfying the Hamming distance bound is given in [**10**], but this code would have failed in the process of ligation and formation of the paths (Step 1 in the procedures) since by slight sliding, the last 8 nucleotides of vertex 5 and the first 12 nucleotides of vertex 3 can be joined by the complement of vertex 2. For the RCP we investigated the codes for the oligos by visual inspection with the aid of the computer program Oligo 4.0 [**18**]. Even for the small graph from Figure 1 this required a considerable amount of time and effort. It is our opinion that much more emphasis needs to be placed on the production of efficient algorithms for encoding oligos used in DNA computation.

We are aware that, due to length, weight and coding barriers, a molecular computation may not be fully applicable in practice to the RCP. However, this question presents the special need for the new operation of **Match** which, to our knowledge, has not been widely discussed.

The application of molecular biological techniques is a fundamentally new approach to computational problems. Even though the massive parallelism in all of the algorithms proposed makes solutions for many of the NP-complete problems attainable, limitations still persist. In essence, none of the proposed approaches eliminate the problem of exponential expansion. There still are unavoidable physical limits to molecular computation. As Hartmanis points out [**12**], if Adleman's experiment were used to solve even a moderately complex Hamiltonian Path Problem, "the biologically encoded set of paths will weight more than $3 \cdot 10^{25}$kg, which is more than the weight of the Earth." Nonetheless, there remain problems which are intractable for the electronic computers (e.g., breaking the Data Encryption Standard code) which could be done in about four months of lab work with molecular computation [**7**].

5. Acknowledgments

The authors wish to thank Erik Winfree for his valuable comments and helpful insights.

References

[1] L. Adleman, Molecular computation of solutions of combinatorial problems, Science **266** (1994), 1021–1024.

[2] L. Adleman, On constructing a molecular computer. In R. J. Lipton and E. B. Baum, editors, *DNA Computers: Proceeding of a DIMACS Workshop,* DIMACS Series in Discrete Mathematics and Theoretical Computer Science, American Mathematical Sociey, 1-21, 1996.

[3] R. Adler, L.W. Goodwin, B. Weiss, Equivalence of topological Markov shifts, Israel J. Math. **27** (1977), 49–63.

[4] M. Amos, A. Gibbons, D. Hodgson, Error-resistant implementation of DNA computations, In L. F. Landweber and E. Baum, editors, *DNA Based Computers: Proceedings of the Second DIMACS Workshop*, DIMACS Series in Discrete Math. and Theoretical Computer Science, Amer. Math. Soc., 151-161, 1998.

[5] F.M. Ausubel, R. Brent, R.E. Kingston, D.D. Moore, J.G. Seidman, J.A. Smith, K. Struhl, P. Wang-Iverson and S.G. Bonitz, Current Protocols in Molecular Biology, Greene Publishing Associates and Wiley-Interscience, New York, NY (1993).

[6] W.M. Barnes, PCR amplification of up to 35kb DNA with high fidelity and high yield from lambda bacteriophage templates, Proc. Natl. Acad. Sci. USA **91** (1994), 2216–2220.

[7] D. Boneh, R. Lipton, C. Dunworth, Breaking DES using a molecular computer. In R.J. Lipton and E. Baum (eds), *DNA Based Computers: Proceedings of a DIMACS Workshop*, DIMACS Series in Discrete Math. and Theoretical Computer Science, Amer. Math. Soc., 37-66, (1996).

[8] R.J. Britten, E.H. Davidson, Hybridization strategy, Nucleic acid hybridization: a practical approach (B.D. Hanes and S.J. Higgins - editors). IRL Press, Washington, DC. (1985) 3–15.

[9] J. Černy, Poznamka K homogenym s Konecnymi automati, Mat-Fyz. cas SAV **14** (1964), 208–215.

[10] R. Deaton, R.C. Murphy, M. Garzon, D.R. Francceschetti, S.E Stevens, Good Encodings for DNA-Based Solutions to Combinatorial Problems. In L. F. Landweber and E. Baum, editors, *DNA Based Computers: Proceedings of the Second DIMACS Workshop*, DIMACS Series in Discrete Math. and Theoretical Computer Science, Amer. Math. Soc., 247-258, (1998).

[11] J. Friedman, On the road coloring problem, Proceeding of the AMS **110**, (1990) 1133–1135.

[12] J. Hartmanis, On the weight of computations, Bulletin of the European Association For Theoretical Computer Science **55** (1995) 136–138.

[13] N. Jonoska, S. Suen, Monocyclic decomposition of graphs and the road coloring problem, Cong. Num. **110** (1995) 201–209.

[14] R. Lipton, DNA solution of hard computational problems, Science **268** (1995), 542–545.

[15] G.L. O'Brien, The road coloring problem, Israel J. Math **39**, (1981) 145–154.

[16] R. Pool, A boom in plans for DNA computing, Science **268** (1995), 498–499.

[17] D. Perrin, M. P. Schützenberger, Synchronizing Prefix Codes and Automata and the Road Coloring Problem, Symbolic Dynamics and its Applications (Contemporary Mathematics **135**, P. Walters - editor) (1992), 295–318.

[18] W. Rychlik, R. E. Rhoads, A computer program for choosing optimal oligonucleotides for filter hybridizations, sequencing, and in vitro amplification of DNA, Nucl. Acids Research **17** (1989) 8543–8551.

[19] W. Smith, DNA Computers in vitro and vivo, In R. J. Lipton and E. B. Baum, editors, *DNA Based Computers: Proceedings of a DIMACS Workshop*, DIMACS Series in Discrete Math. and Theoretical Computer Science, Amer. Math. Soc., 121-185, 1996.

Institute for Biomolecular Sciences, Department of Mathematics, University of South Florida, Tampa FL, 33620

E-mail address: jonoska@math.usf.edu

Department of Biology, University of South Florida, Tampa FL, 33620

E-mail address: karl@chuma.cas.usf.edu

DIMACS Series in Discrete Mathematics
and Theoretical Computer Science
Volume **44**, 1999

DNA Based Molecular Computation: Template-Template Interactions in PCR

Peter D. Kaplan, Guillermo Cecchi, Albert Libchaber

Abstract. Since Adleman performed a computation with molecules of DNA[**1**], there has been theoretical work on parallel computing with DNA[**8, 10, 2**] but no experimental review of the promises and complications of DNA based computing. In this experiment, we focus on heteroduplex formation during the polymerase chain reaction as one critical complication to large scale DNA computing.

1. Introduction

Von Neumann's pioneering work on large-scale computing devices and formal descriptions of living systems[**9**] has led to a two-fold interest in molecular computation: to reach the extreme limits of miniaturization and to understand the complex behavior of life in terms of a symbolic theory of the manipulation of information. Although this work does not discuss the later issue, we hope that by probing the limits of biomolecular computation, we will gain insight into the information handling capacities of the cell. We choose DNA for molecular computation because it is clear that many of the challenges of manipulating information on a molecular level have been solved by biological systems; moreover, computationally intractable problems of molecular design have been solved using nucleic acids by the technique of *in-vitro* evolution[**3, 11**]. By pointing out that today's biotechnology allows us to bring over a billion years of evolutionary development to molecular computation, Adleman has allowed us to foresee a day when molecular computing will be a reality. As experimental scientists, we have been exploring the power and limitations of current technology. In this report, we focus on the DNA amplifier known as the polymerase chain reaction (PCR) and an inevitable complication of using PCR for parallel computations in which many distinct but similar strings are being copied.

Adleman's key innovation was finding a method of encoding a computer style problem in DNA sequences (finding a Hamiltonian Path on a directed graph). A directed graph is a set of vertices connected by edges on which it is only possible to move from one vertex to another if there is an edge connecting them in the correct direction (*e.g.* Fig.1a). A Hamiltonian Path passes through the graph visiting every vertex exactly once; the path must start at vertex S and end at vertex E. The computer is realized by using single stranded 20 nucleotide sequences of DNA to represent each vertex and each edge. The edge sequences are chosen so that they will hold two vertices together by base pairing (Fig.1b). By making a mixture of all edges and vertices, we imagine that all possible paths will randomly assemble; the computational task is to sort through these possible paths for Hamiltonian Paths. Another way to describe this is to call the vertices words and the edges (grammatical) rules which specify which sequence of words are legal. Our algorithm is to generate a complete language and then sort through it for sentences which represent Hamiltonian Paths. This scheme is generally applicable in the sense that many problems can be solved by generating all possible solutions and then sorting through them for the correct answer. In the course of repeating the previous experiment[**4**], we found that the major difficulties were at the beginning, in assembling a complete language of paths and in copying a subset of that language by PCR. In this paper we examine apparently oddly sized DNA produced when copying the language and describe one way this DNA is produced by PCR.

2. PCR as an amplifier

The polymerase chain reaction is an important tool for DNA based computation. In fact, PCR, as a DNA amplifier, will almost certainly be a key component of any large scale DNA computer. In PCR, the sample is heated to separate or denature the double stranded template and then each of the two strands is copied. Each time this thermal cycle is repeated, both the strands and the copies will be duplicated — therefore the number of copies grows exponentially with the number of thermal cycles. The copying is initiated by the hybridization, of one of two "primer" oligonucleotides (Fig.1c). The assumption that by maintaining a surplus of primer to template we can neglect template/template interactions turns out to be incorrect when amplifying heterogeneous populations of DNA.

As part of our efforts to reproduce Adleman's experiment, we did some experiments on a simple graph (Fig. 1a). In the simple graph, the Hamiltonian Path ($S12345E$) is the only path. In the next two graphs, SG+13 and SG+35, we add a single edge making possible a single non-Hamiltonian Path (for SG+13 it is the path $S1345E$, and for SG+35 it is $S1235E$). To begin the calculation, in a 30 μl reaction, we mix together a 50 picomoles of single stranded oligonucleotides representing each edge and each vertex of the graph (Table 1). and then we allow the oligonucleotides to hybridize in the presence of the enzyme T4 DNA ligase which forms permanent paths by linking the sugar phosphate backbone of the oligonucleotides (Fig.2 lanes 4-6)[**1**] [1] After a two hour incubation at 37° C, 5 μl is transferred to a 50 μl PCR reaction and paths stretching from S to E are copied using oligonucleotides S and the compliment to E as primers[2] (see Fig.1c). The PCR product is examined by polyacrylamide gel electrophoresis[3] (Fig. 2 lanes 1-3) On observing the ligation product on this gel (lanes 4-6), we find a continuous

[1] A 94° C mixture of edges and vertices containing 50 pm of each oligonucleotide was cooled to 37° C and Promega DNA Ligase buffer containing 10 units of T4 DNA Ligase was added.

[2] 50 μl PCR reactions were performed in a buffer of 10 mM Tris (pH 8.3), 50 mM KCl, 1.5 mM $MgCl_2$, 0.001% gelatin, 100 μm each of dATP, dTTP, dCTP, and dGTP, 50 pm of both primers, and 2.5 units of recombinant *Thermus aquaticus* DNA polymerase. The complete temperature cycle was: 94° C for 5 min; 35 cycles of (94° C, 30 s; 55° C, 30 s; 72° C, 60 s); 72° C for 5 min. The reaction was then cooled to 4° C in a Perkin Elmer model 9600 thermal cycler.

[3] Native polyacrylamide gels were made by casting a mixture of 6 ml of 40% 19:1 Acrylamide:bis-acrylamide, 30 ml H_2O, 4 ml 10X TBE (1 M Tris, 0.9 M Boric acid, 10 mM EDTA), 270 μl of 10% Ammonium persulfate and 50 μl of TEMED (N,N,N',N'-Tetramethylethylenediamine) between glass plates separated by 1.5 mm. A teflon comb was

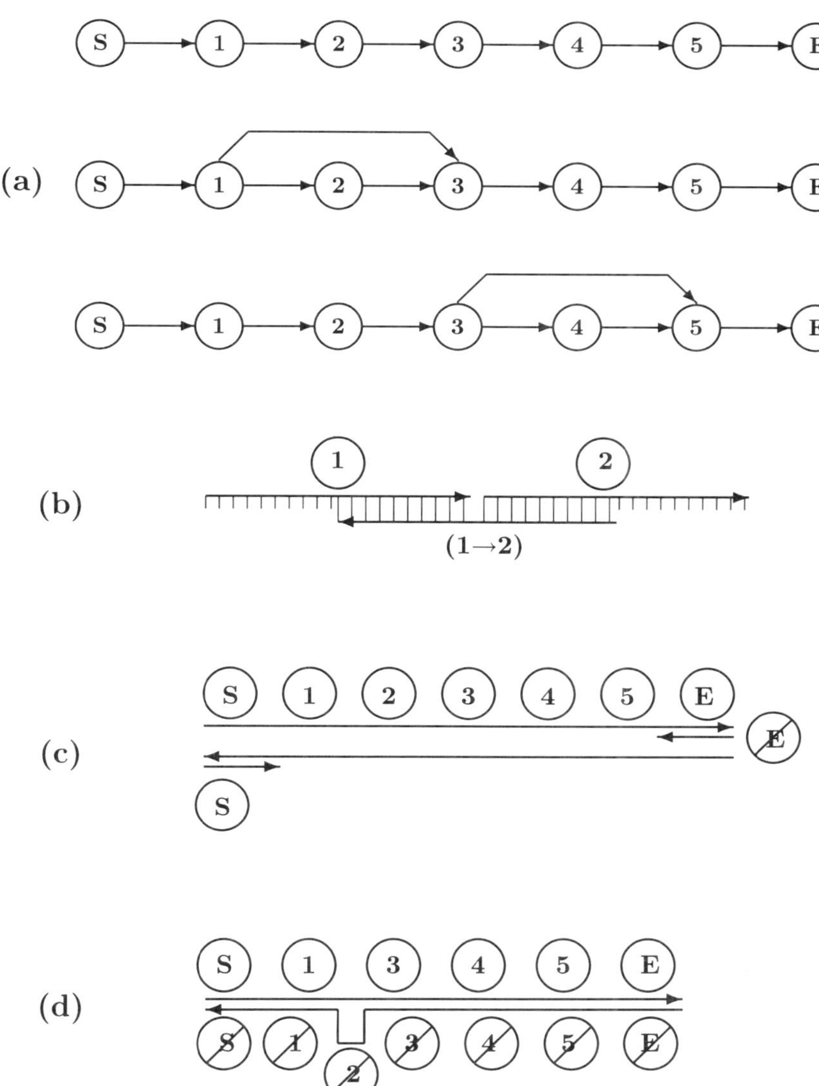

FIGURE 1. (a) By solving the Hamiltonian Path on these three graphs, the Simple Graph (SG), SG+13, and SG+35, we have been able to identify one source of surprising bands in a DNA computation. (b) The vertices of the graph are represented by single strands of DNA. Vertices are held together by annealing to edges whose sequence matches the end of one vertex and the beginning of the next. This structure is the template for ligation. (c) To amplify sequences beginning with S and ending with E, we perform PCR using S and the compliment to E as primers (in the figure the compliment is denoted by a slash through the symbol). (d) The heteroduplex structure formed by 120 and 140 nucleotide strands annealing during PCR. This structure migrates slower than 120 and 140 bp double stranded DNA.

TABLE 1. Sequences used to represent the graph in Fig 1a (* denotes 5' phosphorylated oligonucleotides).

vertex	sequence(5' to 3')
S	GAAGGTAGCAGACCGAAGTA
1	*GCGGAACTGCACAATAAGCA
2	*TGGAGAATACAAGAGCAACC
3	*AAGTCACTTGACCTTCGGCC
4	*CGCGGATGGACCAGCGTCCA
5	*AGCCTGGAATGGTTAGTCGA
E	*AACCTATGTGTGGCGGCGCT

distribution of electrophoretic mobilities, not a set of clear bands corresponding to the discrete numbers of oligonucleotides. The continuous distribution may be due to the formation of complex structures, which may involve folded DNA, complexes between several strands, or incorrect ligation products. On examining the PCR product (lanes 1-3) we note that the simple graph (lane 3) produces a single strong band at 140bp (7 steps times 20 bases/step) as expected. In the other graphs (SG+13 lane 5, SG+35 lane 1), in addition to the expected bands at 140 bp (7 steps) and 120bp (6 steps), we find a striking band at about 190 bp. This band is surprising because the only way to generate a path longer than 7 steps is to have an edge pointing backwards; no such edge is included in Fig.1a. Further, 190 bp corresponds to 9.5 steps. We do not know how to interpret a half step. We decided to explore this 190 bp band as a prototype of the unexpected bands seen in experiments on DNA computation.

We isolated the DNA in the anomalous band by cutting it from the gel, soaking the gel slice in water overnight, and amplifying DNA in the soak water by PCR. Although we expected to see only a band at 190 bp, the PCR product included all three original bands (120, 140, and 190 bp). By varying the amount DNA used as input to PCR, we found that at low concentrations, only the short bands are produced, but at higher concentrations, all three bands appear (Fig. 4, lanes 1-4). To rule out the possibility of mis-priming during PCR, we performed a complete set of optimization reactions, varying the pH from 8.5 to 10, the concentration of Mg^{++} from 1.5 to 3.5 mM with both 0 and 10% DMSO[**6**]. All three bands (120, 140, and 190 bp) always appeared together. To explore the possibility that an illegal or ungrammatical path was responsible for the 190 bp band, we performed PCR using many combinations of vertices and their complements as primers. No coherent pattern emerged.

We then examined the PCR product on a denaturing gel [4] on which DNA separates into single stranded molecules (Fig.3). On the denaturing gel, we found only 2 bands (120 and 140 bp). From this result, we inferred the possibility that

inserted between the plates to make loading wells. Gels were allowed to polymerize for at least 2 hours. Samples mixed with a 6X loading buffer of 10% Ficoll in 50 mM Tris (pH 8.3). After loading gels, equilibrated for at least 30 minutes prior to electrophoresis for 50 min in a running buffer of 1X TBE at 13 V/cm and 25° C. Gels were visualized using a UV transilluminator after a 10 min soak in 1 μM ethidium bromide.

[4]Denaturing gels were run as native gels except that the gel contained 7 M Urea, 12% acrylamide, the loading buffer was a 50/50 mix of sample and formamide, the electrophoresis apparatus was heated to 55° C, the electric field was 21 V/cm and the gel was stained with a 20 min soak in 1μM SYBR Green I (FMC).

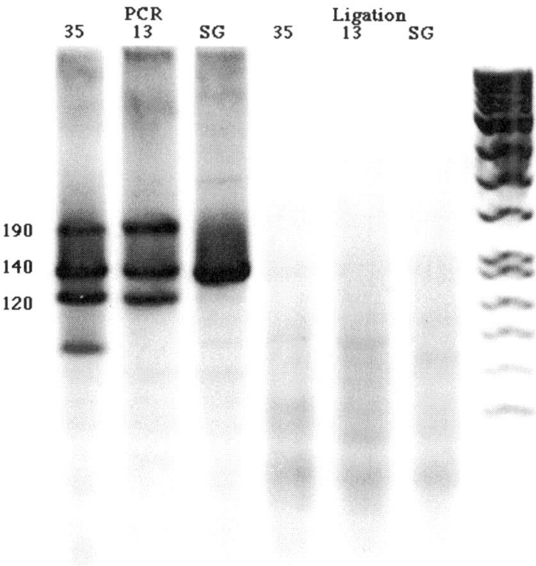

FIGURE 2. Ligation and PCR of the three graphs in Figure 1a. While the PCR product of SG (lane 3) displays only the expected band at 140 bp, the products of SG+35 (lane 1) and SG+13 (lane 2) display bands at 120 and 140 bp as well as a surprising band at 190 bp. Note that the ligation products (lanes 4-6) do not migrate in discrete bands.

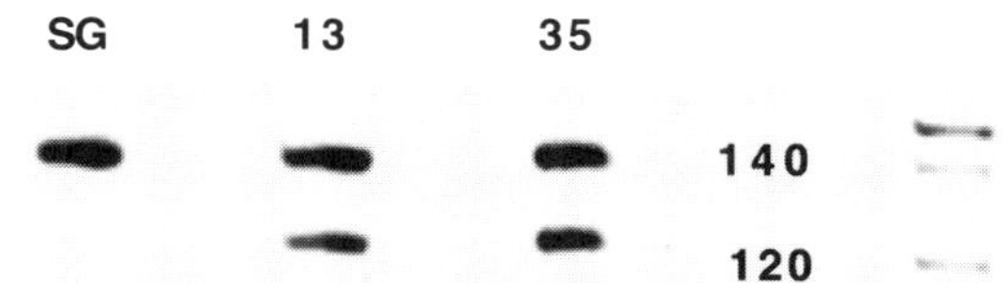

FIGURE 3. Denaturing gel of PCR product displayed in Figure 4 lanes 5-7. When the strands are separated, we see only two sizes of DNA, 140 and 120 nucleotides. This suggests that the DNA in the apparently 190 bp band is actually the heteroduplex structure sketched in Figure 1d.

the DNA in the long band was a hybrid of complimentary 120 and 140 bp DNA (Fig. 1d). To prove that this is the source of the long band, the PCR product was digested with the restriction endonuclease *AvaII* which cuts inside the sequence for vertex 4. The digestion products (Fig. 4 lanes 5-7) are consistent with our structure. In particular, the long and short bands in the graph SG+35 are not cut by *AvaII* because vertex 4 never appears as a *double stranded* object in this DNA while the long and short bands in SG+13 *are* cut by *AvaII.* We conclude that the creation of heteroduplex structures during PCR is a significant source of unexpected bands during computation.

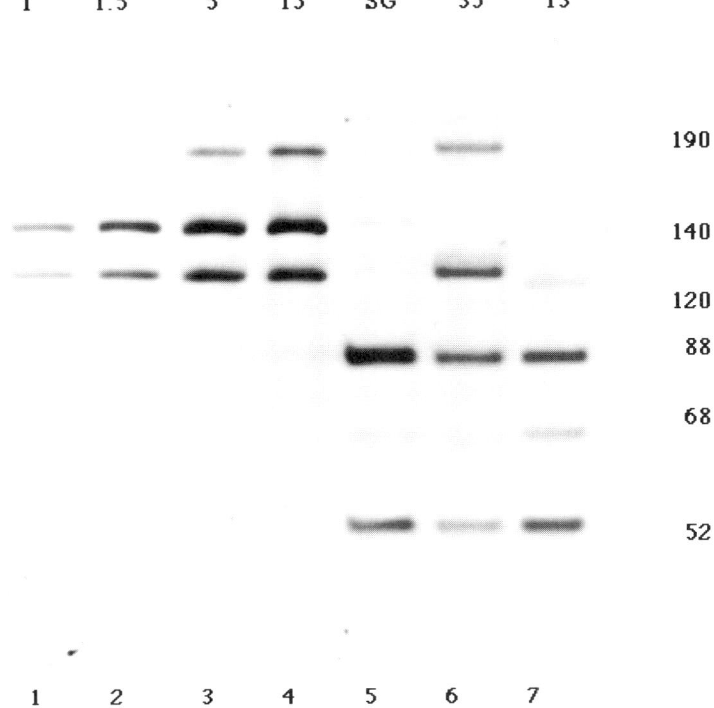

FIGURE 4. Details of the anomalous 190 bp band. (Lanes 1-4) We amplify the 190 bp band with different amounts of 190 bp input (0.5, 1.5, 5, and 15 μl of the water used to soak the gel slice from 190 bp in Figure 2, lane 6 were added to a 100 μl 15 cycle PCR reaction). For small quantities of input DNA, the shorter 120 and 140 bp DNA is produced while at high concentrations, the surprising 190 bp band emerges. (Lanes 5-7) PCR product of the 140 bp amplification of SG and the 190 bp amplification of SG+35 and SG+13 digested by the restriction endonuclease *AvaII* which cuts the sequence G'G(A/T)CC at position 88 of the Hamiltonian Path. This digestion decomposes the 140 bp band into pieces 52 and 88 long, the 120 bp band of SG+13 into 52 and 68, the 120 bp band of SG+35 is not cut. The critical result of this digestion is that the 190 bp band of SG+13 is cut while the 190 bp band of SG+35 is not cut because vertex 4 is part of the single stranded loop in the heteroduplex and *AvaII* will not cut single stranded DNA. *Methods*: 20 μl of the PCR product displayed in lanes 4-6 of Figure 2 are mixed with 24 units of *AvaII* (New England Biolabs) in the manufacturer's recommended buffer with 0.1% Bovine Serum Albumin. The reaction is incubated at 37 C for 16 hours.

3. Heteroduplex Formation

The remainder of this report discusses the significance and implications of heteroduplex formation in computation and in biology. Biologists use heteroduplex formation to identify deletion or insertion mutations[**12, 7**]. PCR amplification of the gene of interest is performed in the presence of a non-mutant allele and the results are analyzed by gel electrophoresis. As the normal and mutant alleles can differ by one or two bases, they are difficult to distinguish directly on a gel. The heteroduplex's single stranded bulge results in a vastly different mobility creating a distinct, distant band which is a clear signature of mutations[**7**].

In the context of computation, the phenomenon of heteroduplex formation requires careful consideration. First, heteroduplex formation will be ubiquitous in DNA based computation in which PCR is used as an amplifier. In this simple graph there is a single heteroduplex; in a more complicated graph (*e.g.* [**1**]) there is an enormous set of possible heteroduplex structures. Next, although it is the best available tool, gel electrophoresis is not a good tool for analyzing DNA. As mobility varies greatly with DNA conformation we should use denaturing or single stranded gels rather than native gels for analysis; we hope that a new technology will replace electrophoresis. Next, there are implications for designing calculations. Heteroduplex formation can occur during the annealing step of each cycle of PCR. If the 3' end of a valid computational string anneals to the middle of another string, DNA polymerase will extend this 3' end producing a new, illegal string while simultaneously destroying a valid string. Extreme caution is required to design protocols so that 3' ends will not hybridize incorrectly and so that broken strings are removed before they contribute their 3' ends to a new, illegal, hybrid string. Finally, it is not necessary to use high concentrations of template or to produce large concentrations of product during PCR. By keeping all concentrations low, we believe that heteroduplex formation and other template/template interactions will be minimized. Finally, good computational design should take advantage of this process to create new algorithms which use relatively infrequent template/template hybridization to *advance* the calculation[**5**].

4. Acknowledgements

We gratefully acknowledge invaluable conversations with Marcelo Magnasco, David Thaler, Shumo Liu, and Nadrian Seeman.

References

[1] L. M. Adleman, Molecular Computation of Solutions to Combinatorial Problems. *Science*, 266:1021-1024, 1994.

[2] D. Beaver, Computing with DNA. *J. of Comp. Bio.*, 2:1-7, 1995.

[3] G. F. Joyce, In-Vitro Evolution of Nucleic-Acids. *Curr. Opin. Struct. Biol.*, 4:331-336, 1994.

[4] P.D. Kaplan, G. Cecchi, A. Libchaber, unpublished.

[5] P.D. Kaplan, A. Libchaber, in progress.

[6] Kit from Invitrogen (San Diego, CA).

[7] D. M. J. Lilley, Kinking of DNA and RNA by base bulges. *Proc. Natl. Acad. Sci. USA*, 92:7140-7142, 1995.

[8] R. J. Lipton, DNA Solution of Hard Computational Problems. *Science*, 26:542-545, 1995.

[9] J. Von Neumann, Theory of self-reproducing automata, A.W. Burke, ed. Univ. of Illinois Press (Urbana, IL 1966). Probabilistic logic and the synthesis of reliable organisms from unreliable components, C.E. Shannon and J. McCarthy, ed. Princeton Univ. Press (Princeton, NJ 1956).

[10] J. Reif, Parallel Molecular Computation: Models and Simulation. *Proc. of 7th Annual ACM Symposium on Parallel Algorithms and Architectures SPAA'95*, 213–223, 1995.

[11] J. W. Szostak and A. D. Ellington, In Vitro Selection of Functional RNA Sequences. In R. F. Gesteland, J. F. Atkins, editors, *The RNA World*, Cold Spring Harbor Press, Cold Spring Harbor, 1993.

[12] B. L. Triggs-Raine and R. A. Gravel, Diagnostic heteroduplexes: simple detection of carriers of a 4-bp insertion mutation in Tay-Sachs disease. *Am. J. Hum. Genet.*, 46:183-184, 1990.

NEC Research Institute, 4 Independence Way, Princeton, NJ 08540
E-mail address: kaplan@athena.rockefeller.edu

(P. Kaplan and G. Cecchi) Center for Studies in Physics and Biology, Rockefeller University, New York, NY 10021

E-mail address: guille@babel.rockefeller.edu

DIMACS Series in Discrete Mathematics
and Theoretical Computer Science
Volume **44**, 1999

Use of a Horizontal Chain Reaction for DNA-Based Addition

Frank Guarnieri and Carter Bancroft

ABSTRACT. We have reported previously the development of a DNA-based algorithm for addition of two non-negative binary numbers, and the experimental application of a simple example involving the addition of two binary digits. We describe here the extension of this algorithm to the addition of two rational non-negative numbers expressed in any base.

1. Introduction

The seminal studies by Leonard Adleman [**1**] demonstrated theoretically and experimentally how DNA can be employed for solving a directed Hamiltonian path problem, thus illustrating the utility of DNA for performing massively parallel operations. A subsequent investigation by Richard Lipton provided the theoretical framework for the generalization of this approach to the use of DNA for the solution of hard, NP-complete computational problems such as the "satisfaction" problem [**6**]. Since then, a number of investigators have proposed additional novel concepts that are continuing to push forward the frontiers of DNA-based computing [**3, 7, 4**].

Much of the current thought in this area is directed to the use of the enormous potential of DNA for carrying out massively parallel operations to solve interesting combinatorial problems. However, it seems to us that an obvious prerequisite for designing a novel computer that will be useful in a wide range of applications is the development of the bit manipulations required to carry out mathematical calculations such as addition. We are therefore exploring the question: *Given the chemical properties of DNA and the enzymes that have evolved to operate on this molecule, is it possible to design and execute controlled biochemical reactions that mimic the bit-flipping operations of a computer?*

To this end, we have recently described the development and application of a DNA-based algorithm for adding any two non-negative rational binary numbers [**5**]. In that work, we first developed a simple DNA-based algorithm for adding two binary digits, and carried out the biochemical execution of this algorithm. We then developed a general algorithm for employing DNA to add any two non-negative rational binary numbers. A central feature of this algorithm is the encoding of input values in single-stranded DNA in such a way that performance of a primer extension reaction results in a process termed a *horizontal chain reaction.* In this reaction, input DNA sequences serve as successive templates for repeated extension of a newly synthesized result DNA strand. The final result strand produced by this algorithm encodes a series of digits, interspersed with "operator" sequences, that represent in the correct order the result of addition of the two input binary! numbers. In the present paper, we demonstrate that this approach is not limited to the addition of two numbers expressed in binary, but can be modified to permit DNA to be employed to add two numbers expressed in any base.

2. General Model

Applying the horizontal chain reaction to perform the addition of two numbers expressed in higher bases requires a straightforward generalization of our DNA base 2 addition algorithm described in [**5**]. For a given base n, the first digit at the n^0 position must be represented by n non-complementary strands of DNA. The first strand, known as the defining strand, encodes the input digit. The other $n-1$ strands each encode one of the other digits so that all possible base n digits are represented. The first strand carries a DNA sequence at its 3' end called $\overline{DEF}(0)$ (an overlined element indicates that its DNA sequence is complementary to the sequence of the corresponding non-overlined element), while the other $n-1$ strands carry a DNA sequence at the 3' end called either $\overline{OP}_1(0)$, $\overline{OP}_2(0)$, $\ldots$, or $\overline{OP}_n(0)$. The second digit at the n^0 position is a single strand of DNA which is: $DEF(0)$ if this input number is a 0, $OP_1(0)$ if this input number is a 1, $OP_2(0)$ if this input number is a 2, and so on. Thus, if the second number is a 0, $DEF(0)$ will use the defining strand of the first digit as a template in a primer extension reaction producing a result strand with the same value as the defining strand. If the second number is a 1, $OP_1(0)$ will use the second strand of the first digit as a template in a primer extension reaction producing a result strand with the correct value of the addition operation. This applies to $OP_2(0)$, $OP_3(0)$, $\ldots$, $OP_n(0)$ in an obvious way.

The strand representing the first digit at the n^0 position carries one of two DNA sequences at its 5' end, called either $\overline{DEF}(0,1)$ or $\overline{OPP}(0,1)$. The function of these sequences is to carry information from the n^0 position to the n^1 position (hence the arguments $(0,1)$). $\overline{OPP}(0,1)$ indicates "carry the 1" to the n^1 position, and $\overline{DEF}(0,1)$ indicates "carry the 0" to the n^1 position. The second digit at the n^1 to the n^n position is one of n pairs of non-complementary single-stranded pieces of DNA chosen solely on the basis of the input value of this digit. Only one member of the pair will be employed for further primer extension based on whether a 0 or a 1 was carried over from the previous position. These strands all contain a DNA sequence called J (for junk) at their 3' end which avoids primer extension of template DNA strands, followed by either $\overline{DEF}(i-1,i)$ or $\overline{OPP}(i-1,i)$ associated with carry the 0 or carry the 1 from the previous position respectively, and a DNA sequence $\overline{DEF}(i)$, $\overline{OPP}(i)$, or $\overline{CR}_1(i)$, $\overline{CR}_2(i)$, $\ldots$, $\overline{CR}_n(i)$, at the 5' end which uniquely directs a further primer extension reaction based on the different paths that different combinations of "carry the 1" may lead to. To be able to handle the "carry the 1" situation, it is necessary to represent the first digit of positions $n > 0$ analogously to the representation of the n^0 position, but with an additional DNA strand, for a total of $n+1$ strands containing the appropriate CRi's to handle all possibilities. In addition, to provide a primer for a final "carry the 1" operation, a placeholder strand representing the $n+1$ position, with the properties described previously [**5**] is included.

3. Base 3 addition

To illustrate how these concepts are used to extend our previously reported base 2 DNA addition algorithm, we now demonstrate how to do base 3 DNA addition of two numbers. The DNA representation of all possible pairs of input two-digit numbers, each either a positive integer or zero and represented in base 3, is shown in Fig. 1. The "first digit" and "second digit" at a given 3^n position refer to the value (either 0, 1, or 2) at that position of, respectively, the first and second numbers to be added. Representation of a given digit is seen to require either 1, 2, 3, or 4 strands of DNA, whose functions are described below. Each of the indicated elements is represented by a unique DNA sequence; and each DNA sequence is non-complementary to all other sequences, except that an overlined element indicates a DNA sequence that is complementary to the sequence representing the element (*e.g.*, $\overline{DEF}(0,1)$ is complementary to $DEF(0,1)$ and $\overline{1}(0)$ is complementary to 1(0)). The representation of digits employs the following convention: a number in parentheses refers to a position, while a number not in parentheses refers to the value of the digit at that position. For example, 0(1) indicates a value 0 at the 3^1 position of a number.

The functions of the elements in the input DNA sequences shown in Fig. 1 are as follows:

1. **The first digit at the 3^0 position** is represented by three strands of DNA each containing the following three elements (all descriptions of the elements in a given DNA strand begin with the 5' end of the strand): a "position transfer operator" (*e.g.*, $\overline{DEF}(0,1)$), which transfers information from the 3^0 to the 3^1 position), an element representing the value of the digit at the 3^0 position, and a "position operator" (*e.g.*, $\overline{DEF}(0)$), which is employed only at the 3^0 position.
2. **The second digit at the 3^0 position** is represented by a single DNA strand containing a single element whose sequence depends upon the value of this digit. This DNA strand represents not only this value, but also an operator. This operator will serve as a primer in a primer extension reaction to direct the extension of only one of the strands representing the first digit at the 3^0 position, thus yielding the first step in the synthesis of the *result DNA strand* that will ultimately encode the result of the addition operation. After this first step, the result strand is designated Result Strand 1 (RS1).
3. **The second digit at the 3^1 position** is represented by two DNA strands, each containing three elements: a position operator, a position transfer operator, and a J (junk) element that will prevent primer extension of the input DNA strand. This representation is designed so that only one of the strands representing this digit will hybridize with the result strand described above, and then serve as a primer extension template for further elongation of RS1 to yield RS2.
4. **The first digit at the 3^1 position** is represented by four DNA strands each containing the indicated four elements, with each element defined as described above. Only one of these strands will hybridize with RS2, and serve as a primer extension template for further elongating RS2 to yield RS3. If the placeholder strand described below does not prime an additional reaction, RS3 is the *final result strand.*

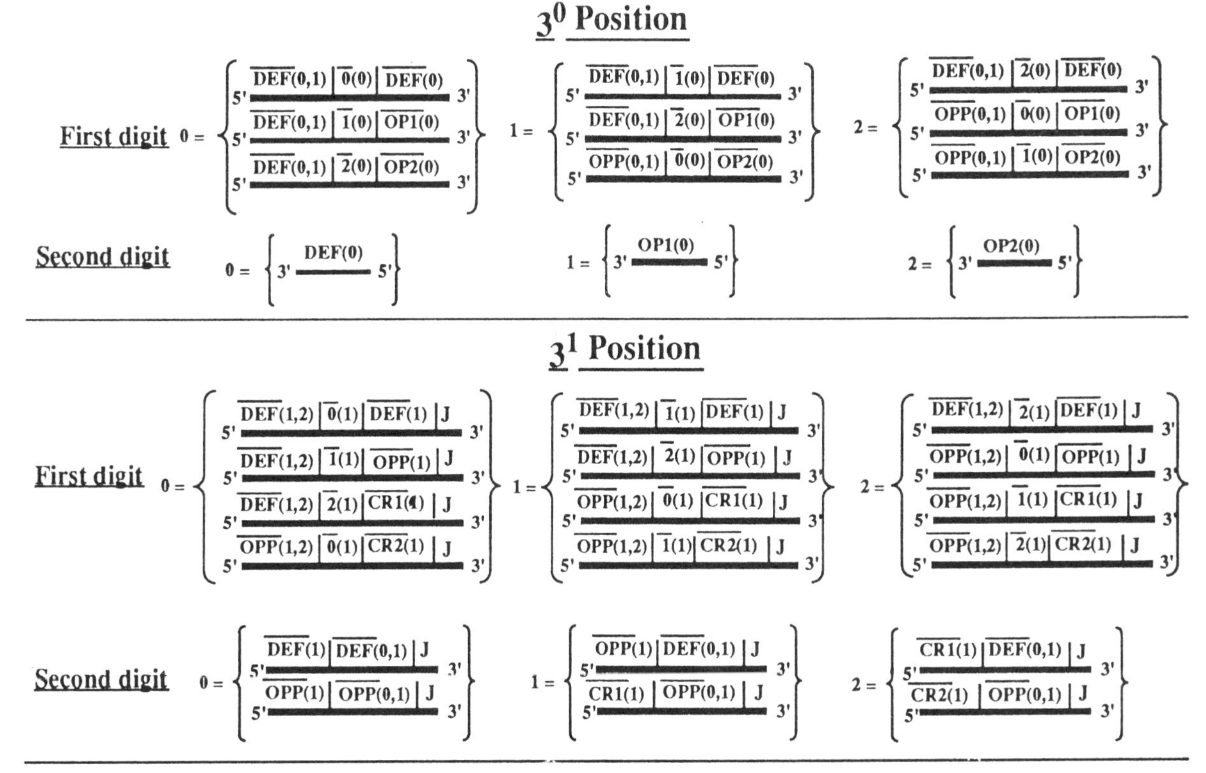

FIGURE 1. Representation of the DNA sequences to be employed in the horizontal chain reaction for pairwise addition of all possible two-digit numbers in base 3, each either a positive integer or zero.

To permit a 1 to be carried to the 3^2 position if the result of the addition operation under the above rules require it, the *placeholder strand* for the 3^2 position illustrated in Fig. 1 is included. This placeholder strand will function only when a 1 must be "carried" to the 3^2 position, when it will act to carry a 1 to the 3^2 position by serving as a template to further extend RS3 to yield the final result strand.

An example of the application of this algorithm for adding two numbers expressed in base 3 is shown in Fig. 2, which illustrates the addition of the two base 3 numbers 12 + 21. The input DNA's for each of the four reactions represent, in order: the first and second input 3^0 digits (respectively, 2 and 1); the second digit at the 3^1 position (2); the first digit at the 3^1 position (1); and the placeholder strand. In reaction 1, the operator (primer) representing the second digit at the 3^0 position initiates a horizontal chain reaction by hybridizing to the appropriate strand representing the first digit at the 3^0 position and, upon primer extension, yields Result Strand 1 (RS1), which encodes a 0 at the 3^0 position. In reaction 2, RS1 primes the appropriate strand of the second digit at the 3^1 position. Following extension of the primer but not of the template (because the input DNA strands contain the J element), RS1 is extended to yield RS2. In reaction 3, RS2 primes the appropriate strand of the first digit at the 3^1 position, resulting in extension of RS2 to yield RS3, which now additionally encodes a 1 at the 3^1 position. Finally, in reaction 4, RS3 primes the placeholder strand for the 3^2 position, resulting in extension of RS3 to yield the final result strand. This final result strand encodes a series of three digits, interspersed with operator sequences, that represent precisely and in the correct order the outcome of this addition operation: 110. In addition, each DNA element representing a result digit encodes both the value and position of the digit. A number of molecular biological approaches could thus be applied to readout of the answer encoded in the result strand; *e.g.*, use of appropriate oligomers as hybridization probes or polymerase chain reaction primers, or direct DNA sequencing [**2**].

4. Extensions and Conclusions

In our previous work [**5**], we described the straightforward extension of our algorithm for adding two two-digit binary numbers, to the addition of two m-digit binary numbers, each either a positive integer or zero. By an entirely analogous procedure, the algorithm described here can be extended to the addition of two m-digit numbers, each either a positive integer or zero, expressed in any base. Similarly, this more general algorithm can be further extended in an obvious way to the addition of any two non-negative decimal numbers expressed in any base. Finally, we note that the addition algorithm at its present early stage of development can add only two numbers at a time, and thus cannot be employed for either iterative or parallel addition. Modification of this algorithm will therefore be required to take full advantage of the massive parallel processing capabilities of DNA.

The *horizontal chain reaction* is seen to play a central role in the DNA-based addition algorithm described here and in our previous studies [**5**]. This reaction yields a successively elongated DNA strand that plays a dual role in the execution of the algorithm: the growing result strand serves both as an operator that participates actively in the information flow during the execution of the algorithm, and as a tape that records the output of the algorithm. It thus seems likely that the

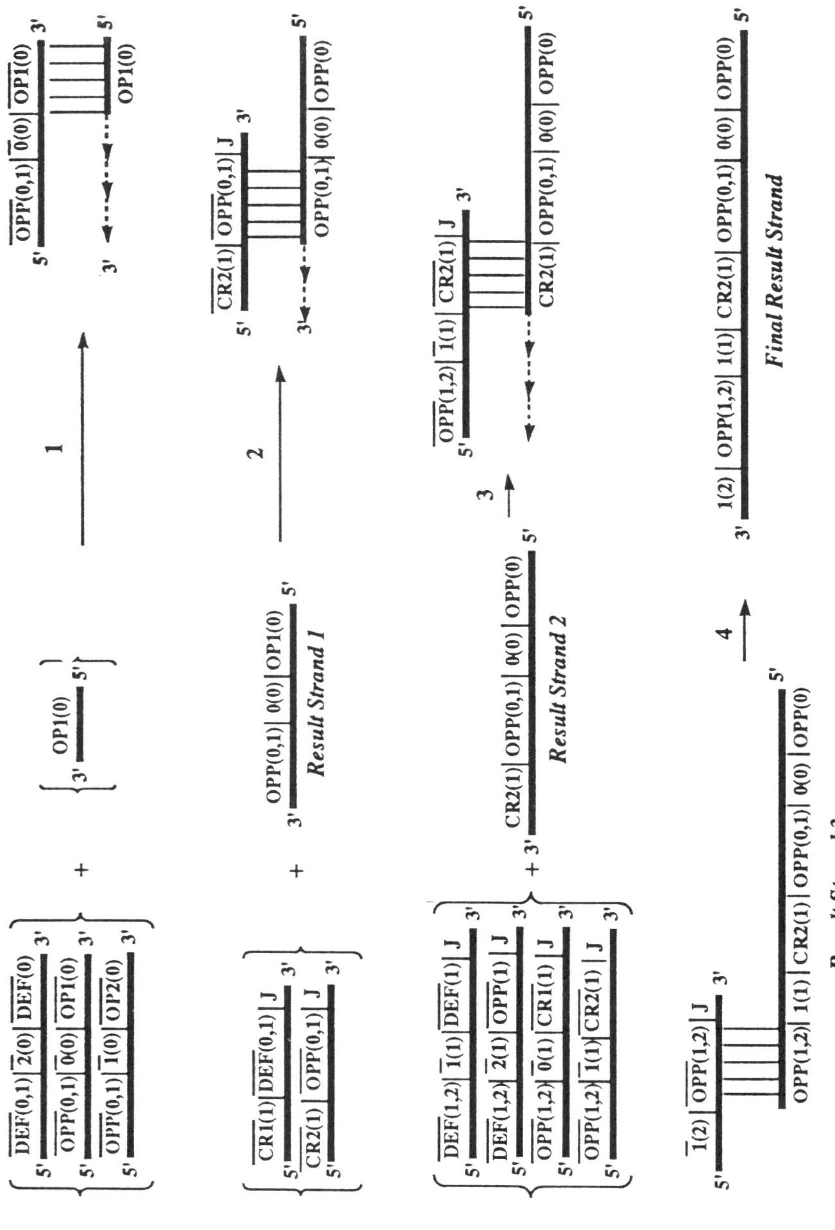

FIGURE 2. Illustration of the operation 12 + 21 as an example of the use of the horizontal chain reaction for addition of two two-digit numbers expressed in base 3. Vertical dotted lines represent hybridization between complementary DNA elements, and reiterated arrows represent primer extension.

horizontal chain reaction will prove to be useful not only in DNA-based algorithms for performing mathematical calculations such as addition; but also in other areas of DNA-based computing, such as solutions to combinatorial problems and possibly also the development and application of an algorithm for employing DNA to simulate a Turing machine.

References

[1] Leonard M. Adleman, Molecular Computation of Solutions to Combinatorial Problems. *Science*, 266:1021-1024, 1994.

[2] Fred M. Ausubel et al, Eds., *Short Protocols in Molecular Biology*, Wiley, New York, ed. 3, 1995.

[3] Eric B. Baum, Building an Associative Memory Vastly Larger Than the Brain. *Science*, 268:583-585, 1995.

[4] Donald Beaver, Computing with DNA. *J. Computational Biol. 2*, 1-7, 1995.

[5] Frank Guarnieri, Makiko Fliss, and Carter Bancroft, Making DNA Add. *Science*, 273:220-223, 1996.

[6] Richard J. Lipton, DNA Solution of Hard Computational Problems. *Science*, 268:542-545, 1995.

[7] John H. Reif, Parallel Molecular Computation: Models and Simulation, *Seventh Annual ACM Symposium on Parallel Algorithms and Architectures(SPAA95)*, Santa Barbara, June, 1995, pp. 213-223.

Department of Physiology and Biophysics, Mount Sinai School of Medicine, New York, NY 10029-6574

DIMACS Series in Discrete Mathematics
and Theoretical Computer Science
Volume **44**, 1999

Computation With DNA: Matrix Multiplication

John S. Oliver

ABSTRACT. If chemical reactions are to be used as the basis for computers, efficient instruction sets will need to be developed. A chemically based computation can not at this time be expected to compete with an electronic computer. However, the potential usefulness of a chemical computer provides a compelling reason to investigate and design procedures for the solution of varied problems. DNA based methods which may be used to calculate the product of Boolean matrices or matrices containing positive, real numbers are presented. This provides a method to perform a quantitative calculation with DNA.

1. Introduction

Two papers report the use of DNA to perform calculations. The first solves a Hamiltonian path [**1, 4**] while the second outlined an extension of these methods to solve the SAT problem [**11**]. These papers have generated significant discussion, both positive and critical. As in any new area many questions immediately arise. What types of problems can we solve with chemical reactants? Are there easier methods for solving these problems than those previously outlined?

In this report, I outline procedures to solve multiplication of Boolean matrices or matrices containing positive real numbers by means of manipulations of DNA. In addition I extend these procedures to the calculation of powers of matrices.

Matrix multiplication can be represented as a directed graph and is a simpler problem than the Hamiltonian path or the SAT problem. However, two novel features will be presented. First, an analysis method different than that already described is presented to determine the successful synthesis of a path. Second, a strategy is presented to perform a quantitative calculation rather than simply to determine the successful formation of a path. It is hoped that the presentation of a new instruction and means of analysis may stimulate DNA based solutions to more challenging combinatorial problems.

2. Boolean Matrices

Two Boolean matrices [**14**] and their product can be represented by a directed graph [**9, 12**]. For example the product of two Boolean matrices, X and Y, and the directed graph representation are shown in Figure 1.

In this graph, the initial vertices 1, 2, and 3 represent the row identifiers for the first matrix and the product matrix. The terminal vertices A and B represent the column identifiers for the second and product matrix. By definition for matrix multiplication, the numbers of columns in the first matrix must equal the number

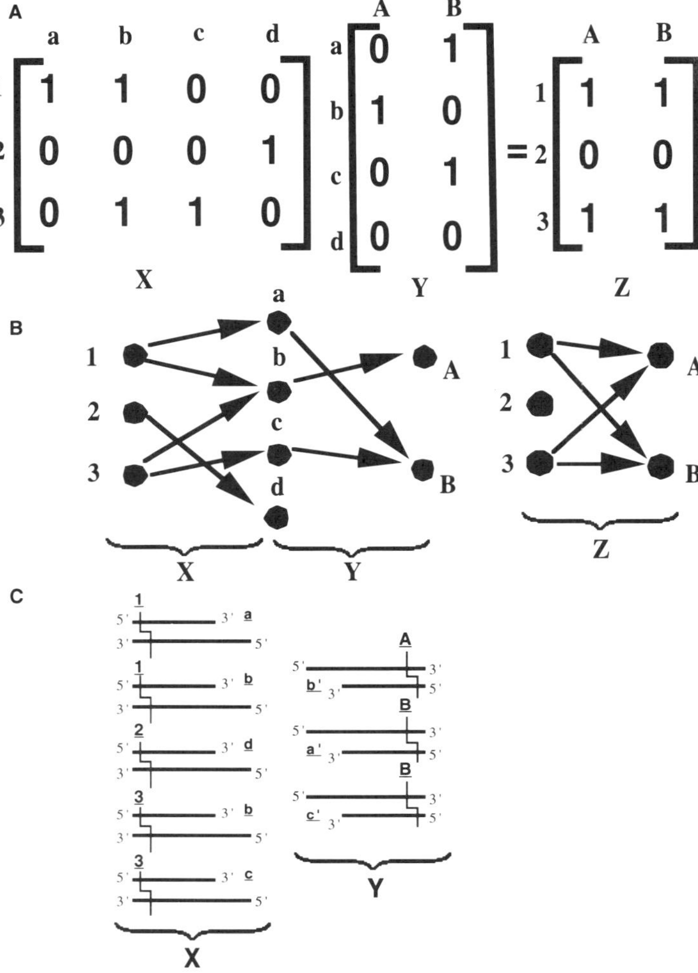

FIGURE 1. Design of DNA sequences for multiplication of two matrices. *(A)* Multiplication of matrices X and Y. *(B)* The graph representation of the operation. The row and column identifiers of the matrices are represented by vertices (circles) in the graph. Non-zero elements in matrices X, Y, and Z are represented by directed edges (arrows) which connect vertices for the appropriate row and column identifying that element. Thus the directed edge connecting vertices "1" and "a" in the graph represents the symbol "1" in row 1 column a of X. The graph representation of Z also can be drawn by inspection of the matrix. Alternatively, the graph of Z can be determined from the graph representing the product of X and Y. The edges in Z represent paths (in this case a path is simply a sequence of edges which connect an initial vertex (1, 2, or 3) to a terminal vertex (A or B)) between initial and terminal vertices in the graph on the left. Z is constructed by replacing the paths in (X)(Y) with edges and removing all intermediate vertices (a, b, c, and d). Thus the edge from 1 to B in the graph of Z represents the path 1-a, a-B in the graph for (X)(Y). *(C)* The DNA strands used to represent the non-zero elements (edges in the graph) of each matrix. The ends of the DNA sequences represent vertices at either end of the respective edge in the graph. Thus 1, 2, and 3 label restriction enzyme sites which represent the initial vertices 1, 2, and 3. A and B label restriction enzyme sites which represent the terminal vertices A and B. The intermediate vertices for which an edge is entering the vertex are represented by the single stranded DNA sequences a, b, c, or d. Intermediate vertices for which an edge is exiting the vertex are represented by the complementary sequences a', b', c', and d' respectively.

of rows in the second matrix. The intermediate vertices a, b, c, and d represent the columns of the first matrix and the rows of the second matrix.

Non-zero elements in the first or second matrix are represented by a directed edge (vector connection) in the graph. Connections between initial and intermediate vertices represent non-zero elements in the first matrix. Connections between intermediate and terminal vertices represent non-zero elements in the second matrix.

For example, in the first matrix the element at the intersection of row 1 and column a is the symbol "1". Thus we draw an edge proceeding from vertex 1 to vertex a. Again, in the first matrix the element at the intersection of row 1 and column b is the symbol *zero*. Thus there is no edge connecting vertex 1 and vertex b. Similarly we can construct edges between vertices a, b, c, and d and A and B to represent symbols of one in the second matrix. Each of the edges that we construct has directionality indicated by an arrowhead.

The elements of the product matrix symbolize the existence or non-existence of paths made from continuous linkages of edges between initial and terminal vertices. Paths can only be constructed by proceeding along edges in the direction indicated by the arrowhead associated with each edge. In essence one is asking if it is possible to form a path from each initial vertex to each of the terminal vertices. The problem of evaluating the elements of Z then reduces to determining a complete set of paths from the initial vertices 1, 2,...to terminal vertices $A, B \ldots$ Now consider a method to perform this calculation (i.e. determine this set of paths) using DNA to represent the edges, vertices, and paths in the graph above. The calculation consists of several steps which we now discuss in more detail. The steps are shown in Figure 2.

1. Design and synthesis of DNA representing vertices and edges in the graph,
2. Reaction of the constructed DNA to form paths,
3. Analysis of the reaction to identify paths between initial and terminal vertices by restriction enzyme digestions. Paths between initial and terminal vertices will be represented by a linear piece of double stranded DNA (ds DNA) containing different vertex-specific restriction enzyme sites at each terminus.

2.1 Design and Strategy: From the graph representing the multiplication of two matrices, it is clear that a path is formed between an initial vertex (1-3) and a terminal vertex (A and B) only when an intermediate vertex (a-d) is linked to both. Pieces of DNA that represent an edge and the two vertices at either end of the edge, and thus the respective symbol "1" in the matrix are constructed. A path is formed by linking pieces of DNA which pass through the same intermediate vertex.

Each initial vertex which is connected to an intermediate vertex is represented by a unique restriction site at the end of a short (0-10 nucleotide) piece of double stranded DNA. Each terminal vertex connected to an intermediate vertex is also represented by a distinct restriction site at the end of a duplex strand of DNA. At the other end of these duplex strands is a single stranded sequence that represents the intermediate vertex to which the initial or terminal vertex is attached. Each intermediate vertex is represented by a unique sequence. In addition, the single stranded overhangs are synthesized so that an edge which ends at an intermediate vertex, i.e. one that proceeds from an initial vertex to an intermediate vertex, has a sequence which is complementary to the single stranded sequence for edges

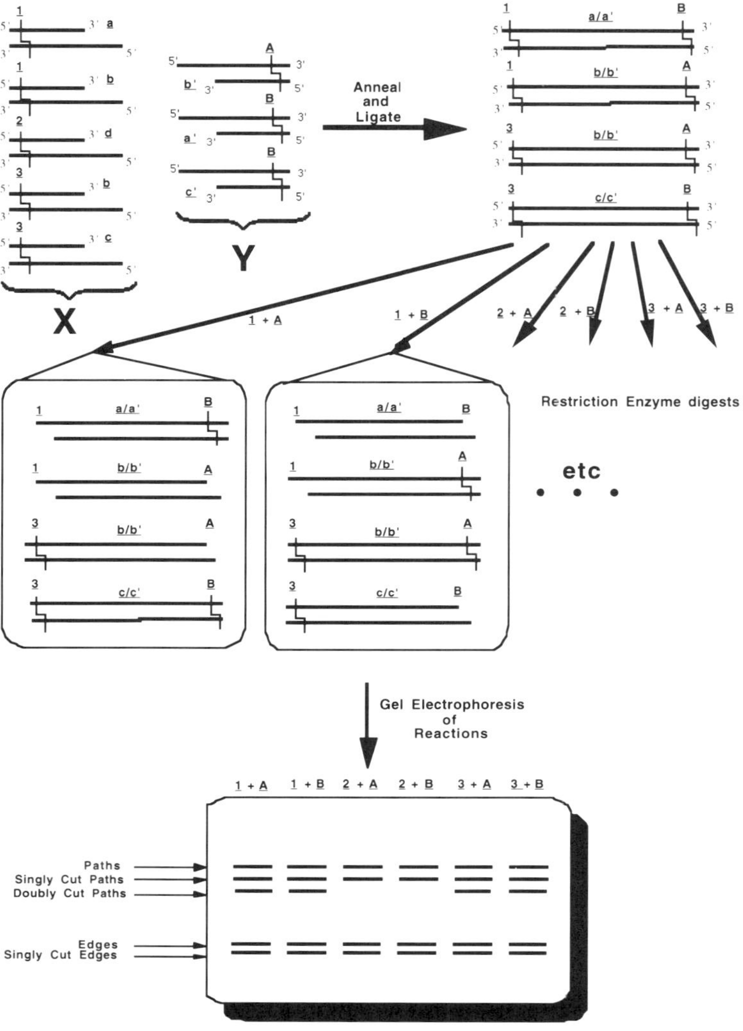

FIGURE 2. Reaction sequence used to multiply matrices X and Y from Figure 1. The desired DNA strands are synthesized such that the sequences representing initial and terminal vertices are double stranded. The intermediate vertices are represented by single strand overhangs. The DNA sequences are mixed, annealed and ligated in a reaction that generates all possible (in this case three) paths. The reaction mixture is divided into six equal aliquots which are used in separate restriction enzyme digest reactions. Each aliquot represents an element in the product matrix. To each aliquot is added two restriction enzymes. One enzyme corresponds to the row of the product matrix that the element occupies and the other restriction enzyme corresponds to the column that the element occupies. The paths in each of these six reactions will either be uncut, cut once, or cut at both ends depending on the restriction sites incorporated in the path. A portion of each of the restriction enzyme reactions is submitted to gel electrophoresis which separates the strands based on size. Paths which have been cut by each of the enzymes in a particular reaction, thus representing the symbol "1" for that element. will appear as bands on the gel.[**15**]

beginning at the same vertex, edges that proceed from an intermediate vertex to a terminal vertex. Thus, each piece of DNA represents an edge with an initial or terminal vertex coded by a restriction site at one end and an intermediate vertex coded by a single stranded overhang at the other end (see Figure 1C).

The single stranded portions of these DNA strands will anneal to a complementary sequence under defined reaction conditions. In order to specify a direction to the edge, the single stranded sequences representing intermediate vertices are non-palindromic. Thus, sequences representing an edge that enters an intermediate vertex will anneal only to sequences representing an edge that exits the same intermediate vertex. Each of the DNA sequences representing an edge and the two vertices connected by that edge are of equal length. As a result all assembled paths in this graph will be the same length.

2.2 Path Formation: The DNA sequences representing all the edges in a graph are mixed in the presence of a DNA ligase (see Figure 2). This enzyme will form a covalent bond between two DNA molecules as long as they have complementary single stranded overhangs. Thus, sequences representing edges entering and exiting a single intermediate node are ligated to form a single piece of duplex DNA which represents a path between an initial vertex and a terminal vertex.

2.3 Analysis: Each element in the product matrix describes the existence or non-existence of a path connecting two vertices. The initial and terminal vertices are represented by DNA sequences which can be cut with a unique restriction enzyme. Thus each element which defines a path (a 1 in a Boolean matrix) is represented by a DNA sequence connecting two different restriction enzyme sites. Connectivity can thus be determined by shortening of the DNA by the restriction enzymes representing the initial and terminal vertex. The reaction mixture is aliquoted into separate tubes. The number of aliquots is the same as the number of elements in the product matrix. The two restriction enzymes corresponding to the initial and terminal vertices (representing, respectively, the appropriate row and column of the product matrix) are added to one of the reaction tubes. If a path exists between the two vertices the DNA sequence representing the path is restricted at both ends and a new DNA fragment is generated which has been shortened at each end by a defined amount. Paths which contain only one of the initial or terminal vertices (in this case only edges used as starting material in the ligation reaction) will be shortened at one end. The size of the DNA generated in these restriction reactions is determined. Those reactions containing DNA shortened at both ends denote an element of 1 in the product matrix. Reactions which do not result in DNA shortened at both ends correspond to an element of 0.

If the initial or terminal vertices are represented by nonpalindromic restriction enzyme sites the product of the ligation reaction can be cut with the required enzymes and used in a subsequent ligation (representing a subsequent multiplication). Analysis after the second ligation reaction would proceed as above.

3. Matrices to N^{th} power

The graph representing the square and the cube of a matrix is shown in Figure 3. It is readily apparent that there is translational symmetry relating the vertices and edges. This symmetry continues when one takes the matrix to a higher power.

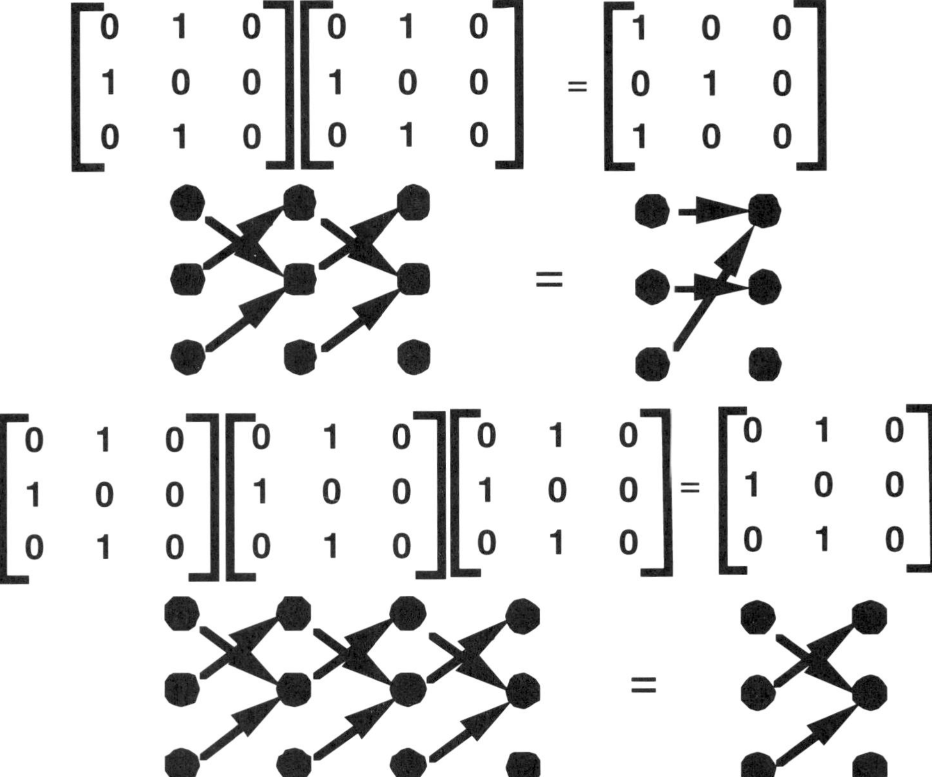

FIGURE 3. The square and cube of a matrix and the graph representations of the operations.

To multiply a matrix by itself several times, DNA sequences are constructed to represent this graph as described above with the exception that no differentiation is made between initial, terminal, and intermediate nodes. The symmetry of the graph allows the calculation to be performed with DNA sequences representing the first set of vertices and edges. All other vertices and edges are the same. All vertices are constructed as single stranded overhangs. Thus in the example shown only three DNA sequences need to be synthesized.

The subsequent ligation reaction will produce an array of products. These will be grouped into multiples of the length of the original DNA stands. The length will correspond to the power to which the original matrix was taken. Thus if the original DNA strands are constructed of length X, products of length $2X$ are the result of squaring the matrix. Products of length $20X$ are the result of taking the matrix to the 20th power. A single reaction will generate the elements for a variety of powers of the matrix. The different path lengths can be separated and the lengths corresponding to the powers of interest separately isolated. Analysis of each of these reactions can be carried out in any of a number ways by recognizing that the initial and terminal vertices remain single stranded. It should be specifically pointed out that the initiation and termination sequences for the same vertex are complementary and thus different as long as nonpalindromic single stranded overhangs were used in the construction of the original DNA sequences. As a result affinity extraction methods similar to those already described by Adleman [**1**] could be used to identify sequences containing particular initial and terminal vertices. This is a rather laborious procedure, however. Alternatively a procedure conceptually similar to the restriction enzyme analysis described above could be carried out. The reaction is divided into aliquots each representing an element in the product matrix. In this case, rather than observing a decrease in the length of the DNA due to restriction at both ends, sequences containing a single stranded region complementary to the initial and terminal vertices could be annealed and ligated to the overhangs. Strands which are capable of having both of the appropriate sequences ligated would indicate the successful synthesis of a path between those two nodes.

At this time it is not clear how large n can be from a single reaction. Considerations such as ligation efficiency, the efficiency of circularization of products which have complementary ends (which would remove material from further reactions), and the resolution and efficiency of separation procedures [**3, 7**] to isolate the different length reaction products will all play a role. It should be noted that a single ligation reaction has been used to routinely ligate 20 blunt ended fragments or 30 fragments with overhangs. [**10, 5, 6**]

4. Real Number Matrices

The product of two matrices whose elements are real numbers can also be represented by a directed graph. In order to encode the magnitude of the element, however, each edge must also have a transmission factor [**13**] associated with it (see Figure 4). The transmission factor for a path is the product of the transmission factors for each of the edges in the path. Each element in the product matrix is the sum of the transmission factors for each possible path connecting the appropriate initial and terminal vertices. In designing a reaction to calculate the product matrix it is no longer adequate to demonstrate transitive closure between the two vertices of interest. The transmission factors must also be represented in the reaction.

Multiplication of matrices containing non negative real numbers can be achieved using DNA if the transmission factors in the original matrices are encoded by the concentrations of DNA representing the edges. Under appropriate reaction conditions the elements of the product matrix will be proportional to the concentration of the associated path formed. Vertices and edges are constructed as described above and placed in a reaction tube at concentrations proportional to the transmission factors associated with that edge. A non-enzymatic ligation reaction is performed under appropriate kinetic conditions such that the amount of each path formed will represent the transmission factor for that path.

Kinetics of path formation: The formation of a path from two edges requires two steps. The first is annealing of complementary overhangs on DNA strands representing two edges. The second step is the ligation of the annealed strands. A kinetic scheme for the reaction sequence is shown below in which A and B represent the two DNA strands, $A \bullet B$ represents the annealed complex and C represents the ligated product.

$$A + B \underset{k_{-1}}{\overset{k_1}{\rightleftharpoons}} A \bullet B \xrightarrow{k_2} C$$

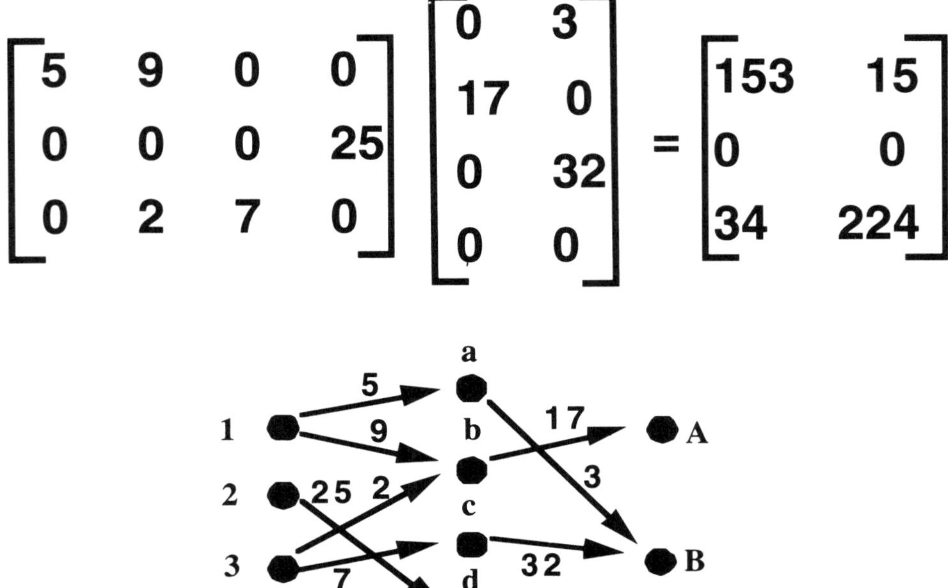

FIGURE 4. The graph representation for the multiplication of two matrices containing real, positive numbers. The numbers over the edges are the transmission factors for each edge.

When the rate constant for ligation is slower than the rate constants for association and dissociation of the strands then the rate of formation of C is:

$$\frac{d[C]}{dt} = k_2[A \bullet B] = k_2\frac{k_1}{k_{-1}}[A][B] = k_{obs}[A][B]$$

As long as the observed rate constants (kobs) for different reactions (formation of different paths) are the same, the relative concentrations of any specific path will be dependent on the *product* of the concentrations of the two original DNA strands A and B as long as the equilibrium constant (k_1/k_{-1}) is small. This is necessary so that at equilibrium the concentrations of A and B are not seriously perturbed from their initial concentrations. In order for the rate constants to be equivalent, the ends to be annealed for the formation of different paths all must have the same T_m. If the melting temperatures are the same, at any temperature the ratio of (k_1/k_{-1}) will be the same for any complementary pair.[1] The rate constant for chemical ligation k_2 is independent of the DNA sequence and thus a constant. Therefore, under appropriate conditions, the rate of formation of a path is dependent on the product of the concentrations of the edges that make up that path.

In the limit we can make k_2 zero and perform the reaction in two separate steps. The first step is the annealing step. At equilibrium the concentration of annealed

[1] In practice the sequences would be designed so that at a particular reaction temperature all the sequences would have the same equilibrium constant. Variations in the shape of the Tm curve for each single stranded sequence will prevent the equilibrium constants from being equivalentat all temperatures.

complexes is dependent on the product of the concentrations of the complementary strands. Once equilibrium has been attained a fast ligation step can be carried out either by addition of a reagent such as cyanogen bromide to initiate ligation [**8**] or by removal of a protecting group. In order that only previously annealed sequences affect the reaction rate, the chemical step must be fast and allowed to proceed for a short period of time. If the ligation step is slow then the rate of formation of ligated product will also depend on the rate constant for formation of annealed duplex k_1.

Alternatively, the reaction can be run under conditions where the equilibrium constant (k_1/k_{-1}) is large. In this case, if we allow the reactants to reach equilibrium before performing the ligation step, the concentrations of the reactants will be perturbed from the initial concentrations and the amount of annealed product will not be a simple product of the initial concentrations. Under these conditions the initial rate of formation of each path will be proportional to the initial concentrations of the reactants as long as the rate of ligation is fast compared to the rate of formation of the annealed complex.

The use of sequences each having the same T_m is also desirable in reactions which do not require consistent reaction rate constants for kinetic analysis in order to control errors due to mismatches. At reaction temperatures much lower than the T_m a sequence can anneal to a sequence containing mismatches. If these mismatches are not at the site of ligation the ligation reaction will be successful and an incorrect path will be formed. On the other hand one must run the reaction at a temperature low enough so that an observable amount of product is made from the lowest melting single stranded sequence. If all complementary overhangs have the same T_m conditions can be adjusted to balance these two requirements. Quantification of path formation is most easily measured by using radiolabeled DNA. Thus, the reaction products which define a path can be quantitated quickly and accurately.

5. Conclusion

A method which may be used to analyze multiplication of Boolean and real matrices using DNA has been presented. DNA is used in natural systems for the solution of different types of problems. For instance, genetic recombination transfers information and genetic rearrangement followed by combinatorial mutations is used to mediate antibody binding. The operations described here utilize many of the same features of DNA which make these recombinations and rearrangements possible. The development of an efficient molecular computer will require the design of a 'good' and 'useful' instruction set. It will also require the design of molecules which have features of molecular recognition and self-assembly and incorporate reporter groups.

References

[1] Adleman, L.M., Molecular Computation of Solutions to Combinatorial Problems. *Science*, 266: 1021–1024, 1994.

[2] Flegg, H.G., *Boolean Algebra And Its Applications*, John Wiley & Sons, Inc., New York, pp. 141–175, 1964.

[3] Fung, E.N.; Yeung, E.S., High-Speed DNA-Sequencing by Using Mixed Poly(ethylene Oxide) Solutions in Uncoated Capillary Columns. *Anal. Chem.*, 67:1913–1919, 1995.

[4] Gifford, D. K., On the Path to Computation with DNA. *Science*, 266:993–994, 1994.

[5] Hagerman, P.J., Sequence Dependence of the Curvature of DNA - A Test of the Phasing Hypothesis. *Biochemistry*, 24:7033–7037, 1985.

[6] Hagerman, P.J., Pyramidine 5-Methyl Groups Infleunce the Magnitude of DNA Curvature. *Biochemistry*, 29:1980–1983, 1990.

[7] Heiger, D.N.; Cohen, A.S.; Karger, B.L., Separation of DNA Restriction Fragments By High-Performance Capillary Electrophoresis With Low and Zero Cross-Linked Polyacrylamide Using Continuous and Pulsed Electric-Fields. *J. Chrom.*, 516:33–48, 1990.

[8] Kanaya, E.; Yanagawa, H., Template-Directed Polymerization of Oligoademylates Using Cyanogen-Bromide. *Biochemistry*, 25: 7423–7430, 1986.

[9] Kim, K.H., *Boolean Matrix Theory and Applications*, Marcel Dekker, Inc., New York, 1–23, 1982.

[10] Koo, H.-S.; Wu, H.-M.; Crothers, D.M., DNA Bending at Adenine/Thymine Tracts, *Nature*, 320:501–506, 1986.

[11] Lipton, R.J., DNA Solution of Hard Computational Problems. *Science*, 268:542–545, 1995.

[12] Robinson, D.F.; Foulds, L.R., *Digraphs: Theory And Techniques*, Gordon And Breach, Science Publishers, Inc., New York, 1980.

[13] Wilson, R.J.; Beineke, L.W. eds., *Applications of Graph Theory*, Academic Press Inc. Ltd., London, 59–80, 1979.

[14] Boolean matrices of two elements using the symbols "0" and "1" are used in this paper. It should be noted that these symbols do not represent the values 0 and 1. Boolean algebra uses the operators "and" and "or" rather than "x" and "+". Thus 1 "or" 1 (sometimes incorrectly shown as 1+1) = 1. To avoid confusion the matrices used as examples in this paper were chosen to give the same products whether considered as Boolean or real matrices. For a more thorough discussion of Boolean Algebra see for instance,[**2**].

[15] Although Fig. 2 implies the use of a non-denaturing gel for simplicity of presentation, in practice a denaturing gel would be used to reduce interactions between DNA fragments.

Department of Chemistry, Brown University, Providence, RI 02912
E-mail address: John_Oliver@Brown.edu

DIMACS Series in Discrete Mathematics
and Theoretical Computer Science
Volume **44**, 1999

A Surface-Based Approach to DNA Computation

Qinghua Liu, Zhen Guo, Zhengdong Fei, Anne E. Condon[†], Robert M. Corn, Max G. Lagally, Lloyd M. Smith

ABSTRACT. A new model of DNA-based computation is presented. The main difference between this model and that of Adleman is in manipulation of DNA strands that are first immobilized on a surface. This approach greatly reduces losses of DNA molecules during purification steps. A simple, surface-based model of computation is described and it is shown how to implement an exhaustive search algorithm for the SAT problem on this model.

Partial experimental progress in solving a 5-variable SAT instance is described, and possible extensions of our model that allow general computations are discussed.

1. Introduction

Adleman [**1**] and subsequently Lipton [**6**] described how genetic engineering tools can be used to solve instances of NP-complete combinatorial problems. Their work has led to hopes of a DNA computer that can outperform the fastest realizable super computers at such problems. Other suggested applications include massive associative memory [**3**] and problems in combinatorial chemistry, such as drug design [**2**].

Adleman solved a tiny instance of the Hamiltonian Path problem using a test tube-based methodology [**1**]. The chemical basis for solution of NP-complete problems is to represent the set of combinatorially possible solutions as a test tube of DNA strands (or oligonucleotides) and to selectively isolate the desired solutions from this set using hybridization, ligation, and other DNA manipulation processes. Problems exist with scale-up of this test tube-based approach for a number of reasons, including the poor efficiencies of the purification and separation steps.

An alternative methodology, in which the DNA strands are immobilized on a surface, offers many advantages for robust DNA computations. The solution set of oligos is initially attached to a surface (glass, silicon, gold, or beads, for example). They are then subjected to operations such as hybridization from solution or exonuclease degradation, in order to extract the desired solution. This method greatly reduces losses of DNA molecules during purification steps. We note that surface-based chemistries have become the standard for complex chemical syntheses such

[†]To whom further communication should be directed.

as solid-phase protein synthesis, solid-phase DNA synthesis, solid-phase protein sequence analysis, and many other chemistries [**7**]. The power of this method was also recognized by Adleman in his use of magnetic support particles in the Hamiltonian path experiment, although his approach did not fully exploit the utility of the support-based approach.

In this paper, we describe a "bare-bones" surface-based model for DNA computation. Our intent is to keep the model simple (incorporating only those operations that we can implement now), yet powerful enough to solve non-trivial exhaustive-search algorithms. We first describe (Section 2) an abstraction of the surface-based model. Briefly, the allowable operations are to selectively mark strands, to destroy either marked or unmarked strands, and to unmark all marked strands. We show that although this model may appear to be more restrictive than Lipton's model, it can be used to implement an exhaustive search algorithm for the Satisfiability problem. In Section 3, we describe in detail how the abstract model can be realized with standard surface-based chemistry. Another feature of our approach is the use of single-base mismatch discrimination in hybridization as a basis for selectively marking DNA strands immobilized on a surface. This method allows us to obtain a high density of information per base.

Partial experimental progress towards our initial goal of solving a 5-variable SAT instance using this model is described in Section 4. One experiment tests our method for selectively marking strands using hybridization. At this stage, we have not achieved a satisfactory level of success with this operation, and we discuss how we plan to improve on this. The second demonstrates successful use of the enzyme Exonuclease I to destroy single-stranded DNA. The third experiment involves both hybridization followed by enzymatic destruction. The results indicate that the hybridized strands are not destroyed, confirming that marking followed by destruction of unmarked strands is feasible.

It is not clear to what degree one can scale up the operations used to realize our surface-based model. In Section 5, we enumerate some of the limitations of our current approach, and suggest possible extensions that we would like to explore in the future.

2. Abstract Model

For simplicity, we let the initial solution space be the set S of binary strings of length n, that is, the set $\{0,1\}^n$. The following operations may be performed on S.

1. **mark**(i, b): this marks all strings of S in which the ith bit has value b.
2. **mark**$((i_1, b_1), (i_2, b_2), \ldots, (i_k, b_k))$: this is an extension of **mark**(i, b) in which a string is marked based on the values of many bits.
3. **destroy-marked**: removes all marked strings from the set S
4. **destroy-unmarked**: removes all unmarked strings from the set S
5. **unmark**: this unmarks all marked strings in S
6. **test-if-empty**: this operation determines whether the set S is empty or not. We assume it is only executed at the end of a computation.

The mark operation here is similar to the select operation of Lipton's model. However, our model is more restrictive than Lipton's, since marked and unmarked strings cannot be physically separated and handled differently. The only operation in our model that allows us to take advantage of marked strings is the destroy operation. Nevertheless, we now show that the Satisfiability (SAT) problem can be solved with these operations.

2.1. Algorithms for SAT. The SAT problem is defined as follows. We are given a Boolean formula in conjunctive normal form, over the set of variables $\{x_1, \ldots, x_n\}$. That is, the formula is the conjunction (logical "and") of clauses, each of which is the disjunction (logical "or") of some variables or their negations. The problem is to determine if such a formula is satisfiable; that is, if there is an assignment of truth values to the variables that sets all clauses to "true."

We present two different exhaustive search algorithms for the SAT problem. The solution space S represents all possible truth assignments, with 1= "true" and 0 = "false".

```
for each clause C of the form (x_{i_1} ∨ ... ∨ x_{i_k} ∨ x̄_{j_1} ∨ ... ∨ x̄_{j_l}) do
    mark((i_1,0),... ,(i_k,0),(j_1,1),... ,(j_l,1))
    comment: all remaining solutions that set C to "false" are marked
    destroy-marked
test-if-empty
```

```
for each clause C do
    for each unnegated variable x_i in C do
        mark(i,1)
    for each negated variable x_i in C do
        mark(i,0)
    comment: all remaining solutions that set C to "true" are marked
    destroy-unmarked
test-if-empty
```

3. A Surface-Based Realization of the Abstract Model

We next describe the chemistry that we are currently developing to implement the above abstract model. We stress that some specific implementation choices that we are making now are not likely to scale to a solution space of size 2^{70}. Issues that arise in scaling up our current approach are discussed in Section 5.

3.1. Solution Space Representation. First, consider the simple problem of representing all possible binary strings of length n as DNA strands. Widely available methods for the solid phase synthesis of DNA molecules can be readily adapted to the synthesis of very complex combinatorial mixtures. In standard solid phase DNA synthesis, a desired DNA molecule is built up nucleotide by nucleotide on a support particle in sequential coupling steps. For example, a support with the nucleotide "A" attached may have the "A" reacted with a "C" to form a dimer, washed and the "C" coupled with "G" to form a trimer (still attached to the surface) and so on. This same chemistry can produce combinatorial sets of molecules by using mixtures of nucleotides at each coupling step. For example, if two nucleotides are used together in five coupling steps, 32 different molecules are made and are present on the support.

For now, we propose to use one base to represent one bit of the binary strings. In addition, all oligos representing a binary string have markers at each end; these will be used as primers in PCR reactions to obtain readout. We can use one base per bit while keeping the GC content of the string constant (say at 50%), by using A or T in half of the positions to represent 0 and 1 respectively, and C or G in the remaining positions to represent 0 and 1 respectively. This is an important aspect of the design of the combinatorial set, as GC content has a very strong effect upon hybridization stability and hence upon hybridization conditions.

3.2. Attachment Chemistry. The attachment chemistry describes the molecules at the interface of the surface and the oligos to be attached to the surface. (Both the surface and one end of the oligos are specially prepared to enable this attachment.) A good attachment chemistry ensures that the properly prepared oligos are immobilized to the surface at a high density, and that other oligos exposed to the surface later (for example, during hybridization) do not bind non-specifically to the surface.

We use a glass surface and an attachment chemistry developed in the Smith laboratory [**4**]. The glass surface is modified with amino-reactive isothiocyanate functionalities in a multi-step process.

3.3. Implementation of Operations. We now describe how each of the operations of our abstract model can be realized on surfaces.

mark(i, b): Let $S(i = b)$ be the set of binary strings of length n in which the ith bit is b. (There are 2^{n-1} strings in $S(i, b)$.) First, the set of DNA oligos that are complementary to the strands in $S(i = b)$ are synthesized. (This is done in the same manner that the initial solution space is synthesized, except that at the ith coupling only one nucleotide is introduced.) Then, each of these hybridizes to its complement on the surface (if present). Thus, the oligos to be marked are now double-stranded whereas those unmarked are single-stranded.

Note that this method assumes that a single-base mismatch is sufficient to discriminate between marked and unmarked oligos. In previous studies [**4**] it was determined that excellent specificity and discrimination of single-base mismatches is obtained using 15mer sequences. In Section 5, we address the fact that this becomes more difficult for longer oligos.

destroy-marked, destroy-unmarked: Either double-stranded or single-stranded DNA molecules may be selectively destroyed using enzymes known as exonucleases, which chew up DNA molecules from the end in, and exist with specificity for either the single-stranded or double-stranded form. We use the enzyme Exonuclease I to destroy single-stranded oligos.

unmark: This is done simply by washing the surface in distilled water. In the absence of salt, which stabilizes the double-stranded pairs, the complementary strands of DNA denature from the oligos on the surface and are washed away, leaving only the original single-stranded DNA attached to the surface.

test-if-empty: One way to implement this is to cleave the attached DNA from the surface, use PCR to amplify this and detect if there is any product as a result. Alternatively, if the strands on the surface are marked at the end of the computation, the complementary strands can be removed as described in the unmark operation and PCR can be applied to those. (In the experiment that we describe in Section 4, the design ensures that there is only one solution to the SAT instance, and so cycle sequencing will be used to actually read out the solution.)

4. Progress To Date

We are working on an implementation of the operations for a solution space of size 32. Our initial computational goal is to solve a 5-variable instance of SAT.

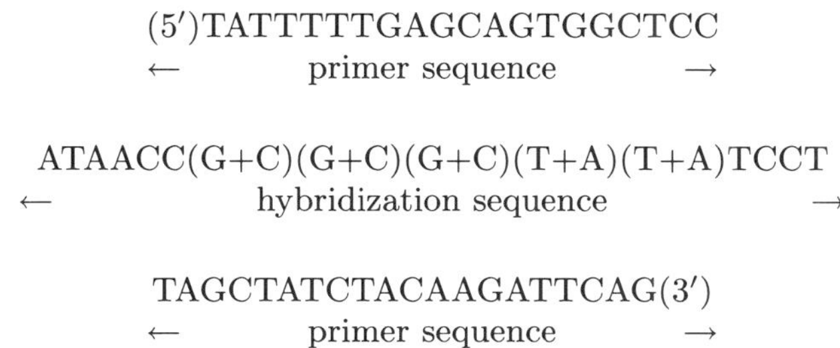

FIGURE 1. Representation of the Solution Space for a 5-Variable SAT Instance

4.1. Solution Space Generation. The solution space, that is, set of 32 distinct molecules, which we have designed and synthesized have the sequences shown in Figure 1. They were synthesized using an Applied Biosystems DNA Synthesizer.

These DNA molecules have an overall length of 55 nucleotides (nt), consisting of a unique 20 nt sequence at each end, and a 15 nt hybridization sequence in the middle. The 20 nt sequences permit both Polymerase Chain Reaction (PCR) amplification (by virtue of the two primer binding sites) of the combinatorial sets of DNA molecules and direct sequence analysis (by virtue of the primer extension site). The 15mer sequence in the middle was chosen to match the length of the oligonucleotides employed in the Smith laboratory for mutation detection [**4**]. The 5 central nucleotides in the 15 mer hybridization sequence were synthesized as a combinatorial set with two possibilities at each position: (G+C)(G+C)(G+C)(A+T)(A+T), yielding $2^5 = 32$ distinct products in the mixture. The adjacent 5 nucleotide sequences at the two ends of the hybridization sequence were synthesized as unique sequences here to limit the size of the combinatorial set to 32 molecules (later we will extend this work to a solution space of size 2^{15}). By keeping the number of positions in the 15mer encoded as A or T roughly equal to those encoded as G or C, the GC content is kept constant at about 50%.

4.2. Implementation of Operations. We next describe three studies that test the mark and destroy-unmarked operations defined above.

4.2.1. *Marking by Hybridization.* In this experiment, we followed previous work [**4**] which successfully achieved single-base mismatch discrimination in hybridization. We note that ultimately we will have do this experiment with the full solution space (32-mixture of molecules) on the surface, but for testing purposes we are currently using just two different molecules on the surface.

We attached two different single-stranded 5'-amino terminated DNA strands to a glass surface. The immobilized single-stranded DNA are both 30-mers with 15 T's as a spacer and the remaining 15 bases as hybridization sequence. The first strand (strand A) is complementary to the center 15-mer on one of the 55mers of our solution space. The other (strand B) is not complementary to any center 15-mer of the 32-mixture occurring in the solution space, and in fact there is a 1-base mismatch between the complement of strand B and some center strand in the solution space. The strands are attached in a group consisting of 4 spots. The two spots in one diagonal are both strand A and the two spots on the other diagonal are both strand B.

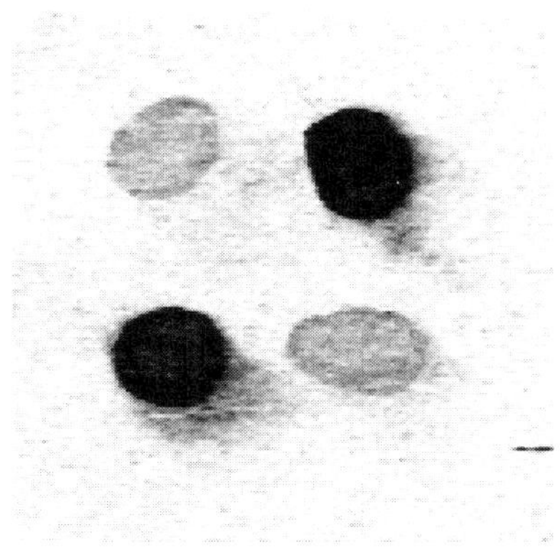

FIGURE 2. Marking by hybridization

The solution space was fluorescently tagged and was hybridized with the A and B spots on the glass for 30 minutes at 37 degrees. Following this, the surface was washed in a buffer, in order to remove excess fluorescently tagged mixture and to improve flourescence detection of the hybridized solution space. Finally, the slide was scanned using a Molecular Dynamics FluorImager 575 for the detection of fluorescence. Hybridization patterns detected in this way show that hybridization occurred to strand A (two darker spots in diagonal) but only to a lesser extent to strand B (two pale spots in the other diagonal). The result is shown in Figure 2.

As can be seen from this figure, some unwanted fluorescence is detected in the mismatched spots. This could be due to (i) non-specific binding of strands on the surface, or (ii) bad discrimination, causing hybridization to occur even in the presence of a single-base mismatch. We have some evidence that (i) is a problem, since fluorescence is still detected after washing with water (which should break the hydrogen bonds of double-stranded DNA). In [**4**], bad discrimination was not a problem, but in their experiment, only a single fluorescently tagged DNA strand, rather than a mixture, was introduced to hybridize to the DNA strand immobilized on the surface. It is possible that the mixture of 32 distinct strands used in this experiment caused the increase in background. We are currently working on alternative surface chemistries to eliminate problem (i) and to increase the effectiveness of discrimination.

4.2.2. *Destroy-unmarked: Destruction of Single Strands via Exonuclease I.* We chose to selectively destroy single-stranded DNA molecules by using Exonuclease. Initially, Exonuclease I and VII were chosen to study the destruction of single strands immobilized on the surface. In preliminary studies, it was observed that Exonuclease VII not only had single-strand-directed $5' \rightarrow 3'$ and $3' \rightarrow 5'$ exonuclease activities, but also had some endonuclease activities (data not shown). Because of this, we chose not to use Exo VII as our enzyme, since endonuclease activities might accidentally destroy a valid solution. On the other hand, Exonuclease I acts specifically on single-stranded DNA, degrading it processively in the $3' \rightarrow 5'$ direction, thus producing $5'$ mononucleotides and leaving the terminal $5'$ dinucleotide intact [**5**].

In our experiments, we found that Exonuclease I could easily destroy single strands in solution without any added proteins or surfactants. However, when

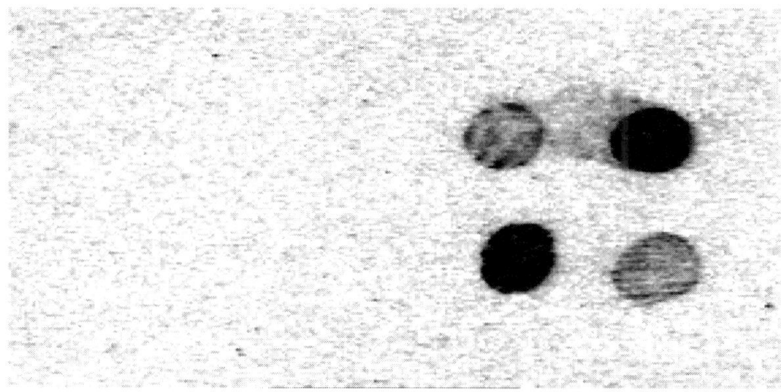

FIGURE 3. Destruction of single strands using Exonuclease I

used to attack single strands attached to a surface, we found that BSA (Bovine Serum Albumin) or non-ionic surfactant Triton X-100 was needed to get the desired nuclease activity.

The experiment shown in Figure 3 demonstrates that Exonuclease I can destroy single-stranded DNA attached to the surface. Strands A and B were immobilized on the glass surface. 8 total spots were applied to the surface in two groups of 4 spots each. In each group, strands were spotted as described in the previous experiment. One group of 4 spots was the experimental group treated with Exonuclease I, while the other group of 4 spots was used as a control (no Exo I). After exonuclease digestion of the strands in the experimental group, the entire surface was hybridized with fluorescent complementary DNA strands. In the control group (right side of Figure 3), we should see the fluorescent signal, since immobilized single-stranded DNA was not destroyed and should hybridize to fluorescently tagged complement. In contrast, the group which was exposed to Exonuclease I (left side of Figure 3) should be destroyed and no fluorescent signal should be detected.

We conclude that Exonuclease I can destroy single-stranded oligonucleotides immobilized on the glass surface with the help of BSA or Triton X-100.

4.2.3. *Destruction of Unmarked Strands Versus Marked Strands.* In this experiment, after attaching strand A in two groups of 8 spots, 4 spots on the left were marked (hybridized) with their complementary strand in the flourescent 32-mixture, and 4 spots of A on the right were left unmarked (still single-stranded). Exonuclease I solution was added to both groups and incubated for 3 hours at 37 degrees. Then, the surface was washed to remove the enzyme and anything it might have digested. Next, the entire surface was hybridized to another portion of the fluorescent complement for 30 minutes at 37 degrees to reveal whether or not any strand A remained on the surface. After washing with washing buffer, the slide was scanned with the FluorImager. As shown in Figure 4, marked strands (on the left) appear to have been protected from destruction by Exonuclease, while unmarked strands (on the right) were destroyed by the enzyme.

5. Scaling up Surface-Based DNA Computation

In our approach, by immobilizing strands to a surface, loss of strands is greatly reduced. Also, our use of single-base mismatches to discriminate between strands

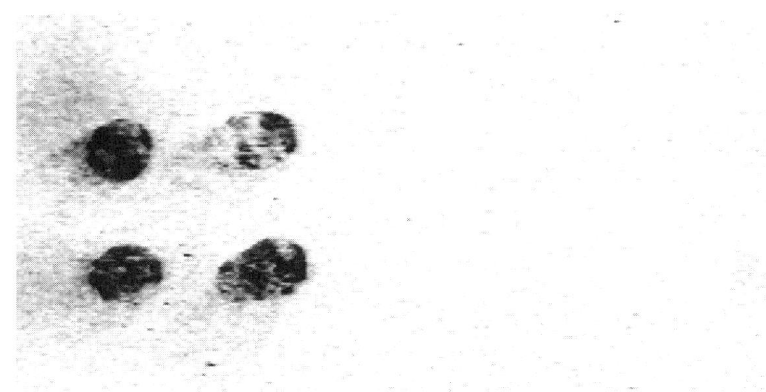

FIGURE 4. **Destruction of unmarked versus marked strands**

to be marked and those to remain unmarked has important implications. First, a high information density - one base per bit or better - can be achieved. Second, this allows synthesis of solution spaces (such as $\{0,1\}^n$) quickly and cheaply using standard technology.

However, there are also limitations of the surface-based approach as described thus far. We enumerate these now, and suggest possible ways of overcoming these limitations.

The scale of computation is severely restricted by the 2-dimensional nature of surface-based computation. With the surfaces currently used, if each DNA molecule occupies an area of 20 angstroms by 20 angstroms, or 400 square angstroms (this is the density found to be optimum in our previous work, [**4**]), then a 1 cm area can accommodate only 2×10^{13} molecules. To go higher one must either a) increase the surface density, b) increase the surface area, or c) build linkage chemistry extending out into solution from which the oligonucleotides can be attached to make a local three-dimensional network on the surface.

Another limitation is that discrimination of single base mismatches in hybridization reactions becomes more and more difficult as the length of the oligos increases from the 15mers that we are now working with. One approach is to design the sequences such that they can be subdivided into independent "words" of, for example, 15 or 20 nt each. For example, a combinatorial set of molecules 60 nt in length would consist of four sequential 15 nt words. Targeting of each word would be accomplished with a complementary 15mer specific for the word being targeted. This places additional constraints upon the sequences, however, and further work is required to design the proper structure for the word elements.

Although our initial studies have employed glass surfaces, we do not anticipate that such glass microscope slides will prove to be the optimal surface for this application. We already noted that one problem with glass is non-specific binding of DNA molecules to the surface (that is, rather than a DNA molecule in solution hybridizing to its complement on the surface, it binds in some other manner to the surface itself). Thermally grown oxides on Silicon wafers or alkanethiol self-assembled monolayers on gold surfaces may be better alternatives.

Since hybridization efficiency is not 100%, some strands which should be marked may not actually be marked, and thus may be destroyed unintentionally. One solution to this problem is to add redundancy to the solution space. Another is to exclusively mark strands that should be destroyed, and use the destroy-marked operation. Now, the opposite problem may occur, in which strands that should be destroyed are not actually destroyed. However, this can be more easily corrected by repeating the mark and destroy-marked operation.

Finally, other operations should be added to the model, to increase its utility. Different methods for labeling desired subsets of the DNA molecules are needed that modify the desired DNA molecules based upon properties of their sequences. This modification must then allow selective manipulation of the labeled subpopulation in successive queries, so that the desired solutions may be found to the mathematical question posed. Many different labeling reactions are possible, based on a) exonuclease degradation of either single-stranded or double-stranded DNA; b) ligation of a marker oligonucleotide onto the ends of selected strands; c) polymerase extension from the ends of selected strands; and d) endonuclease cleavage of selected strands. With the ability to do b) for example, it would be possible to solve the Circuit Sat problem with the surface-based model.

6. Conclusions

We propose a new methodology, namely surface-based DNA computation, that offers many advantages over Adleman's approach. We focus on a very simple set of operations that are standard on surfaces, and show how the Satisfiability problem can be solved with these operations. We describe partial progress on experimental testing of these operations.

In its current form, our approach probably will not scale to manipulation of extremely large solution sets. In spite of this, we believe that further study of surface-based DNA computation will play an important role in the area of DNA computation. First, we predict that in the short term, computations involving several successive operations can be done at a greater scale on surfaces than in test tubes, and these can be used to learn much about the chemical processes underlying any DNA-based computational paradigm. Second, the techniques developed for our current model and the resulting improved understanding of the chemical processes used may well be useful in combinatorial chemistry applications, where the scale of experiments is much less than 2^{70}.

References

[1] L. M. Adleman, Molecular Computation of Solutions to Combinatorial Problems. *Science*, 266:1021-1024, 1994.

[2] L. M. Adleman, On Constructing a Molecular Computer, Manuscript, Computer Science Department, University of Southern California, 1995.

[3] E. Baum, Building an associative memory vastly larger than the brain. *Science*, 268: 542-544, 1995.

[4] Z. Guo, R. A. Guilfoyle, A. J. Thiel, R. Wang, and L. M. Smith, Direct Fluorescence Analysis of Genetic Polymorphisms by Hybridization with Oligonucleotide Arrays on Glass Supports. *Nucl. Acids Res.*, 22:5456-5465, 1994.

[5] I. R. Lehman and A. L. Nussbaum, *J. Biol. Chem.* 239:2628-2636, 1964.

[6] R. J. Lipton, DNA Solution of Hard Computational Problems. *Science*, 268:542-545, 1995.

[7] L. M. Smith, Automated synthesis and sequence analysis of biological macro-molecules. *Analytical Chemistry*, 60:381A-390A, 1988.

(Q. Liu, A. Guo, R. Corn, and L. Smith) Department of Chemistry, University of Wisconsin, Madison, WI 57306 USA

(A. Condon) Department of Computer Science, University of Wisconsin, Madison, WI 57306 USA

E-mail address: condon@cs.wisc.edu.

(M. Lagally) Department of Materials Sciences, University of Wisconsin, Madison, WI 57306 USA

DIMACS Series in Discrete Mathematics
and Theoretical Computer Science
Volume **44**, 1999

Mesoscopic Computer Engineering: Automating DNA-based Molecular Computing via Traditional Practices of Parallel Computer Architecture Design

John-Thones Amenyo

ABSTRACT. How does one go about automating the steps of DNA computing, or what amounts to the same thing, the practical engineering of *hands-free, general-purpose* DNA computers?

The intent of this paper is to indicate how familiar computer design principles for electronic computers can be exploited to build practical computers at the mesoscopic scales of macromolecules and bio-polymers. DNA computing is the most realistic harbinger of such *molecular computers*. Pragmatically, it is expected that DNA computer architectures will be used routinely and not just for solving theoretically *hard* computational problems. The ideas discussed here are akin to the design of a practical programming language for a virtual computer.

The paper shows that all proposed DNA computing algorithms can be run on parallel computer architectures configured from *trellis/lattice banks, filter banks* and *switching banks*. Thus, DNA computations can be re-interpreted as dataflow (or signal flow) networks and subject to conventional treatment.

1. Introduction

Research is proceeding apace on the specifications of algorithms that exploit L. Adleman's method of "DNA-based (molecular) computing," [**1**]. Simultaneously, more work is needed on the molecular biology details that undergird the whole approach. Furthermore, an additional issue that ought to be addressed is how realistic DNA computers can actually be built to serve practical computational needs, just like electronic computers.

Computationally, one can consider L. Adleman's method [**1, 2**] as consisting of a concatenation of two major stages, constituting a waterfall model:

1. Generation of diversity (g.o.d.) stage (G): Intrinsic molecular interactions of DNA molecules are exploited to *code computational problems* and generate a very large space of *solution candidates.*
2. Solution extraction or filtering stage (F): All the generated molecules, representing solution candidates, are subjected to a series of mechanical manipulations and other physical processes in order to identify and extract specific molecules, if available, that are regarded as the exact codings of the

desirable, acceptable or *optimal solutions* of the original computational problem. These manipulations rely on several instrumentally accessible physical properties of the molecules (chemical, mechanical, electrical, magnetic, electromagnetic, thermal, optical, etc.).

Label the g.o.d stage in Adleman's algorithm as (G) and the filtering step as (F). Then, formally, the overall algorithm can be represented as a sequential composition $(G * F)$. One simple formal extension of the original algorithm is that several useful DNA computing algorithms will repeat this basic composition. [For the operations f and g, $f * g(x) = g(f(x))$]:

$$(G * F)^+ = (G * F) * (G * F) * \ldots * (G * F) * \ldots$$

With this view, the close connection between DNA computing and various forms of *darwinism,* including C. Darwin's original formulation [**13**] is apparent. In this case, the g.o.d stage is biological species diversity generated by individual reproduction and mutations, while the filtering stage involves Natural and Sexual Selections. Other modern applications include neural darwinism [**15**]; evolutionary programming and genetic programming [**14**] and [**17**]; and directed molecular evolution [**12**]. So DNA computing can be interpreted as a form of parallel evolutionary programming. DNA computing can also be identified with the (parallelized) *generate/hypothesize and test* paradigm of Artificial Intelligence (AI) research.

Many of the current efforts to extend and understand L. Adleman's seminal work focus primarily on issues connected with the filtering (F) stage. The g.o.d (G) stage is assumed to be straight-forward and treated as a blackbox in the algorithms.

A significant example is the ongoing work by R. Lipton *et al.,* for example, [**20, 5, 9, 10**]. It has been proposed that DNA computational problems be represented by means of "bitmap coding". Namely, for a given problem, consider each element in its solution space as satisfying n 2-valued attributes (predicates). Hence, each element in the solution space is characterized by a *bit map,* a (0,1)-vector of length n. In this view, a DNA computation really starts after the g.o.d. step is presumed to have generated all the bitmap vectors of the solution space and placed them in a mix/"soup" called a *test tube.* The algorithmic specifications then concentrate on filtering and merging operations, to search the space. Eventually, the desirable/acceptable solutions, if any, are extracted from the mix.

In the extraction-oriented algorithms in the open literature, the molecules coding the bitmap are treated as data structures/objects which are to be manipulated by external mechanical/physical processes. An interesting open problem is how to devise DNA computing algorithms in which desirable solutions are *self-extracting* or *auto-selecting.* This means the filtering stage is automated, using only molecular interactions to extract solutions. Ultimately, there is a *simple* answer readout, transduction and then presentation to humans.

In this paper, the g.o.d stage is studied as having interesting and important details in its own right. A set of modules are specified for assembling g.o.d stages for known classes of DNA computing algorithms. For a g.o.d stage, these modules are configured as parallel architectures called trellis/lattice banks. Apart from Adleman's original description, [**1**], existing studies of the details of the g.o.d stage include [**25**] emphasizing computing by annealing and ligation and building universal DNA computers using splicing operations in [**16**] and [**18**].

Another goal of this paper is to introduce concepts, such as *flow,* that enables one to re-interpret some of the computational ideas of DNA/molecular computing in terms familiar from the existing practice of computer architecture design and the computer engineering of semi-conductor based, digital electronic computers.

This *model-based* — "new wine in old bottles" — translation approach to studying DNA computers is merely following an illustrious tradition in mathematics and engineering. Consider, for example, the various definitions of less familiar and abstract types of "numbers" in terms of "natural numbers" by R. Dedekind, E. Galois (imaginaries), Kummer (ideal numbers), G. Cantor, W.J. Gibbs (vectors), H. Grassmann (vectors), W. Hamilton (quartenions), the various researchers responsible for the introductions of such structures as n-dimensional spaces, n-vectors and tensors. An even more pertinent example is H. Poincaré **[21]** studying Euler's conjecture on polyhedra by translating the vector algebra of (0,1) matrices on k-polytopes. Another example is C.E. Shannon's observation, as a graduate student at MIT **[22]**, that electronic relay and switching circuits used in communications and computing can be analyzed using the concepts of Boolean algebra.

This could prove useful as a modest step in moving beyond the invention of the DNA computer to considering engineering issues of physical packaging, manufacturing, fabrication, miniaturization and systems integration for the deployment and practical use of DNA computers, just as has been established for electronic **[7]**.

2. Preliminaries

2.1. Data, Data Structures and Flows. A collection of DNA molecules (oligo-nucleotides and poly- nucleotides) is the basic unit of manipulation in DNA computing. For the purpose of this paper, the fact that the DNA molecules can be oriented, single-stranded or doubly- stranded strings/sequences/words (of oligomers) is abstracted away. The architectural principles described here can handle macromolecules of any topological complexity. To be practical, there must be associated biochemical and biotechnical processes/operations available to manipulate these nonlinear structures[1]

The salient point here is that it is always a *collection* (of molecules) that is manipulated by all operational modules in the architecture. Such collections are represented by X, Y, Z, and indexed when needed. A collection is a set if each individual molecule in it is considered distinguishable. Nonetheless, several of the basic operations of the architecture cannot distinguish among multiple copies of DNA molecules having the same polymeric structure when they are present in a collection. Thus, each collection should be properly interpreted as a multi-set or a hyper-edge (of a hypergraph). When appropriate, operations of union, intersection, etc., refer to multi-set rather than set operations.

It is also useful to visualize the elements of X as being arranged spatially or topologically in some abstract space, as either a vector or as an array, $X = (x_1, ..., x_p)$ or $X = [x_{ij}]$. The dimensionality of this abstract (even nested) array of a collection is a pragmatic choice, dependent on the (human) interpretation of what is being computed. A collection X is interpreted as a *signal,* message, data set, packet, or data cell.

Each operational module of the architecture should be regarded as manipulating all the elements of its input collections *in parallel.* Thus, for collection, every basic operation module (f) is an SIMD operation, (possibly, there may be multiple,

[1]The intriguing, but unproven, possibility of using cell adhesion molecules (CAM) for nonlinear biocomputing has been communicated to the author by Markman Offer.

FIGURE 1. Generic Molecular Flow Manipulation Module

separate but concurrent copies of the basic operation acting on the collections, a quasi-MIMD implementation). This SIMD For example,

$$(fX) \Rightarrow ((fx_1), \ldots, (fx_p)) \Rightarrow (f(x_1), \ldots, f(x_p))$$

This use of SIMD operations is partly responsible for the profound computational power of L. Adleman's approach to molecular computation, in that it is one part where *massive parallelism* occurs in the approach.

Abstractly, a generic operation on a collection X yielding another collection Y is represented as $(X|f|Y)$. For example, from above,

$$Y = (y_1 = f(x_1), \ldots, y_p = f(x_p))$$

The molecular computation embodiment of f is said to have an *input port* to get the collection X and an *output port* to provide the collection Y. The generic molecular collection manipulation is diagrammed as shown in Figure 1.

2.2. Place or Container Operation $(X|p|X')$: **Wires and Vats.** This operation merely formalizes the physical reality that the DNA molecules in a collection have to be contained somewhere, and occupy the volume of something which holds them, between manipulations. A collection occupying a *place* is said to be the *content* of the place. A place with no content is *empty* and has no signal on it. A place with content can be interpreted as computer *memory* or *storage space.*

A place can contain more that one distinguishable collection at the same time. Physically, this means that the elements of each collection X in the place are labelled/tagged or *colored* with an identifying *mark/signature* of the collection. Thus one can talk about collection *types.* This concept of colored collections is used later in Section 7.2 to introduce the ideas of multi-tasking and multi-programming into DNA computing. For example, all collections of a particular computation will be colored to have the same type. Until Section 7.2, it is assumed that only single collection type (single color) is in use throughout the architecture. All collections of the same color are referred to as a *commodity.* Coloring of molecules is also the basis of one general class of inter-molecular communication (for computational purposes).

In the growing literature on DNA computing, a container of DNA molecules is often referred to as a *test tube* [**20**]. (Sometimes, the collection of molecules is itself called a test tube). However, a *place* operation is more general, because so far in current usage, a test tube merely contains passive objects that can be manipulated by filter operations.

The DNA molecules that are the contents of a place can be behave either as a) passive molecules-as-objects (data structures), to be manipulated or as b) active/reactive molecules-as-agents (computational structures/agents) to manipulate others, including themselves. A test tube is an example of the first case. The molecules in a place remain the same until and unless transferred out of the place and/or manipulated in place by other operations. If the molecules remain in place and are not changed/manipulated, then $(X|p|X')$ is equivalent to $(X|p|X)$ and may be written as $(X|I|X)$ or simply as (X). The I-operation later serves as the basis of a temporary store (short-term memory) or "register or cache memory" of the general purpose molecular computer architecture.

In line with circuit terminology, such a place, a mere container of passive molecular collections, can be interpreted as a *wire* or channel or flow/transmission link. In the rest of the paper, the term wire is preferred to that of test-tube, to highlight the correspondence in conventional computer/communications engineering.

In the second case, the molecules occupying a place can self- interact or with other sympatric collections. Therefore, the kind, quantity and number of molecules in the content of the place change over time. That is, the sojourn of a molecule at a place may change its character or even transform or combine it into another molecule. In this case, the place can be interpreted as a *history/memory-sensitive (Markovian) wire.* It may also be termed an *active* or *reactive* place, a reaction container, pot. The term *vat* is used for such a place in the paper. A vat is interpreted as having self- modifying rather than *re-entrant code.* The operations of the g.o.d step occur in vats. Section 5 describes these modules.

2.3. Nodes and Flows. Any operation in the architecture that is not a place (wire or vat) is referred to as a *node.* Places are interconnected by nodes and nodes are interconnected by places. A sequence/path/route of nodes interlinked by places effectively performs a sequential composition of operations.

The requisite architecture of DNA computing, for algorithms published so far, is a pull not a push system. Conventional examples of a *push* system are logic circuits and communication networks. An example of a *pull* system is a Petri Net. In a pull system, contents of a place remain in the place indefinitely until/unless removed and transferred to another place by the action/activity of a node. In other words, the molecules are not self-steering under some power; they are moved about by pick-and-place mechanisms. An in-place manipulation of a collection (in a vat) is also considered a transfer.

The in-place or inter-place transfer/transformation of a collection can be related to traditional computing and communications practice by interpreting it as a *flow.* Formally, given an nodal operation $(X|f|Y)$, the flow through that node is the ordered pair $[X, Y]$. To represent the activity of a nodal operation over time, the collections in its flow can be indexed by a time variable (t). For each t, $[X(t), Y(t)]$ represents a different flow. Sometimes, pragmatically, the input collection of a flow's ordered pair is itself referred to as a flow.

Thus, in general, a DNA computation is manifested as the flows of (multiple) collection(s) (signals) through the places (wires and vats) and nodes of the computer architecture that is implementing it. A DNA computation can be related to an oriented data/signal flow schema, graph or network, with *input* (initial, originating) and *output* (final, destination) nodes

Note also that since places, whether unreactive or active are individually named and labelled, one can talk about repeated visits of a particular molecule (or a collection X) into the same place. This in turn introduces the concept of *feedback flows* into DNA computing, and is used in Section 6 in the construction of general purpose DNA computer architectures.

The count of nodes in a computation imbedded in the architecture gives a measure of its cost and computational complexity. The count of wires used in the computation gives a measure of its layout/spatial costs, in analogy to wire costs in the VLSI design of electronic circuits. The maximum distance (in terms of inter-node hops) between input/output node pairs in the signal graph of the computation gives a measure of the delay performance (computational depth) of the computation. Contrast this macro-level delay cost with the micro-level massive parallelism inherent in the nodal and vat manipulations of collections, as noted earlier.

3. Switching Banks and Flow (re-) Directors

The modules described in this section manipulate flows in the architecture, mostly without regard to the properties of the individual elements in the collections in the flows. These operations *glue* places and nodal operations that implement the g.o.d and filtering stages of the DNA computing paradigm.

Any portion of the architecture that uses only nodes from this section as well as wires is termed a *switching bank* or an *interconnection network.*

In mechanical automations of DNA computing (for example, using micro-robots), the most significant contributors to the (delay and energy) performance costs are the number of switching bank nodes that have to be traversed. It is only when molecular mechanisms are used to automate the switching banks in the architecture that the count of other nodal operations (trellis or filter) become comparable and have to be taken into account in delay calculations. Wires count!

3.1. Splitter or Divider Operation $(X|S.k|Y_1, \dots, Y_k)$**:** A collection X of DNA molecules gets divided by the S-operation and flows into new collections $Y_1, \dots, Y_k$. Each $Y_i, i = 1, \dots, k$, is regarded as a *tap* into the flow. Thus X is the union of $Y_1, \dots, Yk$, $X = \bigcup\{Y_1, \dots, Y_k\}$. A splitter operation can also be used as an inverse multiplexer in special cases.

3.2. Combiner or Aggregator Operation $(X_1, \dots, X_n|C.n|Y)$**:** The combiner is the dual of the splitter operation where collections are merged, *mixed* or joined into a single multi-set flow. Thus, it is a *superposition* operation. Now, $Y = \bigcup\{X_1, \dots, X_n\}$. Combiner nodes can be used to directly implement the **Merge** operations in filter stages as advocated in [**2, 20**], see also [**18**].

Note that both the splitter (S) and combiner (C) operations involved *pouring* of DNA molecules when manual manipulations are used to implement DNA computing. Their operations are analogous to splitters and combiners in cable TV (CATV) and also in lightwave or photonic or fibre optic communications systems.

3.3. Multiplexer Routing Operation $(X|M.[k.i], W|Y_1, \dots, Y_k)$**.** At any time, the value of the variable i is one of $1, \dots, k$, and the collection X flows into Y_i, if this is the case. The node is (re-) *programmed* by changing the value of i over time.

$$\begin{aligned}
Y_j &= \emptyset, \quad \text{if } j \neq i, \quad \forall j = 1, \dots, k. \\
Y_j &= X, \quad \text{if } j = i, \quad j = 1, \dots, k. \\
W &= X, \quad \text{if } j \neq i, \quad \forall j = 1, \dots, k.
\end{aligned}$$

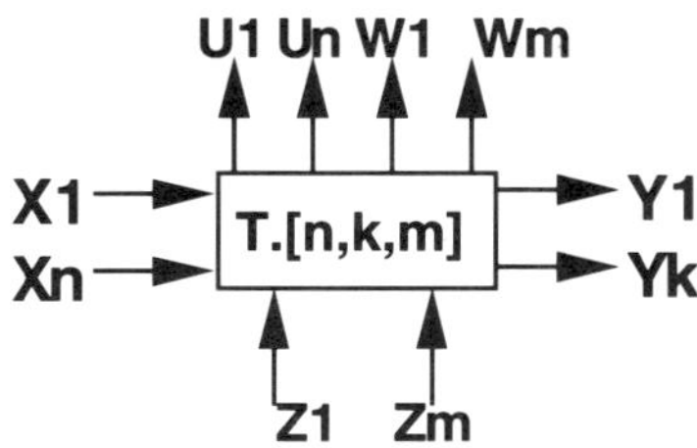

FIGURE 2. "Transistor" Gate Operation Module

3.4. De-multiplexer Routing Operation. This operation is symbolized by $(X_1, \ldots, X_n | D.[n.i], [W_1, \ldots, W_n] | Y)$.

It is the dual of the multiplexer routing node. The selected input flow i is redirected into the output flow.

$$\begin{array}{lll} Y = X_j, & W_j = \emptyset, W_k = X_k, \quad \forall k \neq j, & \text{if } j = i, \quad j = 1, \ldots, n, k = 1, \ldots, n. \\ Y = \emptyset, & W_j = X_j, & \text{if } j \neq i, \quad \forall j = 1, \ldots, n. \end{array}$$

3.5. Context-Controlled Gate ("Transistor") Operation. This is symbolized by,

$$(X_1, \ldots, X_n | T.[n, k, m, [[Z_1, \ldots, Z_m], [U_1, \ldots, U_n], [W_1, \ldots, W_m]]] | Y_1, \ldots, Y_k).$$

Its module is depicted in Figure 2. The general idea here is that the collections $Y_1, \ldots, Y_k$ are built (as recombinations) from the collections $X_1, \ldots, X_n$, provided the molecules in the collections $Z_1, \ldots, Z_m$ satisfy certain specified predicates about their properties. Effectively, the Z-collections are used to "switch" the X-collections into the Y-collections. For the kinds of predicates that can be used, see the definitions of Filter operations in Section 4 below.

$$\begin{array}{lll} Y_j = f_j(X_1, \ldots, X_n), & \text{if } P_j(Z_1, \ldots, Z_m), & \forall j = 1, \ldots, k. \\ Y_j = g_j(Z_1, \ldots, Z_m), & \text{if } \neg P_j(Z_1, \ldots, Z_m), & \forall j = 1, \ldots, k. \\ U_r = E_r(f_1, \ldots, f_k, g_1, \ldots, g_k, X_r), & \forall r = 1, \ldots, n. & \\ W_s = H_s(f_1, \ldots, f_k, g_1, \ldots, g_k, Z_s), & \forall s = 1, \ldots, m & \end{array}$$

The $P_j()$ are predicates on the switching collections, $Z_1, \ldots, Z_m$. The operations $f_j()$ and $g_j()$ are Y-collection builders that use any of the flow operations described here. The $E_r()$ and $H_s()$ operations divert those X_r and/or Z_s collections that are not used to build any new Y-collections into the corresponding U_r- or W_s-channels, appropriately.

A useful special case is the Subsumption Operation $(X | T_S.[1, 1, 1, [Z, U, W]] | Y)$:

$$\begin{array}{llll} Y = X, & U = \emptyset, & W = Z, & \text{if } P(Z) = a. \\ Y = Z, & U = X, & W = \emptyset, & \text{if } P(Z) \neq a. \end{array}$$

Another special case is the Exchange and Cross-over Operation symbolized by $(X | T_X.[1, 1, 1, [Z, -, W]] | Y)$:

$$Y = Z, W = X, U = \emptyset.$$

In essence, for the transistor operation, under certain context-sensitive conditions, a flow Z replaces the flow X that would normally be used as input for the continuation [**24**] of a computation.

Due to the presence of the predicate operations $P(.)$, the controlled gate operation/circuit may be *dirty*. The concept of a dirty circuit is explained in Section 4.2 and solutions are provided in Section 5 below.

4. Filters and Filter Banks

4.1. Extracting Solutions is Filtering is Signal Processing. In a general sense, the filtering stage (F) in DNA computing is completely characterized by noting that it involves an organized collection of signal processing or pattern recognition/analysis ("filters") operations. As noted earlier [**3**],

> ".. [a] little thought shows that most of the *solution extraction* approaches involve determining the optical, mechanical, thermal, magnetic, electrostatic ... and other *physical* properties of oligo-nucleotides and other biomolecules formed...All of these determinations are properly thought of as belonging to the engineering area of signal processing (including pattern/picture/image processing)"

In other words, for the abstract models defined so far in the literature, for example, [**2, 20**], the operations of *Separate*, *Extract* and *Detect* can be described as special cases of *Filter* operations.

In terms of biochemical or biophysical manipulations, all of the following techniques can be interpreted as performing *filtering* operations on the generated candidate solution spaces: magnetic bead extraction; graduated PCR operations; PCR-based extraction; gel-electrophoresis; use of affinity columns; florescence labelling; and sequence determination.

A portion of the computing architecture that uses only wires, switching banks and filter nodes is termed a *filter bank.* This is consistent with current usage in the signal processing community. Thus, the Filter stage (F) of a DNA computing algorithm can be regarded as a multi-stage interconnection network (MIN) or plex structure, where the primary nodes are filter operations.

4.2. Filter Operation $(X|F.P[a_1, \ldots, a_k], W|Y_1, \ldots, Y_k)$**.** From the viewpoint of the end results of a filter step, (essentially an extraction, separation or discrimination), a filter operation can be regarded as a discrete function (generalized logic "gate") F that operates on (one or more) collection(s).

In the simplest case, a basic filter step (operation) is represented as

$$(X|F.[1, P[a], W]|Y_1),$$

with Y_1 and Y_2 being the collections or flows generated by,

$$\begin{array}{rcl} Y_1 & = & \{x \in X : P(x) = a\} \\ W & = & \{x \in X : P(x) \neq a\} \end{array}$$

Here, $P(.)$ is a generalized *predicate* (indicator function) mapping the elements of X into a value domain V, $(a \in V)$. That is, $P(.)$ can be any discrete function defined on the collection X.

Essentially, $P(.)$ is used to separate the elements in the collection/flow into two new collections, according to the properties of the elements. From, the point of view of Y_1 each of its elements can be regarded/interpreted as being *extracted* from the collection X. Concretely, the predicate functions of interest are those which ask questions about instrumentally accessible physical properties of the individual molecules in collections (magnetic, mechanical, optical, etc.)

In reality, there may be errors in the determination of the satisfaction of $P(.)$ by some of the elements. Therefore, some elements may end up in the wrong output collections — for example, the so-called *false positives* and *false negatives.* If such errors occur, the filter operation is said to be *dirty;* otherwise, it is said to be *clean* (which is ideal). Dirty filter operations are properly studied using ideas from either probability theory and statistics or from fuzzy sets [**26**].

For example, we can define the signal-to-noise ratios (SNR) of output collections Y_1 and Y_2, ($Y_2 = W$) of a filter operation. For $Y_k, k = 1, 2$, let Y_k^+ (Y_k^-) be sub-collection of elements where the determination of $P(.)$ is correct (incorrect). Then the SNR(k), is given by:

$$SNR(k) = e^{|(Y_k^+)|-|(Y_k^-)|}/N$$

Where $|Y|$ is the cardinality of a (multi-set) Y considered as a set, with all the elements being distinguishable. N is a normalization/reference constant. The *degree of cleanliness* of a filter circuit/operation is the minimum SNR(k), $k = 1, 2$.

The general basic step of Filter operation can be represented as

$$(X|F.P[a_1, a_2, \dots, a_k], W|Y_1, \dots, Y_k).$$

Here,

$$\begin{aligned} Y_1 &= \{x \in X : P(x) = a_1\} \\ \vdots & \quad \vdots \\ Y_k &= \{x \in X : P(x) = a_k\} \\ W &= \{x \in X : P(x) \neq a_1, \dots, P(x) \neq a_k\} \end{aligned}$$

For convenience, X may be regarded as being a multi-set union of other collections, or what amounts to the same thing, a filter operation can be preceded by a Combiner operation to get multi-port input flows. Thus a filter operation can be interpreted in terms of several metaphors. It behaves as a lowpass/bandpass/hipass filter combination; as a multiplexor; separator; discriminator; selector; classifier; attribute-based labeller; sorter/ranker; content-dependent retriever or as a Böhm-Jacopini conditional branching control structure. A filter operation can also be interpreted as the left-hand side of mapping or signature table. The right-hand side of such a table can be implemented using Context-Controlled Gate Operations. A sequential composition of filter nodes with intervening wires, the simplest form of a filter bank, can be interpreted as performing a multi-attribute classification on the elements of the starting flow/collection. Another interpretation of the filter module is to consider it as implementing fuzzy sets. Here, the predicate results, $P(x) = \alpha_1$, can be interpreted as α-cuts and α-levels, [**26**].

5. Trellis/Lattice Banks: generative/constructive modules

Several operations on vats can be used to implement the details of the g.o.d. stage. Any portion of the architecture that includes only wires, switching banks and operation nodes from this section is called a *trellis* or *lattice bank.*

5.1. Selective Amplification Operation. This is symbolized as:

$$(X[x_1, \dots, x_n]|A.(k_1 \dots k_n)|X'[\{x_1 : k_1 \times \text{times}\}, \dots, \{x_n : k_n \times \text{times}\}]).$$

This module provides the embodiment of the operation where all starting molecules of type $x_i, i = 1, \dots, n$, which are in a (multi-set/flow) collection X are replicated to obtain k_i copies of each. The major bio-chemical process for doing this in DNA computing of course relies on the polymerase chain reaction (PCR).

5.2. Global Replication Operation $(X|A_G.k|X')$. In this case, all the elements of the starting collection are replicated k times. This operation can be used to embody the *Amplify* primitive in the abstract models of [**2**] and [**20**].

5.3. Multiplicative Branching $(X|B.n|X, \ldots, X)$. This is a composition of selective amplification (A) followed by a splitter (S) operation, so that in each of the output places (outgoing branches) there is effectively a copy of the original flow/multi-set (X). This can be interpreted as implementing a broadcast (actually as multicast, multi-point or multi-drop or conference) distribution of the collection X over n new wires.

5.4. Immobilization Operation $(X|I_M.[Z, q, Q]|Y)$: This is a place that permanently hosts a multi-set collection Z of molecules that are capable of interacting with and/or acting upon transient molecules of X flowing in and out of the place. The name comes from the fact that $z \in Z$ are immobilized in that vat, much like resistance/conductivity of an electronic wire or even the EMF of a voltage source. In practical implementations $z \in Z$ are likely to be immobilized enzymes. The total number of copies of any molecule type in Z will never be smaller than q and total number of molecules in the collection will never be greater than Q.

5.5. Self-assembly (Self-Organization) Operation $(X|S_A|Y)$: One result of the self-interactions of the molecules in X may be the self-assembly and self-organization of the x_i molecules to form more spatial/topologically complex structures. The physical examples of self-assembly operations include: annealing/melting; ligation/restriction enzyme cutting; polymerization, mutation and splicing. One should also include *trans-coding* and transformation operations on the molecules, for example, DNA-to-DNA transformations.

5.6. Supervisory Control Operation. One also needs several operations to coordinate active place operations. These will correspond to traditional process control operations of start or stop, pause or resume, redo, delete or edit. At the mechanical level of automation, factors that are typically considered sources of error in DNA manipulation can actually be exploited to obtain these control or coordination operations. Namely, pH level adjustments, solution concentrations, temperature and levels of contamination.

A special case of supervisory control operations is the Purification/Washing Operation. The concept of a dirty operation/circuit was introduced in Section 4.2. Physically, this will be due to inaccuracies of predicate $P(.)$ determinations on collections. A similar issue was identified and resolved in electronic computers, by J. von Neumann, right at the start of the stored-program computer revolution. The essential idea is to use *redundant coding* (of circuits and/or wires).

In a Serial Purification, consider a dirty operation $(X|f|Y_1, Y_2)$. Each flow Y_k, can be passed through a pipeline of multiple copies of the circuit (f). Each time, only the appropriate Y_k-branch is chosen for further manipulation. This results in a pruned of tree of f-circuits (f bank). The Y_k-flows at the leaves of the (washing) tree should have a lower degree of dirtiness than the initial Y_k-flows.

A Parallel Purification version shows an interesting application of genetic programming in DNA computing. Given $(X|f|Y1, Y2)$, make $N \times p$ copies of f to be used in $N+1$ stages. For the initial stage, make p copies of each Y_k. For each succeeding stage, take the p of the Y_k collections from the circuits/operations of the preceeding stage, whose degree of cleanliness exceeds a specified threshold χ. Pass these collections on to be purified by the next stage.

Either form of Purification/Washing is expensive in circuit resources and delay (slow down). The extra circuit costs can be avoided by using a re-circulating switching bank that allows the reuse of the same small number of circuits/operations to implement the purification chaining or stages. The extra delays will be unavoidable.

5.7. Time-gradient Segregation (spatialization) Operation. The formal representation of g.o.d. stage is not complete because each vat always has a mixture of molecular products in various stages of construction. The concept of time- gradient spatialization can be used to provide a better visualization and guarantees of the products of active place operations and g.o.d stages.

Informally, one proceeds as follows. Since the interaction of molecules to produce other molecular structures takes time (that is, does not happen instantaneously), one can introduce the concept of a *generation* (age) to describe the molecules in a vat. The starting molecular materials are said to be of the 0th *generation.* Molecular structures formed as the result of the interactions of the starting molecules are said to be of the 1st *generation.* In general, a molecule formed that involves at least one molecule of the $(n-1)^{st}$ generation and no molecule of a higher generation is said to be of the nth *generation.* The molecules in a vat, considered as members of generations, are regarded as occupying a *time gradient.*

The analytical difficulty of completely formalizing the g.o.d stage and the operations on active places is that they intermingle molecules of many generations. This admixture impacts the predictability of contents of vats and also forces one to use filtering steps that are used merely to locate molecules by their generational attribute.

In time-gradient spatialization, every vat with an uncontrolled admixture of multi-generational molecules, is replaced by a pipeline of vats, called *spatialization vats.* Each spatialization vat contains only molecules of a short specified range of consecutive generations

$$(X|G.g[1] * G.g[2] * \ldots * G.g[q]|Y).$$

In practice, the molecules flow consecutively through the vats in the pipeline of the spatialized time gradient. Thus at any time, one knows the generation (or up to within a small range) of any molecule in a g.o.d step.

One can interpose other operation modules between the vats in a segregated time gradient. Particularly useful will be multiplicative branching, combiner, splitter and filter (selective switching) operations. Formally, let δ represent that minimum interval of time between the interactions that produce molecular generations ("minimum free path"). Let $g(k)$ be a monotonic increasing function of the positive integers, $k = 1, \ldots$ Then each spatialized vat contains only molecules with generations that satisfy $g(k)\delta$, for $r \leq k \leq s$.

6. Parallel Computing Architectures: Flow Networks

The basic operations described above can be combined by means of sequential compositions to form what are effectively DNA computational structures that can be likened to data flow or signal flow networks. Topologically, the resulting flow circuits could take the form of trees, directed acyclic graphs, multi-stage interconnection networks, plex structures or even more complex and dense graphs and (multi-dimensional) polyhedra [**19**].

6.1. Computer Engineering. The flow-oriented architectural model presented can be the vehicle for studying the design of DNA computers using familiar principles used in the design of electronic computers. For example, parallel algorithms [**19**], evolution of the architectures for electronic digital computers [**6**] and ideas analogous to those of VLSI design [**4**]. Other useful ideas come from functional programming and tuple-space programming [**11**]. Issues of characterizing the most suitable biomaterials for practical molecular computers, while very important, are beyond the scope of this paper.

6.2. General Purpose DNA Computer Architectures. The corresponding data/signal flow networks for DNA computing algorithms have arbitrary topologies. In the real world, one is likely to be constrained by upper bounds on resouces of vats, filter and trellis modules. In this situation, one can then design the computer architecture so that its banks have regularized structures/topologies, for example, well-known forms of (augmented), hypercubes, (d, k)- graphs and dense graphs [**19**]. An additional programming step will then be required to map an unfolded, arbitrary dataflow network of any computation into the bounded and topologically regular computer architecture. Portions of the flows of a computation will have to re-circulate through the architecture and repeatedly re-visit and reuse its resources. This introduces in essence the concepts of recursion and iteration in DNA computing.

Formally, consider a bounded architecture (BA) with N filter and trellis nodes. Assume there are at least $p > 0$ of each filter or trellis node type in the architecture. Assume that a switching bank constructed with N diplexer De-multiplexer Routing Operations is available to be used as the Inflow bank (IB). Similarly, assume that a switching bank with N diplexer Multiplexer Routing Operations is available as the Outflow bank (OB). Now consider a recirculating switching bank (RB) that is constructed as a fully connected bipartite graph $K_{N,N}$. RB has N input ports and N output ports. Each input can connect to any output and thus RB is non-blocking [**8**]. As a bipartite configuration RB can be constructed using only Multiplexer Routing (for input) and De- Multiplexer Routing (for output) Operations/circuits. (There are more efficient ways of constructing such RB interconnection networks, see [**8**] and [**23**]).

The banks BA, IB, OB, RB, when properly configured form a general purpose DNA computer architecture (DC). From a "PMSCIO" level viewpoint of computer architecture [**6**], BA serves as "P" (processors), IB+OB+RB serve as "S" (switching), IB and OB serve as "I/O" peripheral interfaces to transduction stages that intercommunicate with the DC. IT is an optional switching bank composed of only identity operations (wires) $(X|I|X)$. Circulation of collections through the IT + IB + OB + RB switching banks can be regarded as flows through delay lines. Hence, IT may be regarded as serving the "M" (memory/storage) function of the DC architecture. The explicit need for externally setting parameter values of the various operational modules (programmability) is indicated by the unit labelled OP in the diagram shown in Figure 3. OP intercommunicates with all the other units. Thus, OP serves as the "C" (control or coordination), stored-program (operating system, firmware, application "software") of the DC. In practice, OP could literally

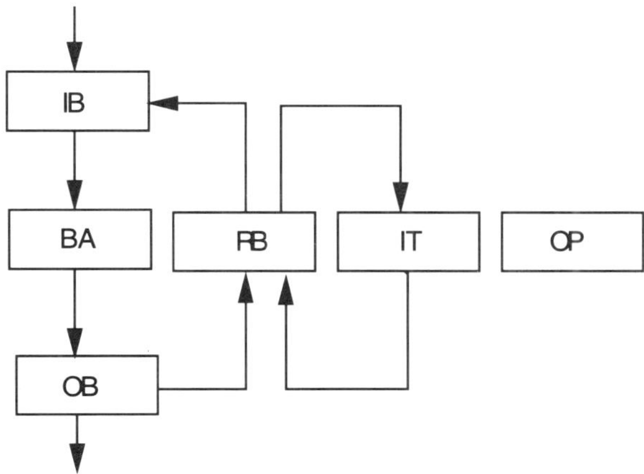

FIGURE 3. Bounded Architecture for Molecular Computer

be driven/controlled by conventional software running on electronic or other digital computers.
The components of the general purpose Ad-Molecular Computer architecture can be re-arranged and re-labelled so that the configuration looks similiar to the more familiar stored-programmed architecture for digital electronic computers. This is depicted in Figure 6.2. Notice that the IB + OB + RB units are now combined into a single switching unit labelled "SW." The BA unit is split into two types of processing units (-PUs), dedicated respectively to Arithmetic/Numeric and Symbolic/Logical processing. Each processing unit will have a fixed collection of g.o.d (G) and filter (F) modules/stages. A unit in the architecture that is now explicitly shown is the Materials Reclamation Unit (MRU). It is the function of the MRU to perform *garbage collection* and recover computational molecules that are no longer in use elsewhere in the computer, and then prepare and reconstitute them for *reuse* in future computations. Thus, the MRU can be regarded as the source of supply of all the computational molecules resources that are employed in the architecture.

7. Architectural Features and Plausible Areas of Applications

The features of such an architecture (the Adleman-style Molecular Computer or Ad-Molecular Computer) can quickly be highlighted. Its computers are *wet* and *fluidic*. Collections of molecules are the fluents. Molecules are carriers of computation, (versatile enough to represent data, functions/operations/relations, operators and combinators, and processes/threads/processors).

All collections are manipulated each time in each "processing unit" of the architecture they go into. The collections are one form of "memory." There is no in-place "storage"; although a re-circulating flow portion can be used as a register/cache. The Ad-Molecular computer is digital, it is capable of manipulating discrete, finite mathematical/formal structures.

Its power resides in the very large-scale micro-parallelism of molecular combinations and manipulations as well as macro-level computational depth. Multiprogramming is a natural fit, and can be achieved by coloring (indexing/labelling)

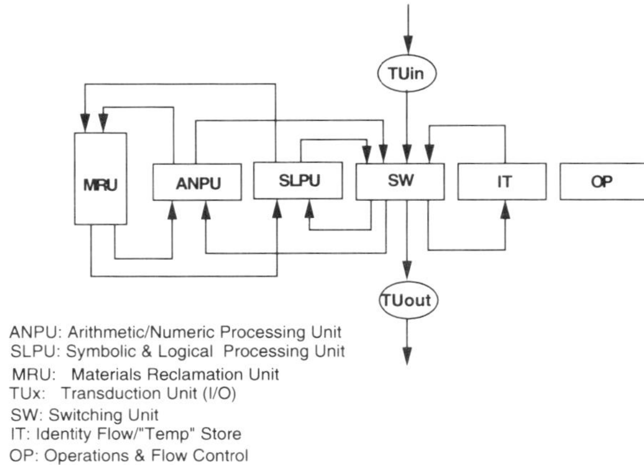

FIGURE 4. A General Purpose Ad-Molecular Computer Architecture

the collections into (virtual containers/categories or aggregations). (Virtual) positional addressing can be achieved by indexing/tagging and associative addressing achieved by pattern matching.

Fault-tolerance all the way to the molecular process level can be achieved using techniques from redundant coding and forward error correction. Equivalent calculations (based on bits or molecular information transfer units (mits)) of performance ratings, bandwidth and capacity show these general purpose computers to be potentially extremely powerful.

From an algorithmic viewpoint, according to the current state-of-the-art of underlying technologies, these general purpose (DNA) Ad-Molecular computers would be most suited for the massively parallel symbolic, non-numerical and logical manipulation of large-scale collections of abstract, discrete, mathematical/logical and data/information structures (entities). These entities and their manipulations are coded using molecules. For the moment, the fundamental molecular operations that would seem to be useful computationally include: tagging/coloring; formation of sum/union domains by merging/superposition and related operation of amplification; combinatorial construction, adjoining, composition or juxtaposition implemented by various (co-) "polymerization" processes and combinatorial product. A programming style well suited to this architecture is combinatorial programming/products followed by pruning.

The notion that DNA computing is the manipulation of large-scale collections brings within the scope of its applications various pragmatic computational collections useful several areas: sets, classes, arrays, images/patterns, vectors, tuples, tables (mapping, translation, signature, decision), signals, data bases, information bases, knowledge bases, distributions, (hyper-) graphs, communicating automata, processor networks, atlases of maps/functions/relations, groups of (situational/reactive) agents, spreadsheets and dependency networks.

Whenever the technology is advanced to cope effectively/efficiently with numeric/arithmetic processing, other types of collections would be brought on board: from the fields of mathematical physics, lattice physics and the physics or science of solid state materials, disordered systems and W. Weaver's organized complexity.

Suitable Ad-Molecular computing schemes can be constructed for structure estimation (a problem that comes from the area of system identification and is related to learning, inductive generalization and prediction of discrete dynamical processes); automation of the symbolic/non-numeric processing of multiple/universal algebras, and thus symbolic computations on linear, bilinear and quadratic forms as well as data parallelism and parallel reduction operations [c.f., K. Iverson and D. Hillis]; relational algebra and thus the implementation and support of very large scale data bases; and both coarse-grained and fine-grained simulations of processor networks, including lattice/trellis/ladder networks/algorithms, dynamics of artificial neuro-networks and related dependency network models for cellular automata and spin glasses. It is also possible to simulate the evolution of board games such as chess.

7.1. Multi-valued logic computers. There exists the possibility of using DNA computing to implement multi-valued (n-valued and variable-valued) logic circuits.

A metaphor which is potentially useful is to interpret the general

$$(X|F.P[a_1, a_2, \dots, a_k], W|Y_1, \dots, Y_k)$$

Filter step as implementing a special case of a multi-valued logic function. Special instances of the Transistor Gate operation can also be interpreted in this way. This points to the exciting possibility that DNA computing could be used for practical realizations of multi-valued (n-valued) logic circuits. The electronic realizations of such generalizations of the Boolean logic switching systems have proved quite elusive. This would seem to be an interesting direction to push the practical applications of DNA-based molecular computing.

7.2. Multi-tasking with DNA computers: Multi-colored architectures. The computer architecture described so far merely functions as a single-tasking computer in that it can handle only one computation at a time.

One way to upgrade the architecture into a multi-tasking system is to tag or color the collections of each specific computation at all times, by using molecular markers/signatures adjoined to the DNA molecules in each collection. The definitions of all the modules need to be modified to account for the color typing of collections. Each operation then manipulates only collections that of the same type. The colored architecture can then be interpreted as a multi-commodity flow network, where each commodity is the computation flowing through. If there is a multi-colored bounded architecture in a DC (Section 6.2) with τ colors then the DC is a general purpose DNA computer with a multi-tasking factor of τ.

7.3. All-Molecular, "One-Soup" Computer Architectures. An intriguing speculation in the literature is the possibility of automating the architecture by using molecular units for every aspect, including the places/containers, switching and filter nodes. Thus, one can place the whole architecture in a single biochemical solution in a "tub" and eventually "fish out" the answers to specific computations. For the flow-oriented architecture, one needs molecular structures to serve as places/compartments and others function as molecular "pick-and-place robots" (*meso-bots*) to transfer collections between places. The only crucial requirement is that the micro-level SIMD character of DNA computing be maintained This is an open problem at the moment.

7.4. A "Virtual" Laboratory for DNA Computing. The author is also currently designing a software-based "virtual lab" for DNA computing. Several of the results presented here came from ongoing attempts to design the accompanying programming language. The motivation for a virtual lab is quite simple. Although several scientists, mathematicitians and engineers from diverse backgrounds are interested in the ideas of DNA computing, very few have the resources, laboratories or the requisite molecular biology experience to actually perform the biochemical protocols themselves.

Thus, a software-based DNA computing "virtual lab" can serve as a starting proxy for research into DNA computing. It can serve as an aid to the imagination and visualization of the processes and results of molecular computing algorithms that are proposed. Ideas can be prototyped in the virtual lab, essentially on what is a small scale "virtual" DNA computer.

The ongoing work involves implementing (small-scale) aspects of Ad-Molecular computing/computers in software to run on electronic computers. Various features would be incorporated include, a "virtual" biochemical lab for computing, providing access to molecular programming resources on electronic computers; software based expert assistants; emulators and molecular programming compilers, translators or assemblers and computer-aided automatic programming to allow mainstream computer scientists and other engineers, scientists and mathematicians to be able to perform prototyping and explorations of Ad-Molecular computing.

7.5. Distributed DNA Computing and Other Models of Computation. The flow-oriented interpretation of DNA computing means it could serve as a *model system* to study other flow-oriented computing systems. This will be useful for logical aspects in which the speeds of operation are not the major focus. The most obvious examples are optical computing and neuro-computing in brains and nervous systems. For example, it is straightforward to model the dendritic arboration and axonal distributions of a biological neuron directly using the operational modules described in the earlier sections. Either a coarse-grained or a fine-grained model is feasible.

The vast amounts of DNA molecules involved means that DNA computing architectures could also be constructed to directly simulate complex systems where cooperative and emergent behaviours are important issues to be studied. Only a brief sketch of this direction is attempted here. Vats can be used to code process cells; collections/flows will model inter-cellular interactions and context-sensitive behaviour will be coded using filter banks. Switching banks will implement the interaction topologies. The resulting arrangement is a dynamic space filter or more familiarly an asynchronous cellular automata. It can be used to study such complex systems as nerve networks, protein conformation, and several problems of condensed matter and lattice physics.

The inter-coupling of successive time stages of two interacting entities (agents) is shown below, with each time stage modeled as an Adleman-style computer. The generalization to k-agents, as well as the introduction of skip level inter-agent couplings is obvious The traditional metaphors for interpreting this distributed model includes, C.A.R. Hoare's communicating sequential processes (CSP), SDL sequence charts for communications protocol models and multi-agent architectures in distributed AI. This coarse-grained model is depicted in Figure 5 as a form of inter-agent (inter-molecular) communication.

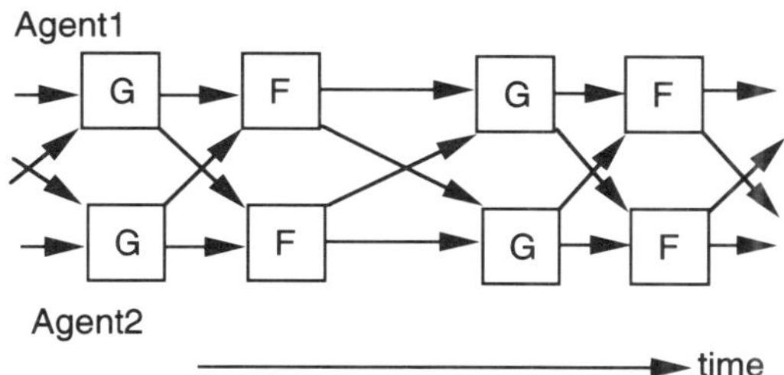

FIGURE 5. Dynamic Process Network — coarse-grained

8. Summary

By introducing the concepts of filter steps, basic flow operations and time gradient spatialization, this report attempts to place DNA computing in a conceptual framework that allows a direct application of the well-established design principles of electronic computer architectures. This is important because it opens the area of DNA computing to a wider audience of scientists and engineers familiar with electronic computers.

Furthermore, the formalization indicates that DNA computing can be regarded as a generalization of the traditional logic circuits (switching) and signal flow networks. In a conventional logic/signal circuit, what flows past a point in a wire at any time epoch is a single unit of value (signal, packet, cell, message). However, according to the ideas discussed here, in a DNA computing circuit, the flow in each wire (place) is composed of concurrent/parallel, multi-colored collections of values (molecules), conceptually arranged as vectors or arrays.

The flow model also suggests several interesting directions for further research, including DNA computing based multi-valued logic, and bio-molecular computing with non-linear structures, going beyond the single and double stranded linear chains of existing DNA computing methods.

The approach also highlights how conventional, not necessarily computationally hard, problems can be exploited on (flow-oriented) Ad-Molecular Computers.

References

[1] Adleman, L., Molecular Computation of Solutions to Combinatorial Problems. *Science,* 266:1021-1025, 1994.

[2] Adleman, L., On Constructing a Molecular Computer. In R. J. Lipton and E. B. Baum, editors, *DNA Computers: Proceeding of a DIMACS Workshop,* DIMACS Series in Discrete Mathematics and Theoretical Computer Science, American Mathematical Sociey, 1-21, 1996.

[3] Amenyo, J.T., BIOCOMPUTER: Architecture, Engineering and Technology. Available as *ftp://ftp.ans.net/pub/misc/biocomputer.txt*, 1995.

[4] Banerjee, P., *Parallel Algorithms for VLSI ComputerAided Design,* PTR Prentice Hall, Englewood Cliffs, NJ, 1994.

[5] Baum, E., Building an Associative Memory Vastly Larger than the Brain, *Science,* 268:583-585, 1995.

[6] Bell, C.G. and Newell, A., *Computer Structures: Readings and Examples,* McGraw-Hill, New York, NY, 1971.

[7] Bell, C.G., Mudge, J.C. and McNamara, J.E., *Computer Engineering: A DEC View of Hardware Systems Design,* Digital Press, Bedford, MA, 1978.

[8] Beneš, V., *Mathematical Theory of Connecting Networks and Telephone Traffic,* Academic Press, New York, NY, 1965.

[9] Boneh, D., Dunworth, C. and Lipton, R., Breaking DES Using a Molecular Computer, In R. Lipton and E. Baum, editors, *DNA Based Computers: Proceedings of a DIMACS Workshop*, DIMACS Series in Discrete Mathematics and Theoretical Computer Science, Amer. Math. Soc., Providence, RI, 37-65, 1996.

[10] Boneh, D., Lipton, R., Dunworth, C. and Sgall, J., On the Computational Power of DNA. Available as: *http://www.cs.princeton.edu/ dabo/papers/biocircuit.ps* (1995).

[11] Carriero, N. and Gelernter, D., *How to Write Parallel Programs: A First Course,* MIT Press, Cambridge, MA, 1990.

[12] Dahir, M., Test-Tube Evolution. *MIT Technology Review,* 14-15, Nov./Dec. 1994.

[13] Darwin, C., *On the Origin of Species by Means of Natural Selection or the Preservation of Favoured Races in the Struggle for Life,* Murray Publ., London. UK, 1859.

[14] Davis, L., *Handbook of Genetic Algorithms,* Van Nostrand Reinhold, New York, NY, 1991.

[15] Edelman, G.M., *Neural Darwinism: The Theory of Neuronal Group Selection,* Basic Books, New York, NY, 1987.

[16] Freund, R., Kari, L., and Paun, G., DNA Computing Based On Splicing: The Existence of Universal Computers. Available as: *http://www.csd.uwo.ca/ lkari/jacm.ps*, 1995.

[17] Goldberg, D.E., *Genetic Algorithms in Search, Optimization and Machine Learning,* Addison-Wesley, Reading, MA, 1989.

[18] Kari, L., DNA Computing: The Arrival of Biological Mathematics. Math. Intell., 19 (2): 9-22, 1997.

[19] Leighton, F. Thomson., *Introduction to Parallel Algorithms and Architectures: Arrays, Trees and Hypercubes,* Morgan Kaufman Publish. San Mateo, CA, 1992.

[20] Lipton, R., DNA Solution of Hard Computational Problems. *Science,* 268:542-545, 1995.

[21] Poincaré, H., Complément a l'Analysis Situs. *Rendiconti del Circolo Matematico di Palermo,* 13:285-343, 1899.

[22] Shannon, C.E., A Symbolic Analysis of Relay and Switching Circuits. *Transactions of the AIEE,* 57:713–723, 1938.

[23] Siegel, H., *Interconnection Networks for Large Scale Parallel Processing: Theory and Case Studies,* Lexington Books, Lexington, MA, 1984.

[24] Stoy, J.E., *Denotational Semantics: The Scott-Strachey Approach to Programming Language Theory.* MIT Press, 1977.

[25] Winfree, E., On the Computational Power of DNA Annealing and Ligation. In R. Lipton and E. Baum, editors, *DNA Based Computers: Proceedings of a DIMACS Workshop,* DIMACS Series in Discrete Mathematics and Theoretical Computer Science. American Mathematical Society. Providence, RI, 199-221, 1996.

[26] (eds) Yager, R.R., Ovichinnikov, S., Tong, R.M. and Nguyen, H.T., *Fuzzy Sets and Applications: Selected Papers by L.A. Zadeh,* John Wiley & Sons, New York, NY, 1987.

ANS CO+RE Systems, Inc., 100 Clearbrook Road, Elmsford, NY 10523
E-mail address: jta@ans.net

DIMACS Series in Discrete Mathematics
and Theoretical Computer Science
Volume **44**, 1999

Error-Resistant Implementation of DNA Computations

Martyn Amos, Alan Gibbons, and David Hodgson

ABSTRACT. This paper introduces a new model of computation that employs the tools of molecular biology whose implementation is far more error-resistant than extant proposals. We describe an abstraction of the model which lends itself to natural algorithmic description, particularly for problems in the complexity class NP. In addition we describe a number of linear-time algorithms within our model, particularly for NP-complete problems. We describe an *in vitro* realization of the model and conclude with a discussion of future work.

1. Introduction

In the past [**7**] and more recently, several papers [**1, 14, 20**] (also see [**10, 17**]) have advocated the realisation of massively parallel computation using the techniques and chemistry of molecular biology. Adleman described how a computationally intractable problem, known as the *directed Hamiltonian Path Problem* (HPP) might be solved using molecular methods. However, a recent attempt [**12**] to repeat Adleman's experiment has cast doubt upon the efficacy of extant models of DNA computation. The researchers performed Adleman's experiment twice; once on the original graph as a positive control, and again on a graph containing no Hamiltonian path as a negative control. The results obtained were inconclusive. The researchers state that *"At this time we have carried out every step of Adleman's experiment, but have not gotten an unambiguous final result."*

The main purpose of this paper is to describe an alternative, but importantly, *feasible* model of DNA computation. The problem with extant proposals is that they assume that certain proposed biological operations are error-free. An operation central to most models is *extraction* of DNA strands containing a certain sequence. There are two problems with extraction. The first, and most important problem is that removal by DNA hybridization of strands containing the sequence is not 100% efficient[1], and may at times inadvertently remove strands that *do not* contain the specified sequence. Adleman did not encounter problems with extraction because only a few operations were required. However, for a large problem instance, the number of extractions required may run into hundreds, or even thousands. For example, a particular DNA-based algorithm may rely upon repeated

This work was partially supported by the University of Warwick Research and Teaching Innovations Sub-Committee.

[1]The actual efficiency depends on the concentration of the reactants.

"sifting" of a "soup" containing many strands, some encoding legal solutions to the given problem, but most encoding illegal ones. At each stage, we may wish to extract only strands that satisfy certain criteria (i.e., they contain a certain sequence). Only strands that satisfy the criteria at one stage survive to the next stage. At the end of the sifting process, we are hopefully left only with strands that encode legal solutions, since they satisfy all criteria. However, assuming 95% efficiency of the extraction process, after 100 extractions the probability of being left with a soup containing (a) a strand encoding a legal solution, and (b) no strands encoding illegal solutions is about 0.006. Repetitive extraction will not guarantee 100% efficiency, since it is impossible to achieve the conditions whereby only correct hybridization occurs. Furthermore, as the length of the DNA strands used increases, so does the probability of incorrect hybridization.

Although attempts [**13**] have been made to simulate highly reliable extraction using a sequence of imperfect operations, it is clear that for any non-trivial problem, reliance on the extraction operation must be minimised, or, ideally, removed entirely. In this paper we describe a novel model of DNA computation that obviates the need for hybridization extraction within the main body of the computation.

The rest of the paper is organised as follows. In Section 3 we describe a DNA implementation for which none of the problems described above exist, and which is therefore truly scalable. In Section 2 we describe an abstraction of the model which lends itself to natural algorithmic description. In addition we describe a number of algorithms within our model and discuss its computational power.

2. The Model of Computation

Here we describe an abstract model of computation which, as we show in Section 3, has a clean implementation in DNA chemistry. We note the initial contribution of Lipton [**14**] to the construction of such models.

Our model is particularly effective for algorithmic description. Moreover, it is sufficiently strong to solve any of the problems in the class NP which includes, of course, the notoriously intractable NP-complete problems [**8**]. As we shall see, these problems naturally have polynomial-time (often linear-time) parallel solutions within the model. This usually comes with the expense of exponentially large data sets.

Within the model, a computation consists of a sequence of operations on finite sets of strings. It is normally the case that a computation begins and terminates with a single set. Within the computation, by applying legal operations of a computation, several sets may exist at the same time. We define legal operations on sets shortly but first consider the nature of an *initial set.*

An initial set consists of strings which are typically of length $O(n)$ where n is the problem size. As a subset, the initial set should include all possible solutions (each encoded by a string) to the problem being solved. The point here is that the superset in any implementation of the model is supposed to be relatively easy to generate as a starting point for a computation. The computation then proceeds by filtering out strings which cannot be a solution. For example, if the problem is to generate a permutation of the integers $1 \ldots n$ then the initial set might include all strings of the form $p_1 i_1 p_2 i_2 \ldots p_n i_n$ where each i_k may be any of the integers in the range $[1 \ldots n]$ and p_k encodes the information "position k". Here, as will be typical for many computations, the set has cardinality which is exponential in the problem size. For our example of finding a permutation, we should filter out all strings in which the same integer appears in at least two locations p_k. Any of the remaining strings is then a legal solution to the problem. We return to this problem (whose solution, incidentally, may be regarded as a very useful standard operation within our model for the solution of other problems) after defining some basic legal operations on strings.

2.1. Basic set operations. Here we define the basic legal operations on sets within the model. Our choice is determined by what we know can be effectively implemented by very precise and complete chemical reactions within the DNA implementation. The operation set defined here provides the power we claim for the model but, of course, it might be augmented by additional operations in the future to allow greater conciseness of computation.

- *remove*$(U, \{S_i\})$. This operation removes from the set U, in parallel, any string which contains at least one occurrence of any of the substrings S_i.
- *union*$(\{U_i\}, U)$. This operation, in parallel, creates the set U which is the set union of the sets U_i.
- *copy*$(U, \{U_i\})$. In parallel, this operation produces a number of copies, U_i, of the set U.
- *select*(U). This operation selects an element of U uniformly at random, if U is the empty set then *empty* is returned.

¿From the point of view of establishing the parallel time-complexities of algorithms within the model, these basic set operations will be assumed to take *constant-time.* This is certainly what the DNA implementation described in Section 3 provides.

2.2. A First Algorithm. We now provide our first algorithmic description within the model. The problem solved is that of generating the set of all permutations of the integers 1 to n. The initial set and the filtering out of strings which are not permutations were essentially described earlier. Although not NP-complete, the problem does of course have exponential-sized input and output.

The algorithmic description below introduces a format that we utilise elsewhere. The particular device of copying a set (as in copy$(U, \{U_1, U_2, \dots, U_n\})$) followed by parallel *remove* operations (as in remove$(U_i, \{p_j \neg i, p_k i\})$) is a very useful compound operation as we shall see in several later algorithmic descriptions. Indeed, it is precisely this use of *Parallel Filtering* that is at the core of most algorithms within the model. The only non-selfevident notation employed below is $\neg i$ to mean (in this context) any integer in the range i=1, 2, ... , n which is *not* equal to i.

- **Problem: Permutations**

  Generate the set P_n of all permutations of the integers $\{1, 2, \dots, n\}$.
- **Solution**
  - *Input*: The input set U contains all strings of the form $p_1 i_1 \dots p_n i_n$ where, for all j, p_j uniquely encodes "position j" and each i_j is in $\{1, 2, \dots, n\}$. Thus each string consists of n integers with (possibly) many occurences of the same integer.
  - *Algorithm*

    for $j = 1$ to n **do**

        begin

        copy$(U, \{U_1, U_2, \dots, U_n\})$

        for i=1, 2, ... , n and all $k > j$

```
                    in parallel do remove(U_i, {p_j¬i, p_k i})
            union({U_1, U_2, ..., U_n}, U)
            end
      P_n ← U
```

- *Complexity*: $O(n)$ parallel-time.

After the jth iteration of the **for** loop, the computation ensures that in the surviving strings the integer i_j is not duplicated at positions $k > j$ in the string. The integer i_j may be any in the set $\{1, 2, \ldots, n\}$ (which one it is depends in which of the sets U_i the containing string survived). At the end of the computation each of the surviving strings contains exactly one occurence of each integer in the set $\{1, 2, \ldots, n\}$ and so represents one of the possible permutations. Given the specified input, it is easy to see that P_n will be the set of all permutations of the first n natural numbers. As we shall see, production of the set P_n can be a useful subprocedure for other computations.

2.3. Algorithms for a selection of NP-complete problems. We now describe a number of algorithms for graph-theoretic NP-complete problems (see [**9**], for example). Problems in the complexity class NP seem to have a natural expression and ease of solution within the model. We describe linear-time solutions although, of course, there is frequently an implication of an exponential number of processors available to execute any of the basic operations in unit time.

2.3.1. *3-vertex-colourability.* Our first problem concerns proper vertex colouring of a graph. In a proper colouring, colours are assigned to the vertices in such a way that no two adjacent vertices are similarly coloured. The problem of whether 3 colours are sufficient to achieve such a colouring for an arbitrary graph is NP-complete [**9**].

- **Problem: Three colouring**
  Given a graph $G = (V, E)$, find a 3-vertex-colouring if one exists, otherwise return the value *empty*.
- **Solution**
  - *Input*: The input set U consists of all strings of the form $p_1 c_1 \ldots p_n c_n$ where $n = |V|$ is the number of vertices in the graph. Here, for all i, p_i uniquely encodes "position i" and each c_i is any one of the "colours" 1, 2 or 3. Each such string represents one possible assignment of colours to the vertices of the graph in which, for each i, colour c_i is assigned to vertex i.
  - *Algorithm*

```
for j = 1 to n do
      begin
      copy(U, {U_1, U_2, U_3})
      for i=1, 2 and 3, and all k such that (j, k) ∈ E
                in parallel do remove(U_i, {p_j¬i, p_k i})
      union({U_1, U_2, U_3}, U)
      end
select(U)
```

  - *Complexity*: $O(n)$ parallel time.

After the jth iteration of the **for** loop, the computation ensures that in the remaining strings, vertex j (although it may be coloured 1, 2 or 3 depending on which of the sets U_i it survived in) has no adjacent vertices that are similarly coloured.

Thus, when the algorithm terminates, U only encodes legal colourings if any exist. Indeed, every legal colouring will be represented in U.

2.3.2. *Hamiltonian path.* A Hamiltonian path between any two vertices u, v of a graph is a path that passes through every vertex in $V - \{u, v\}$ precisely once [**9**].

- **Problem: Hamiltonian path**
  Given a graph $G = (V, E)$ with n vertices, determine whether G contains a Hamiltonian path.
- **Solution**
  - *Input*: The input set U is the set P_n of all permutations of the integers from 1 to n as output from **Problem: Permutations**. An integer i at position p_k in such a permutation is interpreted as follows: the string represents a candidate solution to the problem in which vertex i is visited at step k.
  - *Algorithm*
    for $2 \leq i \leq n - 1$ and j, k such that $(j, k) \notin E$
        in parallel do remove $(U, \{jp_ik\})$
    select(U)
  - *Complexity*: Constant parallel time given P_n.

In surviving strings there is an edge of the graph for each consecutive pair of vertices in the string. Since the string is also a permutation of the vertex set it must also be a Hamiltonian path. Of course, U will contain every legal solution to the problem.

2.3.3. *Subgraph isomorphism.* Given two graphs G_1 and G_2 the following algorithm determines whether G_2 is a subgraph of G_1.

- **Problem: Subgraph isomorphism**
  Is $G_2 = (V_2, E_2)$ a subgraph of $G_1 = (V_1, E_1)$? By $\{v_1, v_2, \ldots, v_s\}$ we denote the vertex set of G_1, similarly the vertex set of G_2 is $\{u_1, u_2, \ldots, u_t\}$ where, without loss of generality, we take $t \leq s$.
- **Solution**
  - *Input*: The input set U is the set P_s of permutations output from the Permutations algorithm. For $1 \leq j \leq t$ an element $p_1i_1p_2i_2 \ldots p_si_s$ of P_s is interpreted as associating vertex $p_j \in \{u_1, u_2, \ldots, u_t\}$ with vertex $i_j \in \{v_1, v_2, \ldots, v_s\}$. The algorithm is designed to remove any element which maps vertices in V_1 to vertices in V_2 in a way which does not reflect the requirement that if $(p_s, p_t) \in E_1$ then $(i_s, i_t) \in E_2$.
  - *Algorithm*
    for j=1 to t **do**
        begin
        copy$(U, \{U - 1, U_2, \ldots, U_t\})$
        for all $l, j < l \leq t$ such that $(p_j, p_l) \in E_2$ and $(ij, i_l) \notin E_1$
            in parallel do remove$(U_j, \{p_li_l\})$
        union$(\{U - 1, U_2, \ldots, U_t\}, U)$
        end
    select(U)
  - *Complexity*: $O(|V_s|)$ parallel time.

For any remaining strings, the first t pairs p_li_l represent a one-to-one association of the vertices of G_1 with the vertices of G_2 indicating the subgraph of G_1 which is isomorphic to G_2. If select(U) returns the value *empty* then G_2 is not a subgraph of G_1.

2.3.4. *Maximum clique and maximum independent set.* A clique K_i is the complete graph on i vertices [**9**]. The problem of finding a maximum independent set is closely related to the maximum clique problem.

- **Problem: Maximum clique**
  Given a graph $G = (V, E)$ determine the largest i such that K_i is a subgraph of G. Here K_i is the complete graph on i vertices.
- **Solution**
  - In parallel run the subgraph isomorphism algorithm for pairs of graphs (G, K_i) for $2 \leq i \leq n$. The largest value of i for which a non- *empty* result is obtained solves the problem.
  - *Complexity*: O($|V|$) parallel time.
    A maximum independent set is a subset of vertices of a graph such that no two members of the set are adjacent [**9**].
- **Problem: Maximum independent set** Given a graph $G = (V, E)$ determine the largest i such that there is a set of i vertices in which no pair are adjacent.
- **Solution**
  Run the maximum clique algorithm on the *complement* of G.
  - *Complexity*: O($|V|$) parallel time.

2.4. Computational power of the model. It is clear that a lower bound on the computational power of the model is that it can solve any problem in the complexity class NP. This follows from the fact that we were able to encode solutions for various NP-complete problems for it and, of course, any instance of any problem in NP may be described as an instance of any NP-complete problem. Our examples also demonstrate that individual NP-complete problems seem to have a natural and direct encoding for the model.

3. Implementation of the model in DNA chemistry

Notice that the algorithms of the previous section work perfectly well if the basic data structure, *set*, is replaced by *multiset*. The **permutation** algorithm now outputs a multiset in which each permutation appears as many times as it was represented in the input set. However, since the *select* operation returns a single element, the output of the other algorithms is exactly as before.

In the proposed implementation outlined below, the algorithms are realized by multisets of single-stranded DNA. In practice, there would be about the same number of copies of each element in any such multiset.

3.1. The structure and enzyme manipulation of DNA. DNA (deoxyribonucleic acid) [**6**] encodes the genetic information of cellular organisms. It consists of *polymer chains*, commonly referred to as DNA *strands*. Each strand may be viewed as a chain of *nucleotides*, or *bases*. An n-letter sequence of consecutive bases is known as an n-mer or an *oligonucleotide* of length n. The four DNA nucleotides are adenine, guanine, cytosine and thymine, commonly abbreviated to A, G, C and T respectively. Each strand has, according to chemical convention, a $5'$ and a $3'$ end, thus any single strand has a natural orientation. The classical double helix of DNA is formed when two separate strands bond. Bonding occurs by the

```
5' A T A G A G T T 3'
           | | |          (a)
3'         T C A   5'

5' A T A G A G T T 3'
   | | | | | | |          (b)
3' T A T C T C A   5'
```

FIGURE 1. (a) Primer anneals to longer template (b) Polymerase extends primer in the $5'$ to $3'$ direction

pairwise attraction of bases; A bonds with T and G bonds with C. The pairs (A,T) and (G,C) are therefore known as *complementary* base pairs. In what follows we adopt the following convention: if x denotes an oligonucleotide, then $\overline{x}$ denotes the complement of x. The bonding process, known as *annealing*, is fundamental to our implementation. A strand will only anneal to its complement if they have opposite polarities. Therefore, one strand of the double helix extends from $5'$ to $3'$, and the other from $3'$ to $5'$.

In order to manipulate DNA strands and thus implement our operations we use specific classes of *enzymes*, each of which performs one or more specific tasks:

1. The DNA *polymerases* perform several functions, including the repair and duplication of DNA. Given a short *primer* oligonucleotide, p in the presence of nucleoside triphosphates, the polymerase extends p if and only if p is bound to a longer *template* oligonucleotide, t. For example, in Figure 1(a)), p is the oligonucleotide TCA which is bound to t, $ATAGAGTT$. In the presence of the polymerase, p is extended by a complementary strand of bases to the $3'$ end of t (Figure 1(b)).
2. *Restriction enzymes* [**21**, page 33] recognize a specific sequence of DNA, known as a *restriction site.* Any *double-stranded* DNA that contains the restriction site within its sequence is cut by the enzyme at that point.[2] In our implementation we use *Sau*3AI, which recognizes site $GATC$.
3. In double-stranded DNA, if one of the single strands contains a discontinuity (i.e., one nucleotide is not bonded to its neighbor) then this may be repaired by DNA *ligase* [**5**]. This allows us to create a unified strand from several bound together by their respective complements.

Another useful method of manipulating DNA is the Polymerase Chain Reaction, or PCR [**15, 16**]. PCR is a process that quickly amplifies the amount of DNA in a given solution. Each cycle of the reaction doubles the quantity of each strand, giving an exponential growth in the number of strands.

3.2. Implementation of the model. Here we first describe how an initial multiset within the model may be constructed in DNA, and then how the set operations may be implemented.

An essential difficulty in the model is that initial multisets generally have a cardinality which is exponential in the problem size. It would be too costly in time, therefore, to generate these individually. What we do in practice is to construct an initial solution, or *tube*, containing a polynomial number of distinct strands. The design of these strands ensures that the exponentially large initial multisets of

[2]In reality, only certain enzymes cut specifically at the restriction site, but we take this factor into account when selecting an enzyme.

our model will be generated automatically. The following paragraph describes this process in detail.

Consider an initial set of all elements of the form $p_1k1, p_2k_2, \ldots, p_nk_n$. This may be constructed as follows. We generate an oligo uniquely encoding each possible subsequence p_ik_i where $1 \leq i \leq n$ and $1 \leq k_i \leq k$. Embedded within the sequence representing p_i is our chosen restriction site. There are thus nk distinct oligos of this form. The task now is how to combine these to form the desired initial multiset. This is achieved as follows. For each pair $(p_ik_i, p_{i+1}k_{i+1})$ we construct an oligo which is the concatenation of the complement of the second half of the oligo representing p_ik_i and the complement of the first half of the oligo representing $p_{i+1}k_{i+1}$. We also construct oligos that are the complement of the first half of the oligo representing p_1k_1 and the last half of the oligo representing p_nk_n. There is therefore a total of $2nk+1$ oligos in solution. The effect of adding adding these new oligos is that the tube will now contain double-strands of DNA; one strand in each will be an element of the desired initial set. The new oligos have, through annealing, acted as "splints" to join the first oligos in the desired sequences. What we really require in solution are only the single strands encoding elements of the initial set. This is achieved as follows. We ligate the double strands and then heat the tube to break the hydrogen bonds between the encoding strands and the splint strands. The splint strands are created with magnetic beads attached to them in order to facilitate their removal at this stage.

It remains in our implementation to describe how the set operations are realized. This is achieved as follows:

- **Remove**

  remove is implemented as a composite operation, comprised of the following:
  - *mark*(U, S). This operation marks all strings in the set U which contains at least one occurrence of the substring S.
  - *destroy*(U). This operation removes all marked strings from U.

  mark is implemented by adding to U many copies of a primer corresponding to $\overline{S}$. This primer only anneals to single strands containing the subsequence S. We then add DNA polymerase to extend the primers once they have annealed, making partially double-stranded only the single strands containing S.

  We may then *destroy* strands containing S by adding the restriction enzyme *Sau*3AI. Double-stranded DNA (i.e. strands marked as containing S) is cut at the embedded restriction sites, with the single strands remaining intact. We may then remove all intact strands by separating on length using *gel electrophoresis* [**5**]. However, this is not strictly necessary, and leaving the fragmented strands in solution will not affect the operation of the algorithm.
- **Union**

  We may obtain the *union* of two or more tubes by simply mixing them together, forming a single tube.
- **Copy**

  We obtain i "copies" of the set U by splitting U into i tubes of equal volume. We assume that, since the initial tube contains multiple copies of each candidate strand, each tube will also contain many copies.
- **Select**

We can easily detect remaining DNA using PCR and then sequence strands to reveal the encoded solution to the given problem. One problem with this method is that there are often multiple correct solutions left in the soup which must be sequenced using nested PCR. A possible solution is *cloning* (see [**2**]).

Although the initial tube contain multiple copies of each strand, after many *remove* operations the volume of material may be depleted below an acceptable empirical level. This difficulty can be avoided by periodic amplification by PCR.

4. Advantages of the model

As we have shown, algorithms within our model perform successive "filtering" operations, keeping *good* strands (i.e., strands encoding a legal solution to the given problem) and destroying *bad* strands (i.e., those that do not). So long as the operations work correctly, the final set of strands will only consist of good solutions. However, as we have already stated, errors can take place. If either good strands are accidentally destroyed or bad strands are left to survive through to the final set then the algorithm will fail. The main advantage of our model is that it doesn't repeatedly use the notoriously error-prone separation by DNA hybridization method to extract strands containing a certain subsequence. Restriction enzymes are *guaranteed* [**4**, page 9] [3] to cut any clean double-stranded DNA containing the appropriate restriction site, whereas hybridization separation is never 100% efficient. Instead of extracting *most* strands containing a certain subsequence we simply destroy *all* of them with absolute certainty, without harming those strands that *do not* contain the subsequence. Even if, in reality, restriction enzymes have a small non-zero error rate associated with them, we believe that it is far lower than that of hybridization separation. Another advantage of our model is that it minimizes physical manipulation of tubes during a computation. Biological operations such as pipetting, filtering and extraction all lose a certain amount of material along the way. As the number of operations increases, the material loss rises and the probability of successful computation decays. Our implementation uses relatively benign physical manipulation, and avoids certain "lossy" operations.

5. Fundamental limitations on DNA computations

Although our implementation is much less error-prone than those previously proposed, there still exists at least one barrier to scalable DNA computation. This concerns the sheer weight of DNA required to solve any problems of large size in the model described in this paper. As Hartmanis points out in [**11**], if Adleman's experiment were scaled up to 200 vertices the weight of DNA required would exceed that of the Earth. DNA computations, like all others, cannot escape the "exponential curse" of the NP-complete problems. The approach of generating all possible solutions to a given problem and then gradually filtering out illegal solutions seems impractical for problem sizes beyond a moderate threshold. It is interesting to note, however, that estimates lead us to believe that our model will perform more rapid computations than conventional computers on a range of feasible problem sizes. One possible way around the weight problem (for problems in the complexity class P, for example) is to look for implementation of existing

[3] "New England Biolabs provides a colour-coded 10X NEBuffer with each restriction endonuclease to ensure optimal (100%) activity."

computing paradigms in DNA. Reif [**18**] points to one possible way forward, describing a method for simulating a CREW P-RAM using DNA. However, Reif uses the error-prone hybridization separation method described earlier, which we believe may be avoided by our methodology. Another development is the proposed [**19**] DNA and restriction enzyme implementation of Turing Machines, which establishes formally that any algorithm can be realised through DNA, although Turing machine emulation is not likely to be of practical significance. Baum and Boneh [**3**] describe an alternative form of DNA computation which is not, in their words, "brute force". Their implied criticism is valid if it is directed at the notion of using exponential-sized volumes of DNA for problems belonging to the complexity class P. However, every erstwhile algorithm that has appeared in the literature has been for an NP-complete problem and no matter what polynomial-time DNA solution is advocated for these, the difficulty of exponential-sized volumes of DNA cannot be avoided. This difficulty ought not to be a feature of good DNA solutions for problems in P.

6. Conclusions

Recent work in DNA computation has cast serious doubt on the reliability of extant models. The use of error-prone operations such as DNA hybridization separation can lead to inconclusive final results. In this paper we described a feasible model of DNA computation that avoids the problems of those previously proposed. An abstract model of computation was introduced, and we showed that it is sufficiently strong to solve any of the problems in the class NP (which includes the NP-complete problems.) We then described algorithms within the model for a selection of graph-theoretic NP-complete problems. A description of the proposed implementation of our model in DNA was then presented before we concluded with a discussion of some of the factors limiting scalable DNA computation. Although this paper is theoretical in nature, we expect to test our model in the laboratory in the near future. Future work is expected to focus on an *in vitro* P-RAM simulation with $error - free$ operations.

7. Acknowledgements

The authors would like to thank Erik Winfree and Paul Rothemund for their helpful comments on an earlier draft of this paper.

References

[1] Leonard Adleman. Molecular computation of solutions to combinatorial problems. *Science*, 266:1021–1024, 1994.

[2] Martyn Amos, Alan Gibbons, and David Hodgson. Error-resistant implementation of DNA computations. Research Report CS-RR-298, Department of Computer Science, University of Warwick, Coventry, UK, January 1996.

[3] Eric B. Baum and Dan Boneh. Running dynamic programming algorithms on a DNA computer. In L. Landweber and E. Baum, editors, *DNA Computers: Proceeding of the Second DIMACS Workshop*, DIMACS Series in Discrete Mathematics and Theoretical Computer Science, American Mathematical Sociey, 31-44, 1998.

[4] New England Biolabs. Catalog, 1996.

[5] T.A. Brown. *Genetics: A molecular approach*. Chapman and Hall, 1993.

[6] James D. Watson et. al. *Recombinant DNA*. Scientific American Books, 1992.

[7] Richard P. Feynman. There's plenty of room at the bottom. In D. Gilbert, editor, *Miniaturization*, pages 282–296. Reinhold, 1961.

[8] Michael R. Garey and David S. Johnson. *Computers and Intractibility: A Guide to the Theory of NP-Completeness.* W. H. Freeman and Company, New York, 1979.

[9] A. M. Gibbons. *Algorithmic Graph Theory.* Cambridge University Press, 1985.

[10] David K. Gifford. On the path to computation with DNA. *Science*, 266:993–994, 1994.

[11] Juris Hartmanis. On the weight of computations. *Bulletin of the European Association For Theoretical Computer Science*, 55:136–138, 1995.

[12] Peter Kaplan, Guillermo Cecchi, and Albert Libchaber. Molecular computation: Adleman's experiment repeated. Technical report, NEC Research Institute, 1995.

[13] Richard M. Karp, Claire Kenyon, and Orli Waarts. Error-resilient DNA computation. In *7th ACM-SIAM Symposium on Discrete Algorithms*, pages 458–467. SIAM, 1996.

[14] Richard J. Lipton. DNA solution of hard computational problems. *Science*, 268:542–545, 1995.

[15] Kary B. Mullis. The unusual origin of the polymerase chain reaction. *Scientific American*, 262:36–43, 1990.

[16] Kary B. Mullis, François Ferré, and Richard A. Gibbs, editors. *The polymerase chain reaction.* Birkhauser, 1994.

[17] Robert Pool. A boom in plans for DNA computing. *Science*, 268:498–499, 1995.

[18] John H. Reif. Parallel molecular computation: Models and simulations. In *Proceedings of the Seventh Annual ACM Symposium on Parallel Algorithms and Architectures (SPAA)*, Santa Barbara, June 1995.

[19] Paul W. K. Rothemund. A DNA and restriction enzyme implementation of Turing Machines. In R. J. Lipton and E. B. Baum, editors, *DNA Based Computers: Proceedings of a DIMACS Workshop.* Ser. in Disc. Math. and Theor. Comp. Sci., 27 American Mathematical Society, 75-119, 1996.

[20] Warren D. Smith. DNA computers in vitro and vivo. In R. J. Lipton and E. B. Baum, editors, *DNA Based Computers: Proceedings of a DIMACS Workshop.* Ser. in Disc. Math. and Theor. Comp. Sci., 27 American Mathematical Society, 121-186, 1996.

[21] J. Williams, A. Ceccarelli, and N. Spurr. *Genetic Engineering.* βios Scientific Publishers, 1993.

(M. Amos and A. Gibbons) Department of Computer Science, University of Liverpool, L69 3BX, England

E-mail address: `martyn@csc.liv.ac.uk`

E-mail address: `amg@csc.liv.ac.uk`

(D. Hodgson) Department of Biological Sciences, University of Warwick, Gibbet Hill Road, Coventry CV4 7AL, England

E-mail address: `dm@dna.bio.warwick.ac.uk`

DIMACS Series in Discrete Mathematics
and Theoretical Computer Science
Volume **44**, 1999

Making DNA Computers Error Resistant

Dan Boneh, Christopher Dunworth, Richard J. Lipton, Jiří Sgall

ABSTRACT. We describe methods for making volume decreasing algorithms more resistant to certain types of errors. Such error recovery techniques are crucial if DNA computers ever become practical. Our first approach relies on applying PCR at various stages of the computation. We analyze its performance and show that it increases the survival-probability of various strands to acceptable proportions. Our second approach relies on changing the method by which information is encoded on DNA strands. This encoding is likely to reduce false negative errors during the bead separation procedure.

1. Introduction

In the short history of DNA (deoxyribonucleic acid) based computing there have already been a number of exciting results. It all started with Adleman's [**1**] beautiful insight that showed that biological experiments could solve the Directed Hamiltonian Path problem (DHP). Then, Lipton [**10**] showed how to use DNA to solve more general problems, namely to find satisfying assignments for arbitrary (directed) contact networks, which includes the important case of arbitrary formulas.

There has since been a series of papers on DNA computation [**4, 5, 15, 3, 12, 13, 11**]. Each of these subsequent results is of the following form: Given enough strands of DNA and certain biological operations; one can simulate some classic model of computation efficiently. Some compare to formulas, some to circuits, others to 1-tape Nondeterministic Turing Machines.

All results assume that the biological operations are *perfect*. Clearly, this is *not* the case and errors must be addressed. However, it is important to know what the ultimate limits are without errors. Clearly, if in the error-free case DNA cannot do exciting things, then there is no hope for the realistic case.

However, DNA operations are *not* error free. Our main result is that it is possible to allow a large class of DNA computations to survive fairly high error rates. This class of DNA computations includes Adleman's original algorithm as well as Lipton's formula SAT algorithm.

Supported in part by NSF CCR–9304718.

J. Sgall was partially supported by grants A119107 and A1019602 of AV ČR. Part of this work was done at Institute of Computer Science, Hebrew University, Jerusalem, Israel and was partially supported by a Golda Meir Postgraduate Fellowship.

All DNA algorithms use strands of DNA to encode many parallel computations. They naturally fall into three classes: Algorithms are *Decreasing Volume* if the number of strands decreases as the algorithm executes; they are *Constant Volume* if the same number of strands remain throughout the computation. A *Mixed Algorithm* is one that fits into neither class. Adleman's and Lipton's algorithms are both of decreasing volume. Note, we use "volume" to denote the total number of strands.

Adleman's and Lipton's algorithms are even more special. In these algorithms each strand is either *good* or *bad.* A good strand encodes a solution to the problem; a bad strand does not. The goal of the algorithms is simple: to keep the good strands around and to eliminate the bad strands. If they succeed, then the final test tube will only contain good strands. The answer can then be found by sequencing one of the remaining strands.

The difficulty with this simple description is that errors can take place. An error may have terrible consequences. If a good strand is damaged or is lost, then the algorithm will fail. If on the other hand, bad strands are not removed, there may be many bad strands left at the end of the algorithm. This also could cause the algorithm to fail.

Errors come from two sources. First, every operation can cause an error. The principal operation in these algorithms is *extraction* [**1, 10**]. This operation removes from a test tube those strands that contain some given pattern. However, extractions are not perfect. They usually only remove about 95% of the strands that match the pattern. In additional, strands that do not match will sometimes be removed anyway: rates are typically 1 part in 10^6. Another source of errors are due to [**9, 14**] the fact that DNA has a "half-life": it decays at a finite rate. If the DNA algorithm takes months, for example, there is a probability that the good strands will disappear.

The reason that errors in operations are so critical is that we envision algorithms that perform thousands of steps. If each extraction removes 99% of the good strands, then the probability that a good strand is there at the end of a 1,000 step computation is tiny (≈ 0.00004). Thus, even at this high extraction rate the algorithm essentially will always fail. This does not even take into account the extraction of bad strands or the half-life problem.

Our first main result is the following: *We can show how to map Adleman's and Lipton's algorithms into new algorithms that are constant volume. These new algorithms are highly resistant to errors.* Moreover, these new algorithms run in almost the same number of steps. Thus, the time penalty for making the algorithms error resistant is small.

However, there is a "space" penalty for making the algorithms constant volume. The number of test tubes remains the same throughout the computation. Thus, the same amount of equipment is needed at the beginning of the computation as later on. This may not be much of a penalty. If one only wishes to perform one computation, then the original algorithm would have more and more equipment sit idle as the volume decreases.

Our second result has to do with decreasing the error rates in the bead separation procedure. Lipton [**10**] describes how binary strings can be encoded as DNA strands in a useful manner. We suggest a modified encoding scheme which is likely to reduce the false negative error rate. We mention that recently Karp, Kenyon and Waarts [**8**] have also presented some techniques for coping with errors in the bead separation procedure.

2. Error resistant computations

We now turn to showing how to make DNA algorithms more error resistant. In order to make the analysis clear we will assume that the algorithm operates as follows: (The more general case in [**10**] is handled in the same way.) Initially, there are 2^n strands of DNA. Just one of these strands is good; all the rest are bad. Clearly, the case of *one* good strand is the worst case.

The algorithm consists of s extraction steps. The good strand always matches the pattern of the extraction. However, a bad strand may or may not match the pattern. A *type I* (or *false negative*) error occurs when a strand is not correctly extracted: let p be the probability that this happens. A *type II* (or *false positive*) error occurs when a strand should not be extracted but is anyway: let q be the probability that this happens. Typical values of p are .95 and q are 10^{-6} (see [**1**]).

Note, just because a strand is bad does *not* mean that at each step it will not match the pattern. The good strand always matches the pattern; the bad strands may or may not. Indeed if x is a bad strand let $M(x)$ be the number of extractions for which it does *not* match the pattern. Clearly, $M(x)$ is at least 1.

We next assume that the computation decreases its volume at a uniform rate. Thus, we will assume from now on that every step decreases the volume at the same rate. Thus, since there are s steps and 2^n strands, the volume goes down by a factor of 2 every s/n steps. If $s = 900$ and $n = 60$, then every 15 steps the volume of DNA halves.

In the case that each step decreases the volume at the same rate, our analysis is simple. However, our method can easily be adjusted for a rate of decrease that is only approximately constant. Our method does not work in the pathological case where the volume goes down in some very complex way. We believe that it may be possible to randomize the order of extractions to "average" out the rate. Finally, such pathological cases may not arise in real problems.

Our modification to the basic algorithms is twofold: First, *every s/n steps double the amount of DNA*. We do this by a PCR operation [**6**]. There are several points: In practice one might wait until the volume decreased by a larger factor. We will do the analysis in this simple case; the generalization to bigger steps is quite easy. Second, the additional cost in time is small. Even if the PCR step is slow, it only happens every s/n steps. In our example, this is every 15 steps. Thus, the additional required time is a small percent of the total time.

Our second modification is to the final detect step. Before we assumed that ideally only one strand remained in the final test tube. Thus, to determine the answer we only had to sequence that strand. Now there may be both good and bad strands in the final test tube. We, therefore, replace this simple detect step by the following: Remove a strand from the test tube and sequence it. Then, check whether or not it is a solution. If it is, the algorithm has found a solution. If it is not, then try another strand. After some fixed number of these steps, say m, stop and report failure. Note, it is easy to computationally check whether or not a sequence does indeed encode a solution to the problem being solved.

The modified algorithm can now fail in two different ways. First, there might be no good strands in the final test tube. Second, there might be good strands in the final test tube, but there might be so many bad strands that the probability that the final step gets a solution is small. This leads naturally to the following: Let P_s be the *survival probability* i.e. the probability that some good strand is in the final test tube; also let P_r be the *selection probability* i.e. the probability that at least δ of the strands in the final test tube are good.

The probabilities P_s and P_r together control the probability that our algorithm succeeds. More precisely, we can use them to easily compute an upper bound on the probability that our algorithm fails. The probability of failure is at most

$$1 - P_s + P_s(1 - P_r) + P_sP_r(1 - \delta)^m$$

where recall that we do m repeated detects and δ is the amount of good strands in the final test tube. Thus, the probability that our method works is at least

$$P_sP_r(1 - \delta)^m.$$

The key is to bound these two probabilities, P_s and P_r. Let us first consider P_s. It should be clear that without the PCR step the probability that the good strand survives is p^s: recall that s extractions are performed and each succeeds with probability p. Now, every s/n steps the surviving strands are doubled in number. This applies equally to the good and the bad strands. Therefore, if a good strand survives s/n steps, it creates 2 new good strands. (Note, we are assuming that PCR step is error-free. However, any error there can be handled by decreasing p.)

This is an example of a *branching process* [**7**]. Branching processes are famous for modelling nuclear reactions and spread of diseases. In our case we are modelling whether or not the single initial good strand or its descendants survive many extractions. An example of such a branching process can be seen in Figure 1. Let r be the probability that a single good strand survives to the next PCR step, i.e. $r = p^{s/n}$. If a strand survives, then it becomes 2 good ones. Thus, the chances have greatly increased that some good strand will survive all the future extractions. Following Feller [**7**] the key quantity is the extinction probability ζ; for this branching process it is the smallest positive solution to the equation

$$1 - r + r\zeta^2 = \zeta.$$

A simple computation shows that the value of ζ is $\frac{1}{r} - 1$. The probability, P_s, that some good strand survives is $1 - \zeta$. Therefore,

$$P_s = 1 - \zeta = 2 - \frac{1}{r} = 2 - p^{-s/n}$$

Table 1 contains some example results. The probability that the original algorithms without PCR improvement succeed is essentially 0 in all cases. The table shows the probability that *some* good strands survive the entire series of steps.

It remains to show that P_r is large, i.e. that it is likely that the ratio of good strands to bad strands is not too small. For this argument we can *ignore* the PCR modification. The reason is simple: we are interested only in the *ratio* of the number

p	probability good survives
.97	.42
.98	.65
.99	.84

TABLE 1. $s = 900$ and $n = 60$

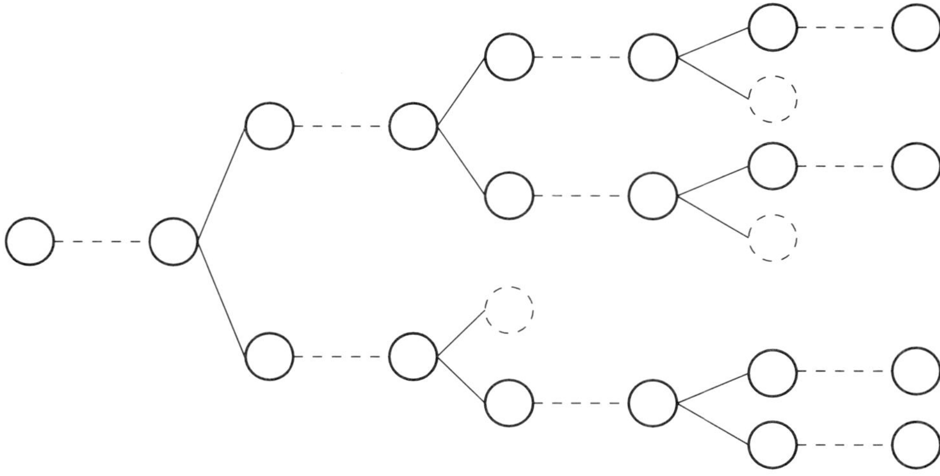

FIGURE 1. An example of a branching process. A good strand is replicated every few steps. Even though some of its descendants die (dashed discs), its prevalence increases exponentially.

of good to bad strands. Since PCR doubles both the good and bad strands equally their ratio is unchanged. Thus, in this case we can assume that there is exactly 1 good strand.

Recall that a bad strand x does not match the pattern of an extract for $M(x)$ extractions. It is therefore the case that a bad x will survive only if $M(x)$ type II errors occur. Thus, the expected number of bad strands that survive is at most

$$\sum_x q^{M(x)}$$

where the sum is over all bad strands since q is the probability that a type II error occurs. (The actual sum is smaller since bad strands still might be subject to type I errors.)

In order to simplify this sum define M_k as the number of bad strands x that have $M(x) = k$. Then, the above is equal to

$$M_1 q + M_2 q^2 + M_3 q^3 + \ldots$$

where the missing terms are insignificant. This follows since $q = 10^{-6}$ and there are at most 2^{70} strands.

The critical point is that $M_1 q + M_2 q^2 + M_3 q^3$ determines how long the final detect step takes. Let this quantity be l. Then, the detect step will likely require order l steps. Clearly, we want this to be as small as possible. However, without any assumption it *is* possible that l is large. Perhaps $M(x) = 1$ for most bad strands.

There are three points. First, in practical problems it seems that it would be rare for many bad strands to satisfy all but a few of the constraints. Typically one would expect that a bad strand might have $M(x)$ fairly large. Clearly, in this case l would be small. Note, even if $M_1 = 10^8$ and $M_2 = 10^{14}$ and $M_3 = 10^{20}$, the value of l is only 300.

Second, we can restructure the computation so that l is made smaller. For example, if l is large we can just repeat the whole computation and reduce l. If one repeats extractions twice, then all values of $M(x)$ are doubled. This is quite expensive and it would be better to avoid this. However, it is interesting that there is a general method to reduce l. Note, we do not have to do this repeating until we fail at the detect step.

Finally, it may be possible to do a more interesting transform on problems that will reduce the number of "almost" solutions, i.e. will reduce l. This seems to be related to the famous Probabilistic Checkable Proofs of [**2**].

The techniques presented above transformed a decreasing volume algorithm into a constant volume algorithm. By doing so we made sure that the prevalence of good strands increases in the test tube. However, this technique can not be applied to an algorithm which is constant volume to begin with. The reason is that the amount of material used by the algorithm would double after every PCR step. Within a few steps the amount of material would be to large to handle.

3. Double encoding of data

We propose a technique which is likely to to reduce the rate of false negative errors in the bead separation step. Our approach will hopefully increase the yield of the extraction procedure.

The basic idea is to modify the way in which binary strings are encoded in DNA. The hope is that the extraction procedure will incur less false negatives when this encoding is used. In Lipton's algorithm for solving SAT [**10**] a binary string $x = x_1 \ldots x_n$ is represented by the DNA molecule

$$3' - B(1, x_1)\ B(2, x_2)\ B(3, x_3)\ \cdots\ B(n, x_n) - 5'$$

where $B(k, b)$ are 20-mers appropriately chosen. We refer to this method of encoding binary strings as the *standard encoding scheme.*

The traditional approach for coping with the false negative problem is to introduce many DNA strands encoding the same binary string into the solution. When there are "many" copies of the same DNA strand, with "high" probability one of them will survive an extraction step despite the effect of false negative errors.

3.1. A modified encoding scheme. We propose to encode the binary string $x = x_1 \ldots x_n$ using the following DNA molecule:

$$3' - B(1, x_1)\ B(2, x_2)\ \cdots\ B(n, x_n)\ B(1, x_1)\ B(2, x_2)\ \cdots\ B(n, x_n) - 5'$$

That is, each 20-mer $B(i, x_i)$ appears twice in the DNA strand in separate positions on the strand. Suppose we need to extract all strands in the solution that contain a specific sequence, say $B(1,1)$. The hope is that since $B(1,1)$ appears twice on the DNA strand, that DNA strand is more likely to be extracted than a DNA strand containing only one copy of $B(1,1)$. Theoretically, this seems to be the case since it becomes easier for the strand to attach itself to some magnetic bead.

We refer to this modified method of encoding binary string as *double encoding.* Note that since the DNA strands in the solution are now twice as long, the test-tube can contain only half the number of strands. As a result only half as many strands can represent a given binary string. It is therefore unclear that using double encoding makes the algorithm more resistant to false negative errors.

A simple analysis can be used to show that double encoding is likely to make the SAT algorithm of Lipton more resistant to false negative errors. Let D_1 be a DNA strand containing the sequence B once. Let D_2 be a DNA strand containing the sequence B twice in separate locations on the strand. Define p_1 (resp. p_2) to be

	T_1	T_2
Type of encoding	Standard encoding	Double encoding
Type of DNA strands	D_1	D_2
Number of copies	$2r$	r
Extraction probability	p_1	p_2
survival prob. after k extractions	$S_1 = 1-(1-p_1^k)^{2r}$	$S_2 = 1-(1-p_2^k)^r$

TABLE 2. Comparison of standard and double encoding schemes

the probability that D_1 (resp. D_2) is extracted during an extraction of all strands in the solution containing the sequence B. We are conjecturing that $p_2 > p_1$ though this still has to be verified by experiments.

Let T_1 be a solution containing $2r$ copies of D_1. Similarly, let T_2 be a solution containing r copies of D_2. Now suppose we perform k extract operations on each of T_1 and T_2. It can easily be seen that the probability that at least one of the strands in T_1 survives the k extractions is

$$S_1 = 1-(1-p_1^k)^{2r} \quad .$$

Similarly, the probability that at least one of the strands in T_2 survives the k extractions is

$$S_2 = 1-(1-p_2^k)^{r} \quad .$$

The above data is summarized in Table 2. The question we need to answer is: how large should p_2 be so that $S_2 > S_1$. Some simple algebra shows that $S_2 > S_1$ when $p_2 > p_1(2-p_1^k)^{1/k}$. Table 3 contains some values for this threshold quantity.

For example, suppose we ran an algorithm which requires 10 extraction steps. That the above table shows that if p_2 is bigger than p_1 by 4.6% then the double encoding method will do better than the standard encoding. Note that the correct value of p_2 can only be determined through experimentation.

3.2. Constructing the initial solution. The first step in the SAT algorithm is the construction of DNA strands which represent all n bit binary strings. For the standard encoding this is done by creating all paths in a certain graph as is described in [**10**].

The generation of the initial solution for the double encoding is a bit more complicated. We can no longer simply form all paths in some graph. Fortunately we can use a technique described in [**5**]. In [**5**] it was shown that it is possible to create a solution of all pairs $(x, f(x))$ for a function f. By taking f to be the identity function we can create the required double encoding solution which consists of all strings (x, x).

k	$p_{2\,\min} = p_1(2-p_1^k)^{1/k}$
5	96.4%
10	94.6%
15	93.6%

TABLE 3. Threshold value for p_2 assuming $p_1 = 90\%$

4. Conclusions

We discussed methods for making the formula satisfaction techniques of Lipton [**10**] resistant to certain types of errors by applying PCR every several steps. This doubling step turns this algorithm into a constant volume algorithm. The analysis demonstrates that the error parameters of the extract operation are sufficiently small to make the formula satisfaction algorithm work in the presence of errors.

We also presented a modified encoding scheme (double encoding) which may be helpful in reducing error rates that arise in the bead separation procedure. The usefulness of this technique will be determined through experimentation.

References

[1] L. Adleman, Molecular Computation of Solutions to Combinatorial Problems. *Science*, 266:1021-1024, 1994.

[2] S. Arora, Probabilistic Checkable Proofs, Ph.D. thesis, U.C. Berkeley, 1994.

[3] D. Beaver, A Universal Molecular Computer. In R. J. Lipton and E. B. Baum, editors, *DNA Computers: Proceeding of a DIMACS Workshop,* DIMACS Series in Discrete Mathematics and Theoretical Computer Science, American Mathematical Sociey, 29-36, 1996.

[4] D. Boneh, C. Dunworth, R. Lipton, Breaking DES Using a Molecular Computer. In R. J. Lipton and E. B. Baum, editors, *DNA Computers: Proceeding of a DIMACS Workshop,* DIMACS Series in Discrete Mathematics and Theoretical Computer Science, American Mathematical Sociey, 37-65, 1996.

[5] D. Boneh, C. Dunworth, R. Lipton, J. Sgall, On the Computational Power of DNA. *Princeton CS Tech-Report CS-TR-499-95.*

[6] H. Erlich, *PCR Technology: Principles and Applications for DNA Amplification*, Stockton Press.

[7] W. Feller, *An Introduction to Probability Theory and Its Application*, Volume I., John Wiley and Sons 1957.

[8] R. Karp, C. Kenyon, O. Waarts, Error-resilient DNA computations. In *Proceedings of Symposium on Discrete Algorithms*, SODA 1996.

[9] Lindahl, Tomas, Instability and decay of the primary structure of DNA. *Nature* 362:709-715, 1993.

[10] R. Lipton, DNA Solution of Hard Computational Problems. *Science* 268:542-545 (Apr. 28) 1995.

[11] C. Papadimitrou, private communications.

[12] D. Rooßand Klaus W. Wagner, On the power of Bio-Computers. Technical report, University of Würzburg, 1995.

[13] Rothemund, A DNA and restriction enzyme implementation of Turing machines. In R. J. Lipton and E. B. Baum, editors, *DNA Computers: Proceeding of a DIMACS Workshop,* DIMACS Series in Discrete Mathematics and Theoretical Computer Science, American Mathematical Sociey, 75-119, 1996.

[14] Saul, R.L., Ames, B.N., Background levels of DNA damage in the population. *Basic Life Sciences* 38 (1986) 529-535.

[15] W. Smith, DNA Computers in vitro and vivo. In R. J. Lipton and E. B. Baum, editors, *DNA Computers: Proceeding of a DIMACS Workshop,* DIMACS Series in Discrete Mathematics and Theoretical Computer Science, American Mathematical Sociey, 121-185, 1996.

(D. Boneh, C. Dunworth, and R. Lipton) Department of Computer Science, Princeton University, Princeton, NJ 08544

E-mail address: dabo@cs.princeton.edu

E-mail address: ctd@cs.princeton.edu

E-mail address: rjl@cs.princeton.edu

Mathematical Institute, AV ČR, Žitná 25, 115 67 Praha 1, Czech Republic

E-mail address: sgallj@mbox.cesnet.cz

DIMACS Series in Discrete Mathematics
and Theoretical Computer Science
Volume **44**, 1999

Active Transport in Biological Computing

Stuart A. Kurtz, Stephen R. Mahaney, James S. Royer, Janos Simon

ABSTRACT. Early papers on biological computing focussed on combinatorial and algorithmic issues, and worked with intentionally oversimplified chemical models. In this paper, we reintroduce complexity to the chemical model by considering the effect problem size has on the initial concentrations of reactants, and the effect this has in turn on the rate of production and quantity of reaction products.

We give a sobering preliminary analysis of Adleman's technique for solving the hamiltonian path problem. Even on the simplest problems, the annealing phase of Adleman's technique requires time $\Omega(n^2)$ rather than the $O(\log n)$ time given by a computationally inspired but chemically naive analysis. On more difficult problems, not only does the rate of production of witnessing molecules drop exponentially in problem size, but the final yield also drops exponentially.

These issues are not objections to biological computing *per se*, but rather difficulties to be overcome in its development as a viable technology.

1. Reaction Rates

We assume familiarity with Adleman's DNA-based technique [**1**] for solving the hamiltonian path problem for directed graphs. We recommend *Molecular Cell Biology* [**5**] as a useful general reference for all chemical and biological concepts not otherwise referenced.

1.1. An Easy Case. We begin by considering a very simple family of directed graphs, the *line digraphs.*

DEFINITION 1. The *line digraph* L_n consists of n-vertices $\{v_0, v_1, \dots, v_{n-1}\}$ and $n-1$ edges $\{e_{0,1}, e_{1,2}, \dots, e_{n-2,n-1}\}$, where $e_{i,i+1} = \langle v_i, v_{i+1} \rangle$.

The line digraphs all have hamiltonian paths, indeed, line digraphs *are* hamiltonian paths.

Adleman's technique begins by constructing molecules that correspond to the edges of the graph. These edge molecules undergo annealing (ligation) reactions

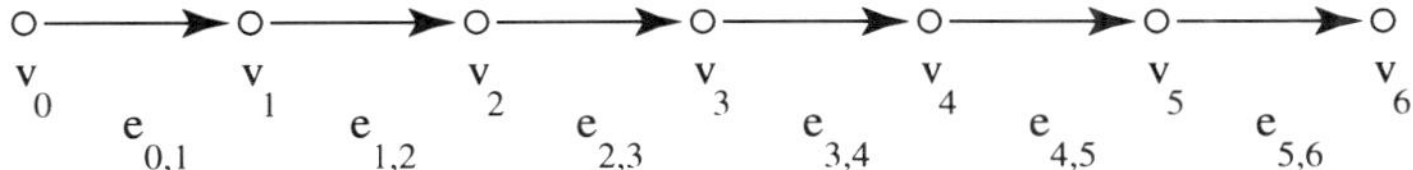

FIGURE 1. The line digraph L_7.

to form molecules that represent paths through the graph. Finally, these path molecules are filtered for length and content, so that only path molecules that represent hamiltonian paths remain.

DEFINITION 2. A path is *useful* if it is a subpath of a hamiltonian path, otherwise it is *useless*.

The molecules that represent useless paths are a potential problem for Adleman's technique, as demonstrated in Section 2 below. Useless molecules are a problem because they reduce the concentration of useful molecules, and therefore reduce the rate of production of witnessing molecules. Worse, they react with useful molecules, making them useless too.

We begin our analysis with line digraphs simply because they are precisely the positive instances of hamiltonian path for which there are no useless paths.

There is a nontrivial path in L_n from v_i to v_j if and only if $i < j$, in which case the path is unique. By abuse of notation, the unique path from v_i to v_j will be denoted by $e_{i,j}$. We will also use $e_{i,j}$ to denote the molecule corresponding to this path.

1.2. Analysis. In this simple case, the ligation reactions take the form

$$e_{i,k} + e_{k,j} \to e_{i,j} \qquad \text{for } i < k < j.$$

The desired end product of these reactions is the molecule $e_{0,n-1}$.

If we ignore chemical considerations, and utilize only our computational intuitions, we would be led more or less directly to consideration of a balanced binary tree of chemical reactions, with path molecules at the leaves and the unique molecule representing the hamiltonian path as its root. Because the depth of this tree is $\lceil \log n \rceil$, we might naively expect that time required to produce a detectable concentration of the hamiltonian path molecule is $O(\log n)$.

In this section, we give a less naive analysis. Our goal will be to give lower bounds on how long it will take to produce the hamiltonian path molecule $e_{0,n-1}$ as a function of the number n. The analysis that follows proves that ligation phase requires time $\Omega(n^2)$, rather than $O(\log n)$. The main purpose behind this section is not to establish the $\Omega(n^2)$ lower bound, even though it was unexpectedly slow, but rather to introduce chemical principles and analytical techniques that demonstrate a far more dramatic loss of performance on slightly more complicated digraphs.

DEFINITION 3. The concentration of a (family of) chemical compound(s) K is denoted by $[K]$.

The reaction rate law (cf. [**5**, p. 28ff]) states that the rate at which a chemical reaction occurs is proportional to the products of the concentrations of the reactants. Thus, e.g., the reaction

$$e_{i,k} + e_{k,j} \to e_{i,j}$$

proceeds at the rate

$$\alpha_{i,j,k}\,[e_{i,k}]\,[e_{k,j}]$$

for some constant $\alpha_{i,j,k}$. For the line digraph L_n, we have $\binom{n}{2}$ different types of path molecules, and $\binom{n}{3}$ reactions to account for. For the sake of simplicity, we assume that the reaction rate constants are all equal, and that the units of time are chosen so that $\alpha = 1$.

This gives rise to the following system of $\binom{n}{2}$ nonlinear ordinary differential equations:

$$[e_{i,j}]' = \left(\sum_{\{k:i<k<j\}} [e_{i,k}]\,[e_{k,j}] \right) - [e_{i,j}] \left(\sum_{\{l:j<l\}} [e_{j,l}] \right) - \left(\sum_{\{h:h<i\}} [e_{h,i}] \right) [e_{i,j}] \,.$$

We will assume, for the sake of simplicity, that units of volume are chosen so that the magnitude of the molar concentration of the solution is one, independent of problem size.[1] This concentration must be partitioned among the $n-1$ edge molecules $e_{i,i+1}$, and so our initial conditions are:

$$e_{i,j}(0) = \begin{cases} \frac{1}{n-1}, & \text{if } j = i+1, \text{ and} \\ 0, & \text{otherwise.} \end{cases}$$

Our analysis focuses on the creation of path molecules that include a given interior vertex v.

DEFINITION 4. Let $\langle v|$ denote the set of molecules that represent paths that end at vertex v, likewise let $|v\rangle$ denote the set of molecules that represent paths that begin at v, and $\langle v\rangle$ denote the set of molecules that represent paths containing v at an interior vertex.

All of the reactions of the form

$$e_{i,v} + e_{v,j} \longrightarrow e_{i,j}$$

can be bundled together into a single reaction of the form

$$\langle v| + |v\rangle \longrightarrow \langle v\rangle$$

All other reactions can be neglected, because they conserve the concentrations of $\langle v|$ and $|v\rangle$. We can explicitly solve the differential equations that arise from the rate law:

$$[\langle v\rangle]\,(t) = \frac{t}{(t+n-1)(n-1)}$$

Because the hamiltonian path molecule is just one of the path molecules that together make up $\langle v\rangle$, it immediately follows that $[e_{0,n-1}] \leq [\langle v\rangle]$, and the $\Omega(n^2)$ lower bound follows immediately.

At this point, we do not have a rigorous analysis that gives a corresponding upper bound, but numerical simulation of the system of differential equations for $n \in [10, 50]$ suggests that the time it takes for $[e_{i,n-1}]$ to reach a given threshold is essentially quadratic in n, as can be seen in Figure 2 below.[2]

---

[1]The *molar* concentration of a compound in solution is expressed in units of molecules per volume. This is somewhat unrealistic, as it does not take into account that for larger instance sizes, the edge molecules will necessarily be longer, and therefore fewer can be accommodated within a given volume.

[2]This plot is based on a threshold concentration of 1.0×10^{-6} in the normalized units in which the initial molar concentration was 1.0.

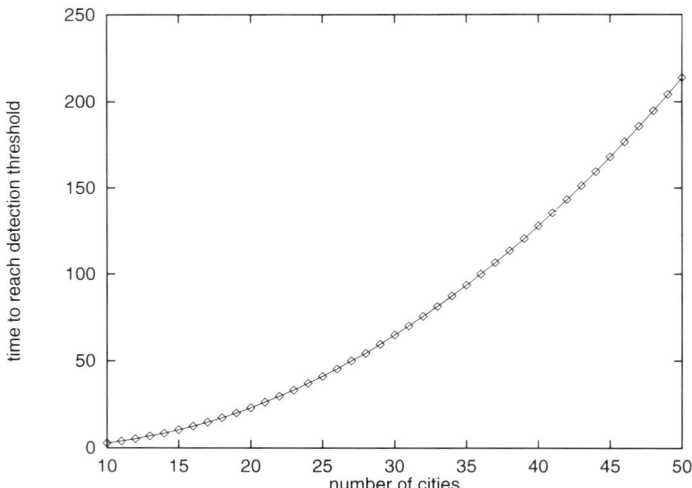

FIGURE 2. Time for $[e_{0,n-1}]$ to reach 1.0×10^{-6}.

2. Concentration

2.1. Line Digraphs Revisited. A somewhat more complete story is told by considering the concentrations of chemicals of a given path length, as a function of time. Let $[k]$ denote the sums of the concentrations of path molecules of length k, i.e.,

$$[k] = \sum_{i=0}^{n-k-1} [e_{i,i+k}] .$$

Figure 3 shows how these concentrations vary as a function of time for the reactions based on the line digraph L_{30}.

Although the plot gives a good "feel" for the data, there are a few points that may not be apparent. The concentration of simple edge molecules, [1], starts at 1.0 and falls steadily during the reaction. Excepting $k = n - 1$, $[k]$ reaches a maximum value in finite time, and falls slowly to zero thereafter. The successive maximums fall quadratically far apart, and because $[k + 1]$ remains below $[k]$ throughout the reaction, it reaches a lower maximum. For the hamiltonian path molecule, $[n - 1]$ tends in the limit to $\frac{1}{n-1}$.

2.2. A Hard Case. Having considered line digraphs, we now move on to consider a more difficult family of digraphs: the *feedback digraphs*[3]. The loss of concentration due to the presence of useless paths makes this a far more realistic example. Useless paths accumulate mass which is not recoverable, and so cannot be used to produce witnessing molecules.

DEFINITION 5. The *simple feedback digraph* F_n has n verticies and $2n-4$ edges. Like the line graph, it has $n - 1$ *forward edges* $e_{i,i+1} = \langle v_i, v_{i+1} \rangle$ for $i = 0 \ldots n - 1$,

[3]This particular family of digraphs is a standard example as to why random walk strategies fail to explore directed graphs efficiently. It is not the worst case graph for diffusion speed.

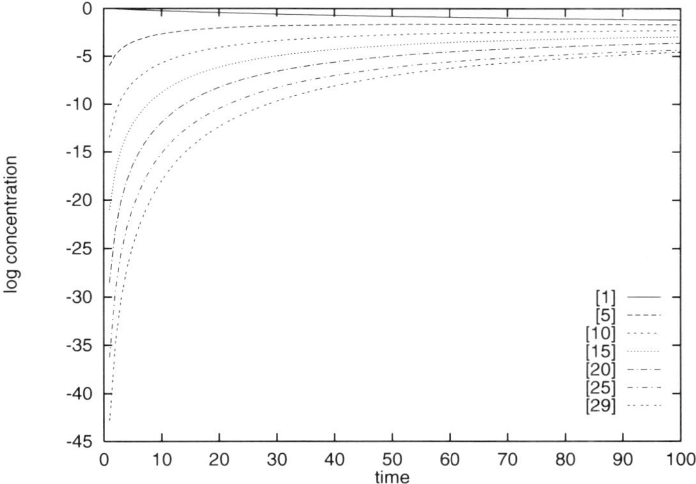

FIGURE 3. $\log_{10}[k]$ as a function of time for L_{30}.

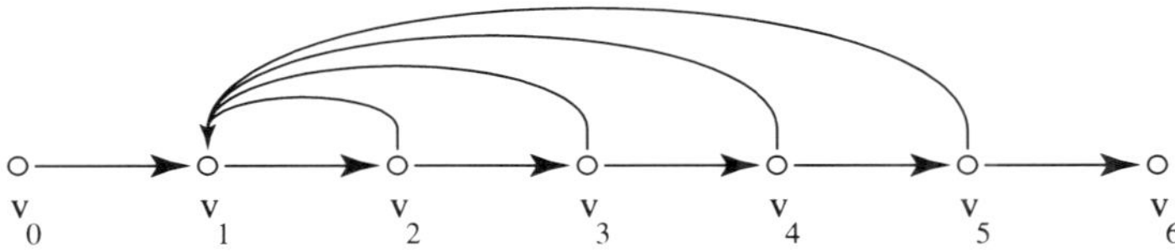

FIGURE 4. The simple feedback digraph F_7.

but it also has $n-3$ *back edges*, $b_k = \langle v_k, v_1 \rangle$ for $k = 2 \ldots n-2$. (See Figure 4 below.)

It should be clear that F_n has a unique hamiltonian path consisting of the edges $e_{i,i+1}$, for $i = 0 \ldots n-2$. Thus, any path that involves a back edge is useless, i.e., it cannot be extended to a hamiltonian path.

DEFINITION 6. We begin by defining $\langle v|$, $|v\rangle$, and $\langle v\rangle$ for F_n as in Definition 4. By extension, if p represents a path through F_n, let $\langle p|$ denote the set of molecules that represent paths that end with p, and define $|p\rangle$ and $\langle p\rangle$ analogously.

If we assume that all edge molecules are present in equal concentrations at the beginning of the ligation phase, then for $i = 2, 3, \ldots, n-2$, $[|e_{i,i+1}\rangle]$ equals $[|b_i\rangle]$ because they have the same initial concentration, and both satisfy the first order differential equation

$$y' = -[\langle v_i|]\, y$$

from the rate law.

Let $e_{i,j}$ denote the simple path molecule from v_i to v_j involving only forward edges. We have

$$[\langle e_{0,i+1}|] = [\langle e_{0,i} b_i|],$$

for $i = 2, 3, \ldots n-2$, because both have the same initial value (0), and both satisfy the differential equation:

$$y' = y\left[|e_{i,i+1}\rangle\right] = y\left[|b_i\rangle\right]$$

By a more complicated version of essentially the same argument $[\langle e_{0,i+1}|] \leq \frac{1}{2}[\langle e_{0,i}|]$. Thus,

$$[e_{0,n-1}] \leq \frac{1}{(2n-4)2^{n-3}}$$

holds throughout the computation (recall that we are using normalized units in which all physical constants have the value one). The ligation phase will never produce the hamiltonian path molecule $e_{0,n-1}$ at or above our detection threshold for $n \geq 18$.

The *complete feedback digraph*, which contains L_n, but has a back edge from v_i to v_j whenever $0 < j < i < n-1$ represents what is presumably the worst case. An analysis similar to that of the simple feedback graph shows that for complete feedback digraphs,

$$[e_{0,n-1}] \leq \frac{1}{(\binom{n-2}{2}+2)(n-2)!}$$

holds throughout the computation. Thus, for the complete feedback digraph, the ligation phase will never produce $e_{0,n-1}$ at or above our detection threshold for $n \geq 10$.

The point behind these two examples is that an exponential loss of concentration occurs even on sparse digraphs (simple feedback graphs are sparse), and that a loss of concentration can become a significant consideration for problems not much larger than that solved by Adleman.

3. Other Algorithms and Issues

While reaction rate and concentration issues suggest that Adleman's technique will not scale as easily as might be hoped, they are devastating for various improvements and/or extensions of his technique.

For example, one version of Adleman's technique considers a "serialized" version of the original graph. For every vertex v_i in the original graph, we have new vertices $v_{i,t}$ for $t = 0 \ldots n-1$; and for every edge $e_{i,j}$ of the original graph we have new edges $e_{\langle i,t\rangle,\langle j,t+1\rangle}$. Hamiltonian paths in the original graph correspond to paths in the derived graph that never visit the same city at two different times.

The advantage of the serialized version of Adleman's technique is that the molecules that correspond to hamiltonian paths of the original graph are nonreactive, and so one never needs to worry about molecules that are "too long," nor about when it is most advantageous to terminate the annealing phase. Unfortunately, we now have quadratically many different reaction families of the form

$$\langle v_{i,t}| + |v_{i,t}\rangle \to \langle v_{i,t}\rangle$$

Because the concentration of reactants in each of these families falls quadratically in the number of vertices, the reaction rates of these families fall quartically.

Reaction rate issues are even more damaging for techniques that attempt to use exponentially many reaction families, such as the algorithms for solving arbitrary PSPACE problems proposed by Beaver [**2**] and Rooß-Wagner [**6**]. Although the vision of exponentially many reactions taking place in polynomially many parallel steps is very attractive, it is now all too clear that each of these parallel reaction steps requires exponential time.

An exponential loss of concentration seems to affect all DNA based algorithms for solving NP-complete problems. For example, the imperfect nature of the "extract" operation makes it possible to construct bad cases for Lipton's satisfiability algorithm in which the relative concentration of molecules representing satisfying assignments falls exponentially with respect to the concentration of nonsatisfying molecules as successive clauses are handled.

4. Active Transport

In a sense, the two problems we've described are different facets of the same issue. Computational systems based on chemical reactions will need to account for, and perhaps actively monitor and adjust, the concentration of reactants involved. Failure to do this leads to bad performance in mild cases of concentration loss, and to pushing witnessing compounds below detection thresholds in more severe cases.

These issues are addressed by biological systems in nature. Cells need to control the concentrations of various compounds, they need to arrange for rare molecules to react, and they need to deal with undesirable byproducts of their own activity.

While many different mechanisms have evolved for managing the concentrations within the cell, the most successful involve membranes to separate a volume (a vesicle) and active systems that transport selected chemicals across that membrane. Since this transportation typically involves moving a chemical from a region of low concentration to a region of higher concentration, it decreases entropy and therefore requires energy. Nevertheless, the benefits to the cell of higher concentration (in terms of higher reaction rates, or under certain circumstances, of sequestering harmful compounds) justifies the expenditure of energy.

4.1. The Microstructure of Biological Computers. Moving from the cell to biological computers, we envision a microstructure of computational organelles that directs nanostructure of molecular computation.

This microstructure is based on a few basic types of vesicles: *collection vesicles* which contain the initial compounds; *reaction vesicles* which contain relatively high concentrations of a relatively few reaction families; *separation vesicles* in which molecules are separated according to the reaction families in which they participate; and *waste vesicles* in which useless molecules are concentrated. These vesicles have impermeable outer membranes, and are connected to one another by microtubes. Reaction products that concentrate in waste vesicles can be digested into shorter path molecules, which then re-enter the computation.

The design of such a microstructure presents some interesting problems.

The components have to fit into a three-dimensional universe. This should not be a significant issue for Adleman's technique, because there are only linearly many reaction families; but it may be significant for other algorithms for which the number of reaction families grows super-cubically. Algorithms that are based on exponentially many reaction families remain impractical.

It is not at all clear how these structures will be fabricated. Undoubtedly there are difficult problems in the domain of materials science and self-organizing systems that will come out of the attempt to build such a structure.

Transporting molecules along microtubes requires time proportional to the length of the tube. This introduces a routing problem reminiscent of VLSI design, but in three dimensions rather than two. The mathematical structure of the resulting system is a family of differential equations with time-delay terms—an impediment to both analysis and numerical simulation.

Finally, the energy requirements of such a system will need to be accounted for—especially at the separation vesicles where a decrease in entropy occurs.

Admittedly, what we are offering here is a crayon sketch of an architecture rather than a detailed proposal. We believe that it is a useful sketch nevertheless. Molecular computers, like Adleman's, that represent their state through the collection of chemicals found in a single well-mixed reaction vessel quickly lose performance as the number of reaction families increase. The key to avoiding this loss of performance is to concentrate the chemicals that make up a single reaction family into a much smaller volume. Because concentrations are bounded above, this implies that many different small volumes are involved—that is, that efficient biological computers must actively transport molecules to separated volumes based on the reaction families they are associated with.

4.2. CNA. The microstructure we discussed in Section 4.1 was first developed for an alternative model of biological computing based on protein synthesis [**4**]. Protein synthesis takes place at a ribosome, where molecules of mRNA are translated into proteins. If we view individual mRNA and protein molecules as strings over a finite alphabet, the process of protein synthesis is simply a finite-state translation.

Traditional models of computation can be viewed as iterated finite-state translation, and so if we ignore the inconvenient fact that the input and output "alphabets" of protein synthesis are different, it seems reasonable to base a computational model on iterated translation. Our model of CNA-based computing is a biologically inspired, biochemically plausible model based on the notion that a ribosome-like structure can be constructed that performs a computationally useful transcription from CNA (a hypothetical class of chemicals such as computationally Complex Nucleic Acids) back to CNA. Lest this seems completely without biological foundation, we point to the literature on the "RNA-world" that likely preceded the present protein world (e.g., [**3**]). The equation CNA = RNA motivates and partially justifies our basic biocomputational model.

CNA-based computing has some significant advantages and disadvantages with respect to DNA-based computing.

Because the key reactions of CNA-based computing depend on only a single low-concentration reactant, our intuition is that the computational performance of CNA-based computers will degrade linearly as concentration decreases, rather than quadratically as in the case of ligation-based computation.

The translation of CNA occurs at membranes and is energetically favorable. Therefore, there do not seem to be any major difficulties associated with the design of separation vesicles; whereas in DNA-based computing, it will be necessary to design molecular "pumps" to move DNA against a concentration gradient.

On the other hand, annealing-based computing is really a triumph of message passing. We do not know of a satisfactory message passing mechanism for CNA, and this remains an area of research.

Finally, recombinant DNA-technology is widely available, and Adleman's successful solution of a seven-city instance of hamiltonian path is a most convincing demonstration of the feasibility of his technique. We have not yet given an equally convincing demonstration; nevertheless, we believe that CNA-based computation is possible. Indeed, we believe that it will be a natural outgrowth of existing research on the RNA-world.

5. Conclusions

Early papers in biological computing did not attempt to analyze concentrations or reaction rates. We have shown that the best known biological algorithm (Adleman's) requires at least quadratic time, even on the simplest instances, which is exponentially worse than our naive expectations would suggest. On more complicated instances, a simple complexity-based analysis shows that even on sparse graphs, the yield of Adleman's algorithm can fall exponentially in problem size. We conclude that these simple chemical notions need to be accounted for in the analysis of biochemical algorithms.

Here we have suggested that a biologically inspired microarchitecture might be constructed based on membrane limited vesicles and the active transport of selected molecules across membranes. Our hope and expectation is that such an architecture may one day circumvent the reaction rate problems inherent in biological computing schemes and may lead to more viable biological computing technologies.

References

[1] L. Adleman. Molecular computation of solutions to combinatorial problems. *Science*, 266:1021–1024, 1994.

[2] D. Beaver. A universal molecular computer. In R. J. Lipton and E. B. Baum, editors, *DNA Based Computers : Proceeding of a DIMACS Workshop*, DIMACS Series in Mathematics and Theoretical Computer Science, pages 29–36. American Mathematical Society, 1996.

[3] E. H. Ekland, J. W. Szostak, and D. P. Bartel. Structurally complex and highly active RNA ligases derived from random RNA sequences. *Science*, 269:364–370, July 21 1995.

[4] S. Kurtz, S. Mahaney, J. Royer, and J. Simon. Biological computing. In L. Hemaspaandra and A. Selman, editors, *Complexity Retrospective, II*. Springer Verlag, 1996. to appear.

[5] H. Lodish, D. Baltimore, A. Berk, S. L. Zipursky, P. Matsudaira, and J. Darnell. *Molecular Cell Biology*. Scientific American Books, third edition, 1995.

[6] D. Rooß and K. Wagner. On the power of bio-computers. Technical report, Universität Würzburg, 1995.

Department of Computer Science, University of Chicago, 1100 E. 58th Street, Chicago, IL 60430

E-mail address: stuart@cs.uchicago.edu

DIMACS, Rutgers Univeristy, P.O. Box 1179, Piscatway, NJ 08855

E-mail address: mahaney@dimacs.rutgers.edu

Dept. of Electrical Engineering and Computer Science, Syracuse University, Syracuse, New York 13244

E-mail address: royer@top.cis.syr.edu

Department of Computer Science, University of Chicago, 1100 E. 58th Street, Chicago, IL 60430

E-mail address: simon@cs.uchicago.edu

DIMACS Series in Discrete Mathematics
and Theoretical Computer Science
Volume **44**, 1999

RNA Based Computing: Some Examples from RNA Catalysis and RNA Editing

Laura F. Landweber

ABSTRACT. This paper will focus on a combined treatment of the fields of RNA evolution and DNA computers. Using examples of *in vitro* selection of functional RNA molecules from large pools of random sequences, we show that these experiments can be viewed as an "RNA computer," in the language of DNA computation. By applying a strict set of criteria to a pool of heterogeneous DNA or RNA sequences, both types of experiments find sequences that contain solutions to hard molecular computational problems. These can even be nondigital problems that an electronic computer cannot solve. For example, a problem we posed to a pool of RNA molecules was one of catalysis, or the ability to ligate a small substrate RNA molecule to itself. *In vitro* evolution searches for the best solution or class of solutions contained within a large library of approximately 10^{15} unique sequences. The advantage of these "molecular computers" is their ability to examine and eliminate billions of possible solutions in parallel. We also describe an example of a cellular paradigm for RNA-based computing: RNA editing.

1. Molecular Selection

In vitro selection is a technique which, like the Polymerase Chain Reaction (PCR), has revolutionized molecular biology, because it allows the study and construction of novel nucleic acid molecules with specific properties or functions. Like DNA computers, *in vitro* selection is a sieving process of collecting desired molecules and discarding unwanted ones from large combinatorial libraries of 10^{13} to 10^{16} unique sequences (Figure 1).

Independently invented by three laboratories in 1989 and 1990 [**6, 4, 16**], *in vitro* selection allows amplification and isolation of rare molecules based solely on their specific functional properties and not on their primary sequence. All random molecules in molecular selection experiments are flanked by identical primer binding sites for PCR amplification; therefore it is essentially "PCR by function." The ability to pass a functional assay and to survive to the next step depends on either binding a target ligand or undergoing a self-modification reaction. This leads to

Key words and phrases. *in vitro* selection, directed evolution, RNA editing, RNA ligase, ribozyme, SELEX.

Thanks to DIMACS for stimulating this research. The research of the author was supported in part by a Burroughs Wellcome New Investigator Award in Molecular Parasitology and by NSF MCB-9520253.

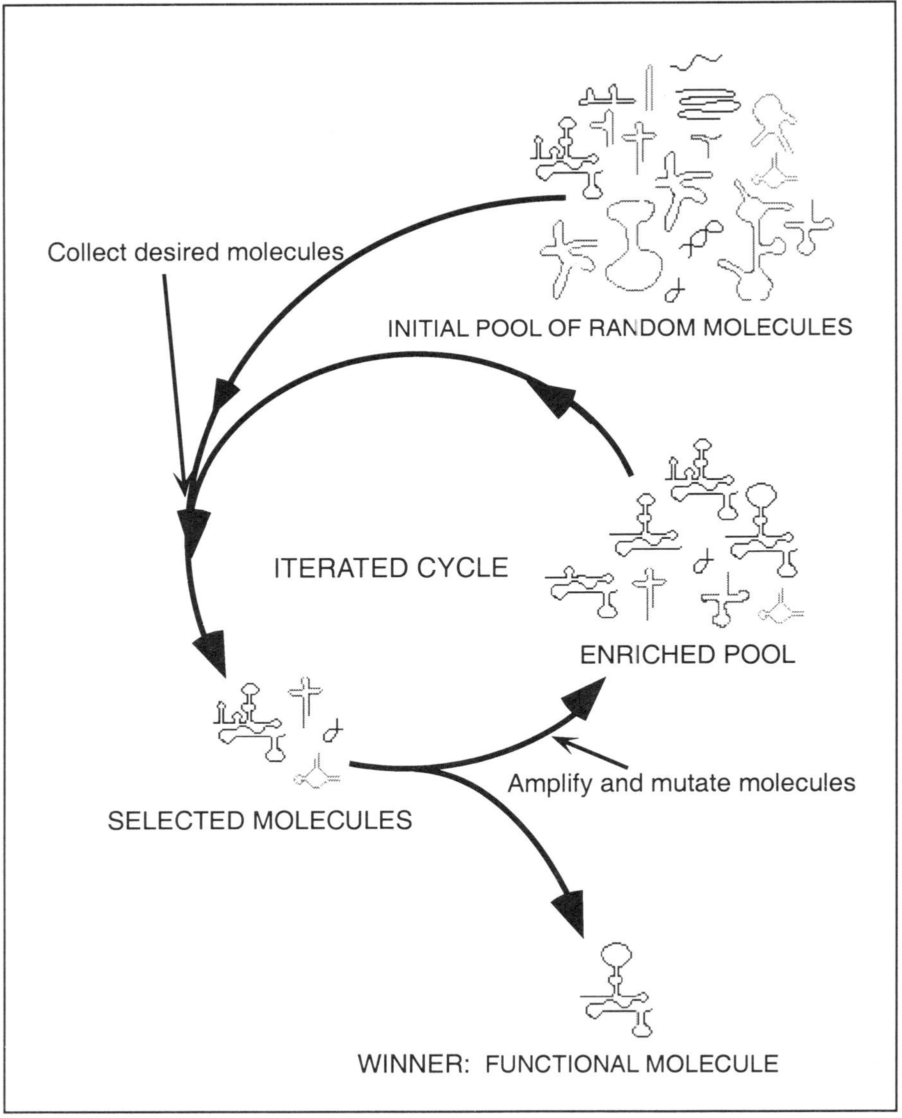

FIGURE 1. Outline of *in vitro* evolution. The "winning" molecule or class of molecules at the end are the most efficient at surviving the selection cycle [**11**].

differential amplification of the "fittest" molecules. In a catalytic assay for RNA enzymes, or ribozymes, the active molecules themselves must become modified by the reaction in such a way as to provide a "fishing hook" to distinguish them from inactive ones and to allow their capture from a large sequence pool. This limits the types of RNA catalysts (ribozymes) which can be retrieved by *in vitro* selection to those that can act in *cis*, allowing "self-tagging" of active ribozymes. Any molecule which catalyzes the reaction only in *trans*, meaning it acts on another molecule but not on itself, will be lost during the filtering process.

Through multiple rounds of selection and amplification, *in vitro* evolution allows amplification, mutation and isolation of single desired nucleic acid molecules from large pools of heterogeneous sequences, leading to purification of molecules present in as few as one in 10^{16} copies. Since these molecules can have novel catalytic properties and myriad applications-such as binding a specific ligand or catalyzing a reaction, the ability to harness combinatorial chemistry to isolate new ribozymes has launched a whole new industry of generating designer RNA tools. By expanding our knowledge of RNA's functional repertoire, the technique has also revealed the possibility of uncovering early catalysts in RNA evolution and even resuscitating active RNA molecules that have been "extinct" for over 3.9-billion years.

Normally the genetic carrier between DNA - informational molecule - and protein - functional molecule, RNA is an ideal unit for *in vitro* evolution, because it can combine both genotype and phenotype into one molecule. Therefore selection can act directly on RNA or DNA, because the primary sequence is contained within the actual molecule. In addition, the sequence is easily mutated and recovered, allowing experimental evolution on test-tubes of nucleic acids.

The basic concept of *in vitro* selection is the differential amplification of molecules with desired properties from pools of random sequences (Figure 1). Amplification of selected molecules increases their frequency in subsequent rounds, or generations, leading to survival of the fittest molecules. Multiple rounds of *in vitro* selection plus evolution allow rapid amplification, mutation, and isolation of a single desired nucleic acid molecule that can bind a novel target or perform a specific function, from a pool of as many as 10^{16} heterogeneous sequences [**11**].

Each iterated cycle mimics several aspects of evolution. Variation is supplied in the initial pool by randomization during oligonucleotide synthesis, and additional variation can be introduced in each cycle by mutagenic PCR, which is amplification with reduced fidelity of replication [**1**]. This allows optimization of the selected molecules. Selection is imposed by the stringency of the protocol used to capture or "trap" desired molecules - usually a binding or a functional assay. Most binding approaches rely on standard molecular isolation steps such as affinity chromatography (in which the specific RNA sticks to a column loaded with target), filter binding (in which the RNA does not pass through a filter if it is complexed with target), gel mobility shift (in which the RNA bound to target migrates as a larger complex), or immunoprecipitation (in which an antibody recognizes the target plus RNA) [**4, 16**]. An example of a direct functional assay is ligation to an oligonucleotide containing a unique sequence tag (Figure 2) or biotin. The sequence tag provides a recognition sequence for affinity chromatography, with the complementary strand serving as "bait," and biotin allows capture of ligated molecules to a streptavidin support; either strategy effectively converts the functional assay into a binding one. In addition, only those molecules that become covalently linked to the substrate RNA can be amplified by selective PCR using the sequence tag as a primer [**2, 1**].

In a landmark experiment [**1**], Bartel and Szostak used *in vitro* selection to isolate new RNA ligase ribozymes from a large pool of random-RNA sequences (Figure 2). Their discovery proves that one can retrieve molecules present in as few as one in 10^{15} copies which have novel enzymatic properties.

The pools of nucleic acid molecules used in molecular selection experiments are generally constructed with 5' and 3' constant regions that contain an RNA polymerase promoter (to allow RNA synthesis) and primer binding sites for reverse transcription and PCR (to recover and produce more copies of the selected

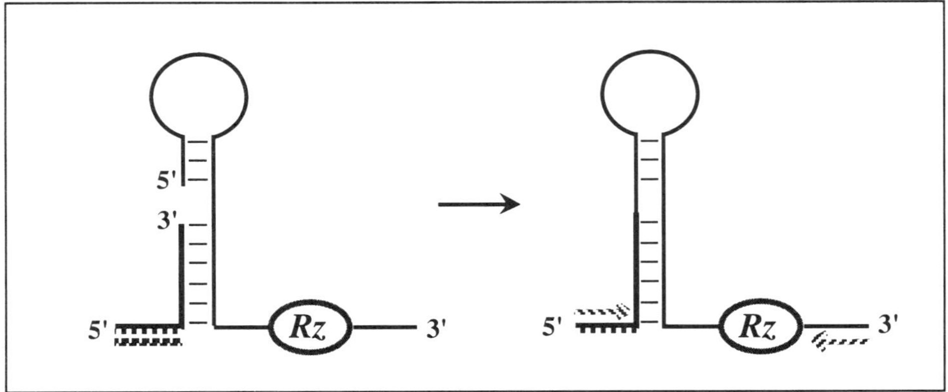

FIGURE 2. Selection for a ligase ribozyme (Rz). Arrows (right) indicate specific PCR primers that enable recovery of novel ligase ribozymes from large pools of random sequences [**1**]. This design allows differential amplification of molecules which ligate the substrate RNA (bold) onto their 5' end, thus incorporating a new DNA sequence tag (dotted) which provides both an affinity tag (shown complexed with the complementary strand on the left) and a selective PCR primer site [**11**].

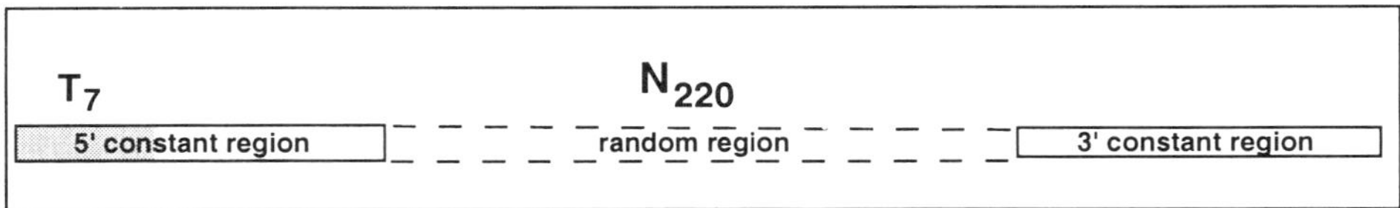

FIGURE 3. Construction of a random pool for *in vitro* selection.

sequences) [**11**] (Figure 3). The constant regions therefore allow PCR amplification of the pool, in addition to the design of the specific selection experiment. The pool in Bartel and Szostak's experiment contained 220 nucleotides of random sequence (N_{220}), flanked by a 5' constant region which could anneal to an oligonucleotide substrate RNA, forming a small hairpin loop which positioned the 3' end of the substrate RNA adjacent to the 5' end of the pool RNA by base-pairing (Figure 2). Ligated molecules become specifically tagged with the new sequence at the 5' end of the substrate RNA. This sequence then provides both an affinity tag for chromatography and a primer for "selective PCR"-amplifying only molecules with a covalently attached 5' primer. Bartel and Szostak were able to select for RNA molecules that could catalyze ligation of the substrate RNA to the pool RNA after just four cycles of enrichment by affinity chromatography and selective amplification. We performed a similar experiment which led to the discovery of a minimal RNA ligase motif (Landweber, unpubl.), and experiments in other laboratories have led to the discovery of many classes of novel nucleic acid catalysts, including DNA enzymes isolated from pools of single-stranded DNA [**11**].

Nucleic acid-based computers and selection both search through a test-tube of molecules representing some subset of "sequence space" to find a molecule or class of molecules which satisfies the requirements of a strict selective filter. DNA computing mimics the field of *in vitro* selection by applying a strict set of functional criteria to a pool of heterogeneous DNA sequences; however, a DNA computer generally has to perform an exhaustive search of all possible solutions. *In vitro* selection, on the other hand, empirically finds the best possible solution in a limited sampling of sequence space. For example, a library containing 10^{15} sequences with a 100 nucleotide random region represents only a very negligible fraction of all possible sequences of length 100 ($4^{100} > 10^{60}$, which is approximately the number of protons in the most massive star in the universe!). The success of *in vitro* selection experiments using pools of fewer than 10^{16} unique sequences already suggests an enormous wealth of functional molecules distributed in sequence space, many of which contain "solutions" to hard molecular computational problems.

The purification of desired sequences to homogeneity usually requires ten or more rounds of *in vitro* selection. The functional molecules which survive a strict selective filter ultimately become the dominant species in the pool. Thus *in vitro* evolution reduces the pool complexity to one (or more) functional class(es) of sequences, much like searching for solutions to hard bio-computational problems [**12**]. The technique has quickly enjoyed popularity, mainly because of its power to examine in parallel billions of possible solutions to important biochemical problems such as catalysis or binding.

A number of parameters in molecular selection experiments greatly affect the nature of the evolutionary process being simulated and therefore the likelihood of finding a "winning" molecule within a large collection of sequences. These parameters include the stringency of the selection, the amount of initial variation in the pool, the amount of variation introduced at each successive round of amplification, the choice of salt concentrations and buffers, and numerous considerations about the design of the pool for the individual experiment [**11**].

The RNA pools can be one of two types. The first is completely random; all bases have an approximately equal probability (i.e. 25% each A, C, G & U) in a region of defined length, such as 220 bases (Figure 3). The pools are either synthesized as long DNA oligonucleotides or assembled from smaller ones. They are then made doubled stranded and amplified by large-scale PCR [**1**] and transcribed into RNA by T7 RNA polymerase. The sequence composition of such pools must be carefully checked to avoid biases introduced during DNA synthesis. Random pools are typically used to isolate novel ribozymes or aptamers - molecules that bind specific ligands. The second type of pool is a "doped," or variant, pool in which an existing sequence is mutated at a defined frequency (e.g. each consensus nucleotide with probability 85% and each non-consensus nucleotide with probability 5%). Doped pools are often used to select variants of existing ribozymes with improved efficiency or to obtain molecules that function with different substrates or in different conditions [**11**]. They can be synthesized either by mutagenic PCR [**2**] or by controlled oligonucleotide synthesis [**4**].

In vitro selection allows the exploration of the sequence space of catalytic RNA molecules, sometimes referred to as "catalytic task space" [**7**]. We and others have selected RNA molecules with desired properties from combinatorial libraries containing over 10^{15} unique sequences. The steps always involve an iterative procedure of artificial selection and PCR amplification of the rare molecules. There are several opportunities for interaction between computational molecular biology and these experiments. For example, one can use the data gathered from *in vitro* selection experiments to model *in vitro* evolution of the pool of RNA molecules. The patterns

```
DNA      G    G     GTTTTGG   AGA         G ATTTGG  A
RNA     uGuuuuGuuuuuGUUUUGGuuuAGAuuuuuuuuGuAU  GGuuAuuu
```

FIGURE 4. Comparison of a representative region of the *H. mariadeanei* COIII mRNA sequences with the corresponding region of the mitochondrial DNA. DNA sequences in upper case; uridines in mRNA that are added by RNA editing in lower case; encoded thymidines deleted from the mRNA indicated by gaps.

of gene-frequency changes in such a system resemble patterns in natural populations under strong directional selection. In addition, *in vitro* selection experiments allow one to map out both the underlying ruggedness of the adaptive landscape as well as the distribution of functional RNA molecules in sequence space. These are problems which have hitherto been treated theoretically by Eigen, Schuster, Kauffman and others (e.g. [**14**]); however *in vitro* selection now provides a means to test numerous predictions and also to refine the theory. Indeed, the molecular "solutions" to computational problems imposed by *in vitro* selection can be treated as satisfiability problems, within the framework of nucleic acid-based computers.

2. RNA Editing

RNA editing is the modification of RNA sequences by addition, deletion, or substitution of bases. Found in a wide variety of eukaryotes, ranging from parasitic protozoa to humans, this remarkable process alters the sequence of messenger RNA before translation into protein such that the resulting protein sequence often differs dramatically from the original gene sequence, which sometimes does not encode a protein at all. For example, in *Trypanosoma brucei*, RNA editing by addition and deletion of literally hundreds of uridine residues creates initiation and termination codons, alters the structural features of transcripts, and creates over 90% of the amino-acid codons [**9**]. The main features of RNA editing are identified by differences between RNA and genomic DNA sequences.

Figure 4 illustrates an extreme example of RNA editing of the mitochondrial cytochrome oxidase subunit III (COIII) transcript in *Herpetomonas mariadeanei*. 91% of the 288 amino acid codons in *T. brucei* [**5**] and *Herpetomonas* [**9**] are created by editing, with this effect restricted only to uridines (Figure 4). The other bases - A, C, and G - are completely conserved between the DNA and the RNA sequence.

The edited proteins also have a fast molecular clock, which accumulates amino acid substitutions at least twice as fast as the unedited versions. This is due to frameshift mutations that are introduced by fixed changes in the number and position of U's inserted or deleted by editing and then compensated by editing at another site that restores the reading frame (Figure 5). Editing therefore allows combinatorially diverse protein products to be generated from a single gene, either over evolutionary time [**9**] or within an individual cell [**15**]. The surprising discovery that RNA editing is a source of sequence mutation, rather than a proper "editing" or correcting mechanism, underscores the importance of the question of why it is still used by some organisms to generate a sequence encoding a single conserved protein [**10**].

The genetic information for editing is stored in the form of "guide" RNA molecules (gRNAs), very small (50-70 nucleotides) transcripts that mediate editing

```
Hsa DNA   G AA      A        G G   GG   AGA      G  G     G G     G
Hsa RNA uuGuAAuuuuuuAuuuuuuu GuGuuuGGuuuAGAuuuuuuGuuGu  uuGuGuuuuuGu
Hme DNA   G AA         ATTTT G G   GG   AGA      G  GT    G A     G
Hme RNA uuGuAAuuuuuuuuuAUUUU GuGuuuGGuuuAGAuuuuuuGuuGU  uuGuAuuuuuGu
Hmm DNA   G AA         ATTTTTG G   GG   AGA      G  GT    G A     G
Hmm RNA uuGuAAuuuuuuuuuAUUUU GuGuuuGGuuuAGAuuuuuuGuuGU  uuGuAuuuuuGu
Hma DNA    GAA    G    G      GTTTTGG   AGA       G ATTTGG  A   G
Hma RNA uuuGAAuuuuGuuuuGuuuu uGUUUUGGuuuAGAuuuuuuuGuAU  GGuuAuuuGuuu
Tbr DNA  G  AA       G    G  GTTTTTGG   AGG       G  GTTTTG  G    G
Tbr RNA uGuuAAuuuuuuuGuuuuGu GUUUUUGGuuuAGGuuuuuuuGuuG  UUGuuGuuuuGu
```

FIGURE 5. An example of fixed frameshifts between species with editing. Hsa, *H. samuelpessoai*; Hme, *H. megaseliae*; Hmm, *H. muscarum muscarum*; Hma, *H. mariadeanei* [**9**]; Tb, *T. brucei* [**5**]. DNA sequences in upper case; uridines in mRNA that are added by RNA editing in lower case; encoded thymidines deleted from the mRNA indicated by gaps.

```
                        gRNA
  3'-UaaUUaUagaaCaUagagaCUgUaaaUaaAUAAACAACCAAAUAUA-5'
     :|||||:|:||||||:|:|||::|||||||||||||||||||||:|
5'...GuuAAuGuuuuGuAuuuuuGAuGuuuAuuuAuuuGuuGGuuuAuGu...3'
                        mRNA
```

FIGURE 6. Representative pairing between a guide RNA (top) and the edited mRNA (bottom). Lowercase u's are added by editing. Boldface u's are edited by base pairing to lowercase a's and g's in this gRNA [**13**].

by base-pairing with specific regions of the edited transcript, allowing some G:U base-pairs [**3**]. Each gRNA contains the sequence information to edit approximately 30 nucleotides of edited RNA [**9**] and pairs more efficiently with the final product than with the pre-edited substrate. For every inserted U in the messenger RNA sequence, there is a corresponding A or G in the gRNA which pairs with the fully edited product (Figure 6). Complete editing proceeds 3' to 5' on the mRNA and requires a full set of overlapping guide RNAs. Editing by each guide RNA creates an anchor sequence for binding the next guide RNA (Figure 7; [**3, 13**] leading to an ordered cascade of editing events. It is also a very slow but faithful mechanism, considering that one can find hundreds of partially edited and even incorrectly edited intermediates at steady state [**9**] although these presumably lead to correct product formation. Thus RNA editing is a surprisingly inefficient cellular computing mechanism for producing messenger RNA sequences, although directed editing 3' to 5' ensures that only fully edited transcripts are translated, since the AUG or UUG [**9**] translation initiation codon is edited last.

3. Conclusions

The ability to isolate unique RNA molecules through *in vitro* selection has given rise to the field of *in vitro* genetics, which can be viewed either as a fantastic tool for the synthesis of novel catalysts through combinatorial chemistry or as a simple

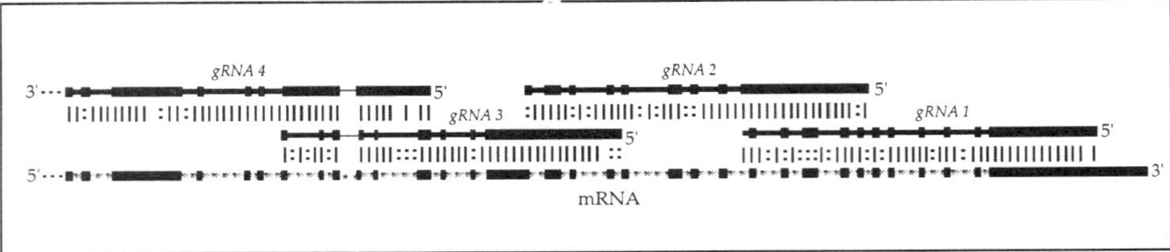

FIGURE 7. Editing by four overlapping gRNAs. Thick lines in the mRNA are encoded in the mitochondrial DNA. Thin shaded lines are inserted U's; the two asterisks are deleted U's [**13**]. Thin lines in the gRNA's are guide nucleotides (A or G) that pair with inserted U's. Vertical lines indicate Watson-Crick base pairs; colons indicate G:U wobble base pairs, illustrating formation of well-paired "anchors" between the 5' ends of gRNA's shaded boxes and the corresponding region of the mRNA.

system for probing the stepwise assembly of processes involving RNA. Together, the results of *in vitro* selection experiments have revealed not only RNA's ability to catalyze a much wider range of reactions than previously thought but even DNA's cryptic ability to operate as an enzyme. Thus, within laws of thermodynamics, the possibilities are virtually endless for the selection of future nucleic acid catalysts, perhaps catalyzing reactions for which no known catalyst exists, such as the breakdown of toxic substances. As a tool for drug discovery, *in vitro* selection offers a new direction for the synthesis of potential therapeutic agents. As a tool for studying evolution, *in vitro* selection allows us the first opportunity to study experimentally the biological origins of catalysis and the emergence of order from random sequences. RNA editing is an example of a cellular process which uses RNA sequences as guides to convert seemingly disordered RNA sequences into functional messenger RNA: a truly RNA-based computer.

References

[1] D. P. Bartel and J. W. Szostak. Isolation of New Ribozymes from a Large Pool of Random Sequences. *Science*, 261: 1411-1418, 1993.

[2] A. A. Beaudry and G. F. Joyce. Directed Evolution of an RNA Enzyme. *Science*, 257: 635-641, 1992

[3] B. Blum, N. Bakalara, and L. Simpson. (1990). A model for RNA editing in kinetoplastid mitochondria: "Guide" RNA molecules transcribed from maxicircle DNA provide the edited information. *Cell*, 60: 189-198, 1990.

[4] A. D. Ellington and J. W. Szostak. In vitro selection of RNA molecules that bind specific ligands. *Nature*, 346: 818-822, 1990.

[5] J. E. Feagin, J. M. Abraham, and K. Stuart, Extensive Editing of the Cytochrome c Oxidase III transcript in Trypanosoma brucei, *Cell*, 53: 413-422, 1988.

[6] G. F. Joyce, Amplification, mutation and selection of catalytic RNA. *Gene* 82: 83-87, 1989.

[7] S. Kauffman, *The origins of order: self-organization and selection in evolution*, Oxford University Press, New York, New York, 1993.

[8] L. F. Landweber, A. G. Fiks, and W. Gilbert, The Boundaries of Partially Edited Cytochrome c Oxidase III Transcripts are Not Conserved in Kinetoplastids: Implications for the Guide RNA Model of Editing. *Proc. Natl. Acad. Sci. USA*, 90:9242-9246, 1993.

[9] L. F. Landweber and W. Gilbert, RNA editing as a novel source of genetic variation. *Nature*, 363: 179-182, 1993.

[10] L. F. Landweber and W. Gilbert, Phylogenetic analysis of RNA editing: A primitive genetic phenomenon. *Proc. Natl. Acad. Sci. USA*, 91:918-921, 1994.

[11] L. F. Landweber, P. J. Simon, and T.A. Wagner, Ribozyme Design and Early Evolution, *BioScience*, 48:94-103, 1998.

[12] R. J. Lipton, DNA solution of hard computational problems. *Science*, 268: 542-545, 1995.

[13] D. A. Maslov and L. Simpson, The polarity of editing within a multiple gRNA-mediated domain is due to formation of anchors for upstream gRNAs by downstream editing. *Cell*, 70: 459-467, 1992.

[14] P. Schuster, W. Fontana, P. F. Stadler, and I. Hofacker, From sequences to shapes and back: a case study in RNA secondary structure. *Proceedings of the Royal Society of London*, 255: 279-284, 1994.

[15] B. Sommer, M. Kohler, R. Sprengel, and P.H. Seeburg, RNA editing in brain controls a determinant of ion flow in glutamate-gated channels. *Cell*, 67: 11-19, 1991.

[16] C. Tuerk and L. Gold, Systematic Evolution of Ligands by Exponential Enrichment: RNA Ligands to Bacteriophage T4 DNA Polymerase. *Science*, 249: 505-510, 1990.

Department of Ecology and Evolutionary Biology, Princeton University, Princeton, NJ 08544

E-mail address: lfl@princeton.edu

DIMACS Series in Discrete Mathematics
and Theoretical Computer Science
Volume **44**, 1999

Universal Computation via Self-Assembly of DNA: Some Theory and Experiments

Erik Winfree[†], Xiaoping Yang, Nadrian C. Seeman

ABSTRACT. In this paper we examine the computational capabilities inherent in the hybridization of DNA molecules. First we consider theoretical models, and show that the self-assembly of oligonucleotides into linear duplex DNA can only generate sets of sequences equivalent to regular languages. If branched DNA is used for self-assembly of dendrimer structures, only sets of sequences equivalent to context-free languages can be achieved. In contrast, the self-assembly of double crossover molecules into two dimensional sheets or three dimensional solids is theoretically capable of universal computation. The proof relies on a very direct simulation of a universal class of cellular automata. In the second part of this paper, we present results from preliminary experiments which investigate the critical computational step in a two-dimensional self-assembly process.

1. Introduction

A fundamental property of DNA is that, under the right conditions, Watson-Crick complementary regions of single-stranded DNA will hybridize and form a double helical structure. This property, in vitro and in vivo, can lead DNA to assume a remarkable diversity of geometric forms[1]. Under certain simplifying conditions, the behavior of hybridization is sufficiently predictable to be considered as a computational primitive; i.e., a function from initial oligonucleotides to final supramolecular structures is computed. The computational aspects of self-assembly were exploited for the first time in [**2**], where linear self-assembly was used as a step in solving the Hamiltonian Path Problem. When the self-assembly of tree-like structures takes place, due to the presence of branched junctions, a slightly more powerful computation results. We review a two dimensional generalization capable

[†] To whom correspondence should be addressed. E. Winfree has been supported in part by National Institute for Mental Health (NIMH) Training Grant # 5 T32 MH 19138-06; also by General Motors' Technology Research Partnerships program and by the Center for Neuromorphic Systems Engineering as a part of the National Science Foundation Engineering Research Center Program under grant EEC-9402726. The experimental portion of this research has been partially supported by grants N00014-89-J-3078 from the Office of Naval Research and GM-29554 from the NIH (to NCS).

[1] In vivo, not only is there single-stranded and double-stranded DNA, but branched junctions are formed during recombination, and trypanosomes maintain complex networks of circular DNA within which RNA editing occurs.

of universal computation, as suggested in [**12**], and also suggest a concrete three dimensional self-assembly process.

2. Computational Power of Abstracted Hybridization

In order to understand the computational implications of DNA hybridization, we will first consider a highly abstracted mathematical model. The physical system we would like to model can be described as follows:

> Synthesize several sequences of DNA. Mix the DNA together in solution. Heat it up and slowly cool it down, allowing complexes of DNA to form. Chemically or enzymatically ligate adjacent strands. Denature the DNA again, and ask, what single-stranded DNA sequences are now present in the solution?

A proper answer to this question is beyond our capability, and realistically detailed models might not be enlightening regarding the logical essence of self-assembly. We therefore investigate very simple models, which, nonetheless, are sufficiently realistic that translation into real world scenarios should be direct. We will consider a number of properties which DNA self-assembly may be postulated to obey, and we will analyze the computational capability and the limits of any self-assembly process which obeys those properties.

Informally, the properties we consider are:

1. **Constant Temperature.** The number of base-pairs required for the stability of DNA complexes does not change during the course of the self-assembly. We thus don't consider annealing, where at high temperatures only long regions will hybridize but later at lower temperatures even short regions can hybridize, but rather we model a "constant temperature" process.
2. **Perfect Watson-Crick Complementarity.** Hybridization only occurs between sequences with perfect Watson-Crick complementarity. Hybridization of mismatching sequences, or that which creates bubbles, branched junctions, triple helices, and other unusual structures, is not considered.
3. **Permanant Binary Events.** All self-assembly interactions occur between two complexes at a time, and no more. These interactions are exclusively hybridizations, joining two complexes together. Furthermore, in the model once two complexes join, they never dissociate.
4. **No Intramolecular Events.** A DNA complex which has self-assembled will not interact with itself, for example by cyclizing. Note, however, that some physically intramolecular interactions can be modeled as a part of a binary event, as discussed below.
5. **Single vs Multiple Binding Regions per Event.** We will consider two cases: either (a) each binary hybridization event creates a single contiguous Watson-Crick region, else (b) the binary events may result in the formation of several physically separated hybridized regions between the two complexes. The latter case is meant to model physical situations where an intermolecular hybridization is immediately followed by an intramolecular hybridization. The case we are interested in is discussed in Section 2.5 (see Figure 7).
6. **Specified Classes of Initial Complexes.** Because of our constant - temperature assumption, it becomes useful to assume that some complexes

have already formed prior to the stage of self-assembly which we will consider. Later in the paper, we will consider initial complexes which consist of (a) oligonucleotides, (b) duplex DNA with sticky ends, (c) hairpins with sticky ends, (d) three-armed junctions with sticky ends, (e) double crossover molecules with hairpins and sticky ends, and (f) arbitrary complexes.

Properties (1), (2) and (3) are used primarily for logical simplicity. If Property (4) were changed to allow intramolecular events, it is possible that some of our results would be slightly modified. We will analyze how our results change under different choices for Properties (5) and (6). In Section 2.5, we impose an additional property in order to incorporate geometrical considerations for lattice self-assembly.

2.1. Language Theory and Grammars. Before we present our model of DNA self-assembly, we should comment on what it means to compute by self-assembly. As mentioned above, the typical case is that one starts with a small variety of synthesized oligonucleotides, and one ends with great variety of self-assembled strands. The resulting strands are not random; they have certain properties that derive from being formed from the original oligonucleotides according to certain rules of hybridization.

An analogous situation arises in formal language theory, which has been well understood for many years. There, rather than test tubes of strands, one is interested in sets of symbolic strings, and in methods of generating them. We will sketch the basics here; for a full development see [**7**].

An *alphabet* is a finite set of symbols, for example $\{A, C, G, T\}$ or $\{0, 1\}$ or $\{x, y, z, (,), +, *\}$. A *string* over an alphabet is a finite sequence of symbols from the given alphabet, for example $TATAA$ or 101011 or $(x + y) * z$. A *language* is a well-defined, possibly infinite set of strings, for example

$$\{\textit{all strings over } \{C,T\} \textit{of length } 70\} \text{ or } \{\textit{all prime numbers, written in binary}\}$$

$$\text{or } \{\textit{all well-formed formulas over } \{x, y, z, (,), +, -\}\}.$$

Although one cannot write down each and every string in an infinite language, one can ask the membership question: is string x in language $\mathcal{L}$? Note that if the language $\mathcal{L}$ contains all bit strings x for which function $f(x) = 1$, the the membership question is equivalent to boolean function evaluation. The membership question may be harder or easier to answer, depending on x and $\mathcal{L}$. Formal language theory goes to great pains to classify languages according to how fancy the computer must be to answer the membership problem. We sketch the fundamental result due to Noam Chomsky, known as the language hierarchy. This requires formalizing the specification of languages by generative rules.

A *rewriting rule* $x \to y$, where x and y are strings, specifies that a string $s = axb$ can be rewritten to produce the new string $s' = ayb$. A *grammar* G is a collection of rewriting rules together with a division of the alphabet into two groups: *terminal symbols* and *nonterminal symbols*, where only nonterminals appear on the left hand side of rewriting rules. Each grammar uniquely defines a language $\mathcal{L}_G$ as follows: the string of terminals s is in $\mathcal{L}_G$ iff it can be obtained from the special nonterminal S by the repeated application of rewriting rules in some order (called a *derivation*).

Grammars may be classified by what kinds of rules they use. We give examples of the three main classes below:

Regular grammars: use rules of the form $A \to pB$ and $A \to p$ where A and B are nonterminal symbols and p is a string of terminals. Languages generated by regular grammars are called regular languages. For example, consider the regular grammar $G_E = \{S \to 0S, S \to 1T, S \to 0, T \to 0T, T \to 1S, T \to 1\}$ where 0 and 1 are terminals. This grammar gives rise to all bit strings with an even number of 1's. $101011 \in \mathcal{L}_{G_E}$ because $S \to 1T \to 10T \to 101S \to 1010S \to 10101T \to 101011$. Note that during the derivation we always have a single nonterminal at the right, where all the action takes place. Despite their apparent simplicity, regular languages have found extensive use in pure and applied computer science, perhaps because their membership question can always be answered by an exceedingly simple abstract computer known as a finite state machine.

Context-free grammars: use rules of the form $A \to P$ where again A is a nonterminal symbol, but now P is an arbitrary string of terminals and nonterminals. Languages generated by context-free grammars are called context-free languages. Consider the grammar $G_F = \{S \to S + S, S \to M, M \to M * M, M \to (S), M \to x, M \to y, M \to z\}$ where the terminals are $\{x, y, z, (,), +, *\}$. This grammar gives rise to well-formed formulas. $(x + y) * z \in \mathcal{L}_{G_F}$ because $S \to M \to M * M \to M * z \to (S) * z \to (S + S) * z \to (S + M) * z \to (S + y) * z \to (M + y) * z \to (x + y) * z$. Note that whereas it is impossible to generate regular languages whose strings all have long-range structure, one can generate long-range "nested" structure in a context-free language – for example, every parenthesis must be matched in the formulas above. Context-free languages include regular languages. The membership question for context-free languages can be answered by a slightly more complex machine known as a nondeterministic pushdown automaton.

Unrestricted grammars: use rules of the form $A \to P$ where now A may be an arbitrary strings of nonterminals, and P is an arbitrary string of terminals and nonterminals. Languages generated by unrestricted grammars are called *recursively enumerable* languages because they include every language which can be generated (enumerated) by any computational process (recursion). Recursively enumerable languages include context-free languages, regular languages, and much more. They are as fancy as you can get. A very simple example: consider the alphabet $\{S, L, R, \overleftarrow{B}, \overrightarrow{B}, \overleftarrow{W}, \overrightarrow{W}, 0, 1\}$ and the grammar $G_P = \{S \to 1,\ S \to LR,\ L \to L\,\overrightarrow{B},\ L \to 1,\ R \to \overleftarrow{B}\,R,\ R \to 1,\ \overrightarrow{B}\overleftarrow{B} \to \overleftarrow{W}\overrightarrow{W},\ \overrightarrow{B}\overleftarrow{B} \to 0,\ \overrightarrow{B}\overleftarrow{W} \to \overleftarrow{B}\overrightarrow{B},\ \overrightarrow{B}\overleftarrow{W} \to 1,\ \overrightarrow{W}\overleftarrow{B} \to \overleftarrow{B}\overrightarrow{B},\ \overrightarrow{W}\overleftarrow{B} \to 1,\ \overrightarrow{W}\overleftarrow{W} \to \overleftarrow{W}\overrightarrow{W},\ \overrightarrow{W}\overleftarrow{W} \to 0\}$ where the terminals are 0 and 1. This gives rise to the rows of Pascal's triangle mod 2. The third row $101 \in \mathcal{L}_{G_P}$ because

$$S \to LR \to L\,\overrightarrow{B}\,R \to L\,\overrightarrow{B}\overleftarrow{B}\,R \to 1\,\overrightarrow{B}\overleftarrow{B}\,R \to 10R \to 101.$$

Later, we will make use of a subclass of unrestricted grammars equivalent to blocked cellular automata, which generalize the example and which are still capable of generating all recursively enumerable languages; that is, they are *universal*. A surprising consequence of universality is that the membership question for recursively enumerable languages is sometimes impossible to answer!

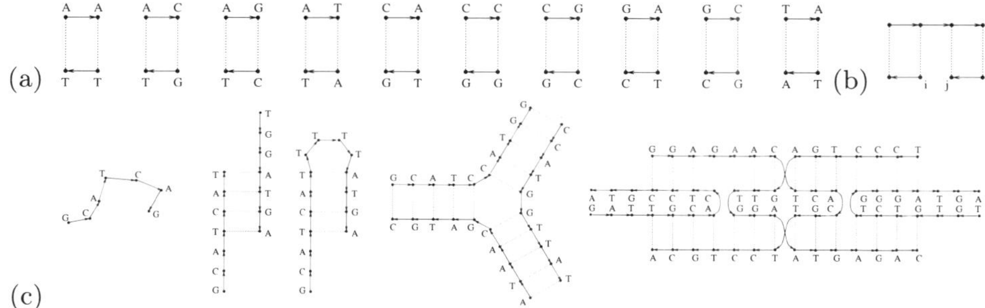

FIGURE 1. Some DNA complexes. Solid lines represent backbone edges; each dotted line represents a pair of reciprocal basepair edges. (a) The 10 Watson-Crick subgraphs. (b) The valid ligation site. (c) A strand, a duplex with sticky ends, a hairpin with a sticky end, a 3-armed branched junction, and a DAO double crossover (DX) unit with sticky ends.

2.2. DNA Complexes and Self-assembly Rules. Grammars turn out to have a close relationship to the self-assembly models we discuss here. However, to make this relationship precise, we define our model formally.

A *DNA complex*[2] is a connected directed graph with vertices labeled from $\{A, C, G, T\}$, edges labeled from $\{backbone, basepair\}$, with at most one incoming and one outgoing edge of each type at each node (thus at most four incident edges total), and where for every basepair edge $x \to y$ there is a reciprocal basepair edge $y \to x$. Furthermore, all base-pairing in a DNA complex must be *Watson-Crick*, that is, every basepair edge must be within a subgraph isomorphic to one of the 10 given in Figure 1a.

A DNA complex (just *complex* for short) represents several DNA polynucleotides bound together by Watson-Crick hybridization. Note that this representation supports a rich variety of DNA structures, but structures such as triple helices are missing; similarly, it is lacking notions of geometry and topological linking. Also, we must be careful because it is possibly to specify physically impossibly structures.

It will be useful to introduce a few examples of DNA complexes, shown in Figure 1c. A *strand* consists of a chain of backbone-connected nodes, with no basepair edges. Strands may be either linear or circular. A *duplex* consists of two strands with contiguous basepair edges between them. A duplex may optionally have a *sticky-end* on either end. An *n-armed (branched) junction* consists of n duplex arms arranged around a central point. A *double crossover unit* (DX unit) consists of two adjacent duplexes with two points of strand exchange[3]. For formal reasons, the empty graph ϵ is a DNA complex.

We now define some operations on complexes. In our model, hybridization is indicated by $C_1 +_B C_2 = C_3$, where $+_B$ denotes the formation of basepair edges B between nodes of C_1 and nodes of C_2. If the graph consisting of both C_1 and C_2 and the edges B is a DNA complex, then C_3 is that graph; else $C_3 = \epsilon$ (for example,

---

[2]Similar to Beaver's *cluster* [**3**].

[3]Real DX molecules [**6**] come in a number of geometric varieties (we use "DAO" here), each of which put constraints on the symmetry and the number of nucleotides between crossover points. We ignore these constraints in the theoretical section.

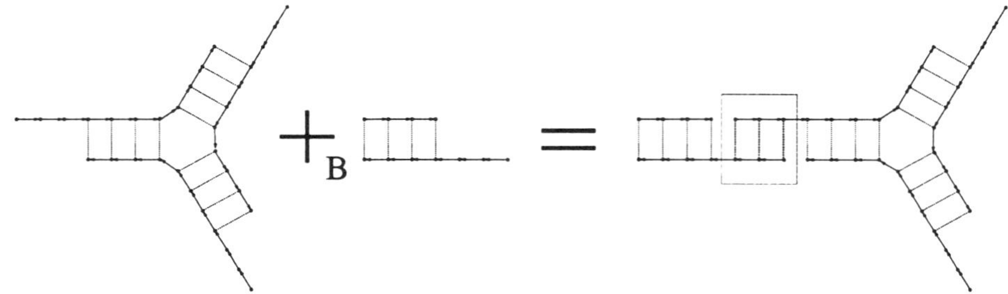

FIGURE 2. A hybridization $C_1 +_B C_2 = C_3$ allowed by R_1^3. The edges of B are emphasized in C_3, and the subgraph induced by B has a dotted box around it.

if a new edge joins two T's). The hybridization operation will be used to describe self-assembly, below.

To analyze the complexes present after self-assembly, we introduce two other operations based on ligation and denaturing. $C' = ligate(C)$ is obtained by adding a backbone edge from node j to node i in every occurrence of the subgraph shown in Figure 1b, so long as nodes i and j have no other incident backbone edges.

To model the denaturing of a complex, we define $\{C_i\} = denature(C)$ to be the set of all *strands* in C, i.e., each C_i is a backbone-connected component of C (with no basepair edges). Note that if C contains topologically linked circular strands, then *denature* will "magically" unlink them from each other[4].

In analogy to formal language theory, we define a *language of DNA complexes* to be a well-defined, possibly infinite set of DNA complexes. We can generate a language of complexes $\mathcal{L}_{R,A}$ by applying *self-assembly rules* R to an initial language A, usually finite[5]. The rules R specify which hybridizations $C_1 +_B C_2 = C_3$ are allowed. Let $\hat{\mathcal{L}}_{R,A}$ be the transitive closure of A under all allowed hybridizations. In other words, (a) $A \subset \hat{\mathcal{L}}_{R,A}$, (b) if $C_1, C_2 \in \hat{\mathcal{L}}_{R,A}$ and $C_1 +_B C_2 = C_3$ is allowed, then $C_3 \in \hat{\mathcal{L}}_{R,A}$, and (c) no other complexes are in $\hat{\mathcal{L}}_{R,A}$. Now let $\mathcal{L}_{R,A} \subset \hat{\mathcal{L}}_{R,A}$ consist of those complexes for which no further hybridization is allowed; these are called the *terminal complexes*. Loosely, $\mathcal{L}_{R,A}$ is meant to model the DNA structures which would form given an infinite volume of DNA and infinite time, presuming that only the hybridizations allowed by R are physically relevant, and ignoring transient structures.

We will be especially interested in the self-assembly rules[6] R_1^T which allow $C_1 +_B C_2 = C_3 \neq \epsilon$ iff (1) the subgraph of C_3 induced by B contains exactly two T-mer (or longer) strands and (2) at most two edges lead to or exit from this subgraph. Thus, R_1^T allows only hybridization of sufficiently long sticky-ends, as illustrated in Figure 2.

Both *ligate* and *denature* can be generalized to set operations by applying the operation to each complex in the original set, and taking the union of all complexes that result. Since single-stranded DNA can be identified with its sequence, written

[4]Circular strands are not necessary in our constructions, but they must be considered in Theorem 2(2).

[5]Logicians may think of A as "axioms" while R may be thought of as "inference rules".

[6]The subscript "1" is used because these rules give rise to essentially one-dimensional complexes.

5' → 3', we can consider *denature* to be a function from sets of complexes to sets of strings over $\{A, C, G, T, \cdot\}$, where $\cdot$ is used to indicate a circular DNA strand.

Finally, we note that to represent strings in alphabets Σ other than $\{A, C, G, T\}$, we may use a *prefix-free codebook* $\mathcal{C}$ which assigns to each symbol σ in Σ a string $\mathcal{C}_\sigma$ over $\{A, C, G, T\}$ such that no string is a prefix of another string. A DNA sequence $s = s_1 s_2 \ldots s_n$ can then be translated into a string $\mathcal{C}(s)$ over Σ by scanning through s from left to right: if s_i begins a subsequence of s which exactly matches some $\mathcal{C}_\sigma$, then s_i is replaced by σ, else s_i is erased; then s_{i+1} is processed, and so forth. For example, if $\Sigma = \{0, 1\}$, $\mathcal{C}_0 = CAG$, and $\mathcal{C}_1 = CTC$, then $\mathcal{C}(AAACTCTCAGTCAG) = 1100$.

In summary, given a finite set of complexes A, self-assembly rules R, and codebook $\mathcal{C}$, we can obtain a language of complexes $\mathcal{L}_{R,A}$ as well as a language of strings $\mathcal{L}_{R,A,\mathcal{C}} = \mathcal{C}(denature(ligate(\mathcal{L}_{R,A})))$.

We now turn to our results. The theorems are stated, explained, and examples are given. Full proofs will appear elsewhere.

2.3. Linear Self-assembly is equivalent to Regular Languages. In this section we address the question of what can be computed by the self-assembly DNA which obeys Properties (1-4), (5a), and (6a) or (6b). This is the familiar case of the self-assembly of long duplex DNA from many small oligonucleotides or sticky-ended fragments. That is, self-assembly begins with oligonucleotides or duplex DNA with sticky ends, and proceeds at a constant temperature, allowing only permanent binary events with a single perfectly complementary hybridization site and no intramolecular hybridization. We make this question precise in our model by asking, what languages of strings $\mathcal{L}$ can be achieved as $\mathcal{L}_{R_1^T,A,\mathcal{C}}$ for some choice of T, $\mathcal{C}$, and A where A contains only linear duplex complexes?

The following[7] can be proved by construction:

THEOREM 1. *(1) For all regular languages $\mathcal{L}$, there exists a positive integer T, a codebook $\mathcal{C}$, and a set of linear duplexes A such that $\mathcal{L} = \mathcal{L}_{R_1^T,A,\mathcal{C}}$. (2) For all positive integers T, codebooks $\mathcal{C}$, and sets of linear duplexes A, $\mathcal{L}_{R_1^T,A,\mathcal{C}}$ is a regular language.*

We will sketch the construction used in the proof of (1) – see Figure 3 for an example. Consider a regular grammar G for $\mathcal{L}$. We design sufficiently dissimilar sequences S_i (we call their Watson-Crick complements S_i') for all the terminal and nonterminal symbols in G. For each rule $A \rightarrow p_1 \ldots p_n B$, we design a duplex with a sticky end S_A', and internal duplex region $S_{p_1} \ldots S_{p_n}$, and a sticky end S_B if B is present. We also design a duplex with one blunt end and a sticky end S_S, to represent the start symbol S. These duplexes make up the initial set of complexes A. T is chosen to be the length of the nonterminal sequences S_i. After self-assembly, the terminal complexes in $\mathcal{L}_{R_1^T,A}$ will correspond to derivations in G. After ligation, each complex will be a blunt-ended duplex whose sequence consists of terminal sequences interspersed with nonterminal sequences. A codebook with $\mathcal{C}_i = S_i$ for each terminal symbol i will "erase" the nonterminal sequences; thus $\mathcal{L}_{R_1^T,A,\mathcal{C}}$ will be exactly $\mathcal{L}$. □

A sketch of the proof of (2) is as follows: we construct a regular grammar G which generates exactly the strands in $denature(ligate(\mathcal{L}_{R_1^T,A}))$. This requires creating a nonterminal symbol for each sticky end of a duplex in A, and considering

[7]We note that this theorem still holds when "duplexes" is replaced by "strands".

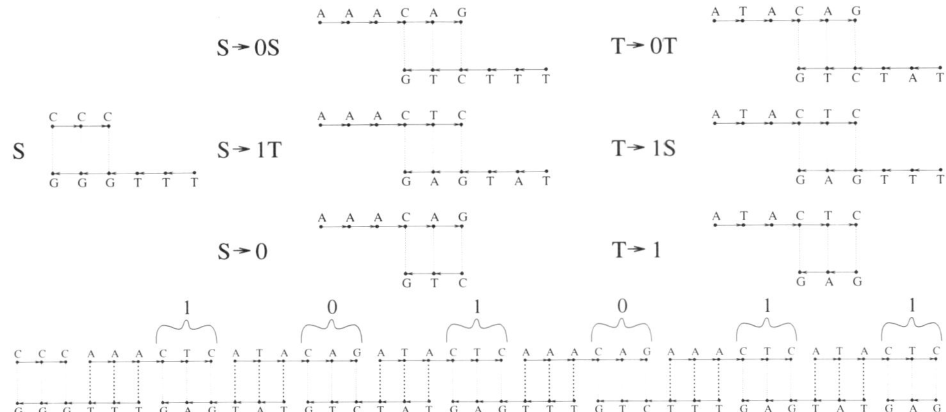

FIGURE 3. The initial complexes A corresponding to the regular grammar G_E, and an example derivation. Note that the self-assembly of the derivation could have occurred in any order. Subsequent ligation and denaturing will produce two strands (top and bottom) from this terminal complex. The codebook $\mathcal{C}$ defines $\mathcal{C}_0 = CAG$ and $\mathcal{C}_1 = CTC$. We use R_1^3.

all (finitely many) T-or-more base overlaps of these sticky-ends; a grammar rule is provided for each such interaction. Care must be taken for gaps and for sticky ends which have no interactions – both lead to termination of the strand sequence, and may require a rule using the start symbol S. Translation by the codebook can be effected by applying a nondeterministic finite state transducer [**7**] to $\mathcal{L}_G$, yielding a regular language equal to $\mathcal{L}_{R_1^T,A,\mathcal{C}}$. □

Thus, our model for linear self-assembly does not permit very interesting computations. It should be emphasized that simple extensions might allow for more complex computations. For example, suppose hairpins appear in A in addition to duplexes. Then, for example, we could replace the duplex for S (Figure 3) by the hairpin, and change the codes for 0 and 1 to the Watson-Crick palindromes $CCGG$ and $CGCG$. Now both the top and bottom strands code the 0 and 1 sequences; furthermore, after ligation the top and bottom strands are joined together by the hairpin. Consequently, we generate the set of all palindromes in which the number of ones is a multiple of four – which is not a regular language! How far can we push this idea?

2.4. Equivalence to Context-free Languages. It has been observed [**1**] that dendrimer self-assembly looks formally identical to context-free grammars. This observation translates very nicely into DNA self-assembly of branched junctions into tree-like complexes. Therefore, in this section we address the question of what can be computed by the self-assembly of DNA which obeys Properties (1-4), (5a), and (6b-d). That is, self-assembly begins with duplexes, hairpins, and 3-armed junctions with sticky ends, and proceeds at a constant temperature, allowing only permanent binary events with a single perfectly complementary hybridization site and no intramolecular hybridization. We note that this form of self-assembly has not been widely studied in the lab, and that full self-assembly would be limited

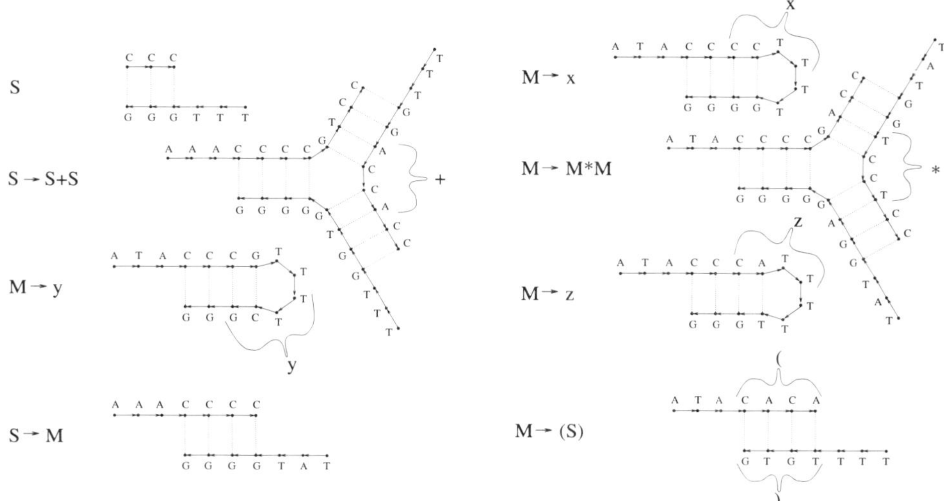

FIGURE 4. The initial complexes A corresponding to the regular grammar G_F. The codebook $\mathcal{C}$ defines $\mathcal{C}_x = CCTT, \mathcal{C}_y = TTCG, \mathcal{C}_z = CATT, \mathcal{C}_(= CACA, \mathcal{C}_) = TGTG, \mathcal{C}_+ = ACCA$, and $\mathcal{C}_* = TCCT$. We use R_1^3.

not only by material but also by geometric (steric) interference and volumetric constraints[8]. Nonetheless, our abstract model allows us to ask the following precise question: what languages of strings $\mathcal{L}$ can be achieved as $\mathcal{L}_{R_1^T,A,\mathcal{C}}$ for some choice of T, $\mathcal{C}$, and A where A contains only duplexes, hairpins, and 3-armed junctions?

An extra complication that immediately arises is the possibility that circular strands may form. Recall our convention that *denature* returns "dotted" sequences to represent circular strands, but didn't specify which permutation of the circle to use. It becomes convenient to work with equivalence classes of sequences, where $\cdot S \cong \cdot T$ if the sequences S and T are circular permutations of one another. Languages $\mathcal{L}_1$ and $\mathcal{L}_2$ are deemed *equivalent* if for every sequence S in one language, there is an identical or equivalent sequence T in the other language.

The following[9] can be proved by construction:

THEOREM 2. *(1) For all context-free languages $\mathcal{L}$, there exists a positive integer T, a codebook $\mathcal{C}$, and a set of duplexes, hairpins, and 3-armed junctions A such that $\mathcal{L} = \mathcal{L}_{R_1^T,A,\mathcal{C}}$. (2) For all positive integers T, codebooks $\mathcal{C}$, and sets of duplexes, hairpins, and 3-armed junctions A, $\mathcal{L}_{R_1^T,A,\mathcal{C}}$ is equivalent to a context-free language.*

We will sketch the construction used in the proof of (1) – see Figures 4 and 5 for an example. The construction is similar to that in Theorem 1. Consider a context-free grammar G for $\mathcal{L}$. Note that there is an equivalent grammar $\hat{G}$ which uses rewriting rules of the form $A \to pBqCr$ where p, q, and r are (possibly null) strings of terminal symbols, and A, B, and C are single nonterminal symbols (or

[8]Consider a tree which branches at every opportunity. It has 2^n nodes within n steps of the center; but the volume of space within n steps grows only as n^3.

[9]We note that this theorem still holds when "duplexes, hairpins, and 3-armed junctions" is replaced by simply "complexes". That is to say, this is a fully general theorem for self-assembly under R_1^T.

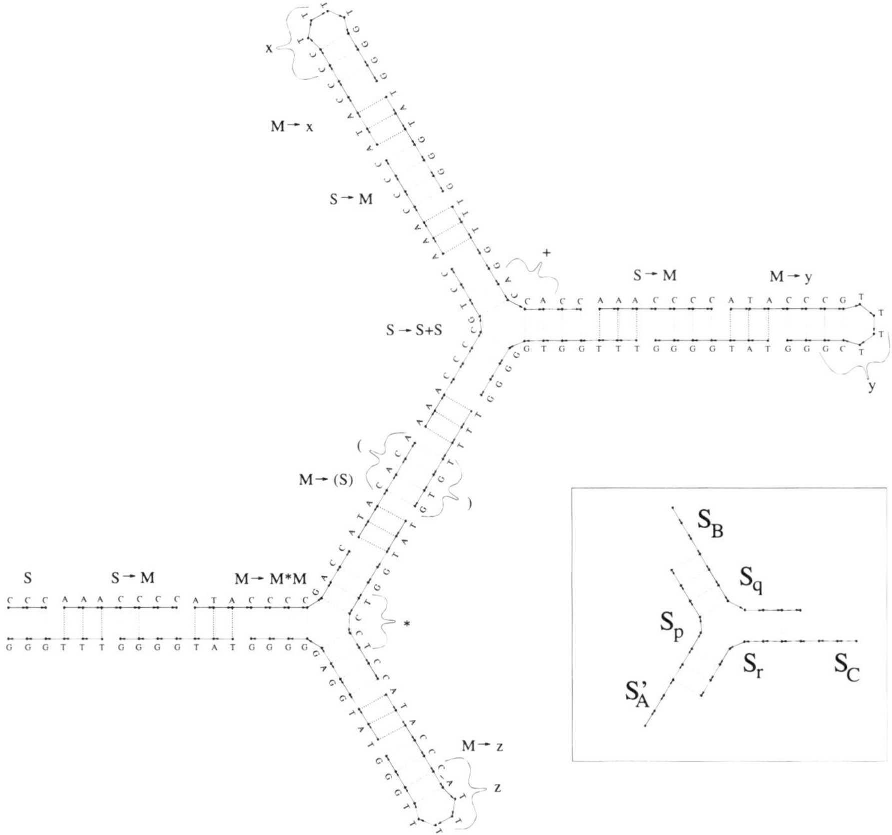

FIGURE 5. An example derivation by self-assembly of the complexes A corresponding to the regular grammar G_F. Note that the self-assembly of the derivation could have occurred in any order, using R_1^3. Subsequent ligation will produce a single strand from this terminal complex. Inset: the three-armed junction corresponding to the generic rewriting rule $A \to pBqCr$.

null). Again, we design sufficiently dissimilar sequences S_i for all the terminal and nonterminal symbols used in $\hat{G}$. For rules of the form $A \to pB$ or $A \to Bp$ (B not null), we design a duplex as before. For rules of the form $A \to p$, we design a hairpin with the sequences for p in the stem. We design a 3-armed junction for each rule of the form $A \to pBqCr$ (B and C not null); it has sticky ends for S'_A, S_B, and S_C, and the sequences for p, q, and r are placed on the arms. As before, we design a blunt-ended duplex for the start symbol S. These complexes make up the initial set of complexes A. As before, at the appropriate "temperature" T, the terminal complexes will correspond to derivations in $\hat{G}$, and ligation will convert each complex into a single strand which encodes the derivation. Processing with the codebook for the terminal symbols will "erase" the nonterminal sequences, and $\mathcal{L}_{R_1^T,A,\mathcal{C}}$ will be exactly $\mathcal{L}$. □

The proof of (2) also follows the form of the proof of Theorem 1, only now we construct a context-free grammar G which, loosely speaking, generates sequences corresponding to backbone paths through complexes in $ligate(\mathcal{L}_{R_1^T,A})$, where gaps are filled in with the symbol $\diamond$, and where several (but not necessarily all) permutations of each circular strand are given using $\cdot$. This language is then passed through a nondeterministic transducer which returns the strand sequences in $\{A, C, G, T\}$ and circular strand sequences in $\{A, C, G, T, \cdot\}$. As before, the final strings are produced by another nondeterministic transducer, which this time translates using the codebook. Thus the final language is context-free, and is equivalent to $\mathcal{L}_{R_1^T, A, C}$. □

More intuitively, we can reason that because no intramolecular hybridizations are allowed by R_1^T, the initial complexes can aggregate only into tree-like structures. No matter how convoluted the original complexes are, paths through the resulting tree-like structures are well modeled by context-free languages.

Our model of self-assembly of DNA into tree-like structures has strictly more computational power than the model of linear self-assembly. However, it is still a far cry from universal computation. It turns out that when we attempt to model intramolecular interactions, in the form of cooperative binding sites, a much more powerful model results. We consider a particular case in the following section.

2.5. Two Dimensional Self-assembly is Universal. To prove that two dimensional self-assembly can be universal, it suffices to demonstrate a restricted class which is universal. We review the class of structures introduced in [**12**], which are geometrically based on a lattice of double crossover (DX) units of DAO type [**6**]. It was shown in [**12**] that the self-assembly of DX units can directly mimic the operation of an arbitrary one dimensional cellular automata system. An example is shown in Figure 2.5, where a simple blocked cellular automaton rule (corresponding to the unrestricted grammar G_P of Section 2.1, but without the termination rules) is used to generate a Sierpinski triangle pattern.

The model of self-assembly used here follows Properties (1-4), (5b), and (6e), and it is motivated by additional physical concerns. As shown in Figure 2.5, the hybridization events may now involve *two* binding sites arranged as a *slot*. Geometry becomes important; only sticky ends which are close to each other and arranged properly may form a slot where binding can occur. Physically, one sticky end of an unattached DX unit would hybridize to one side of the slot, followed shortly by (the now intramolecular) hybridization of the DX unit's other sticky end to the slot's other binding site. For full computational generality, it is critical that a DX unit which matches one site in a slot, but not the other site, will *not* hybridize to the lattice. Under appropriate conditions, DX units which bind to only one site in a slot would soon dissociate, while fully matching DX units would bind nearly irreversibly. We therefore model slot-filling as a single permanent binary event involving two binding regions, and T is chosen so that single-site binding will not occur.

We emphasize that this form of DNA self-assembly has not yet been demonstrated experimentally, although we report some preliminary results in Section 3.

We must define new self-assembly rules: R_2^T allows hybridizations allowed by R_1^T, and additionally allows two-region slot-filling hybridizations between complexes containing the subgraphs shown in Figure 7, so long as the total number of basepair edges in B is at least T. This rule is meant to model local geometry in complexes; it will be a good model only for certain structures, including (we believe) the ones used in our construction.

The following can be proved by construction:

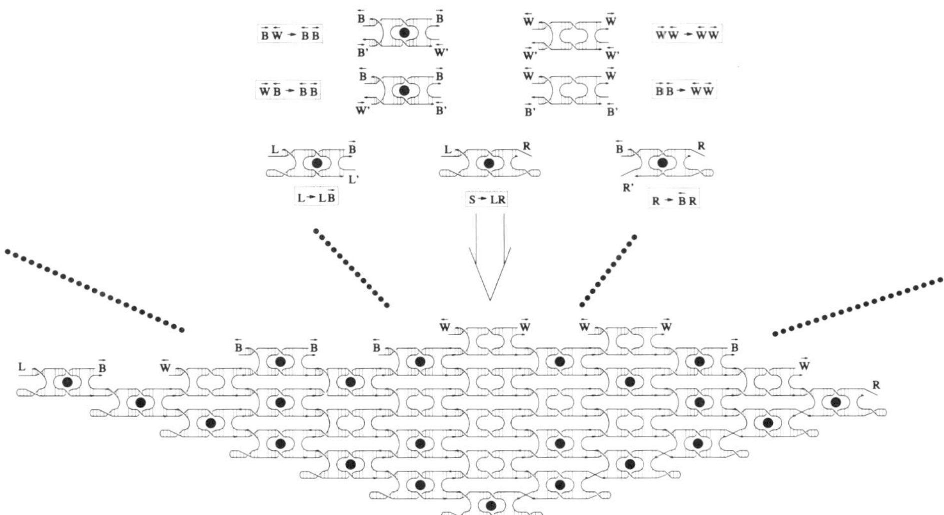

FIGURE 6. An algorithmic pattern in a self-assembled lattice. At the top, the seven initial DX units in A are shown (the black dot is a visual aid to identify "black" complexes), involving 22 oligonucleotides. The corresponding rewriting rules from G_P are presented in boxes. The units use 12 unique sticky end sequences, denoted by $\{L, R, \overleftarrow{B}, \overrightarrow{B}, \overleftarrow{W}, \overrightarrow{W}\}$ and their complements L' etc. The L and R sequences are both length T; the other sequences are length $T/2$. Upon self-assembly according to R_2^T, a V-shaped chain of the lower three units is formed due to hybridization of L and R, while the open slots in the initial chain are filled by the unique unit whose sticky ends match those on both sides of the slot. In this example, the process continues indefinitely. Each strand in $ligate(\hat{\mathcal{L}}_{R_2^T,A})$ represents one or two columns of Pascal's triangle mod 2.

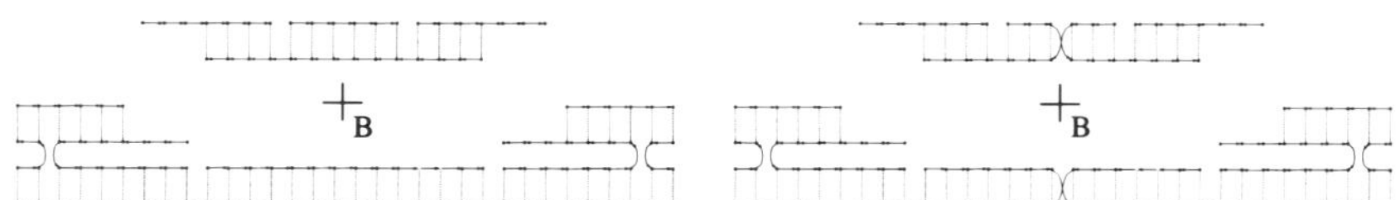

FIGURE 7. Two allowed slot-filling hybridizations in R_2^6. These graphs represent requires subgraphs of the complexes C_1 and C_2 in $C_1 +_B C_2 = C_3$. Other positions of nicks are also allowed, as are other lengths of the duplex regions.

THEOREM 3. *(1) For all recursively enumerable languages $\mathcal{L}$, there exists a positive integer T, a codebook $\mathcal{C}$, and a set of duplexes and DX units A such that $\mathcal{L} = \mathcal{L}_{R_2^T,A,\mathcal{C}}$. (2) For all positive integers T, codebooks $\mathcal{C}$, and sets of duplexes and DX units A, $\mathcal{L}_{R_2^T,A,\mathcal{C}}$ is equivalent to a recursively enumerable language.*

The proof of (1) is based on the constructions in [**12**]. As cellular automata are capable of universal computation, for example by directly simulating Turing machines, we conclude that two dimensional self-assembly is universal. (2) follows because there is an algorithm for generating all the complexes in $\mathcal{L}_{R,A}$ so long as R is computable: keep trying new hybridizations of complexes known to be in the language, and remember the resulting complex. □

Although universal, one dimensional cellular automata are not often a convenient model for computing functions of interest, although they are faster and more efficient than 1-tape Turing Machines, due to their parallelism.

2.6. Solving the Hamiltonian Path Problem. As a concrete example of using two dimensional self-assembly for computation, we will solve the same Hamiltonian Path Problem (HPP) used in [**2**]. Recall that the problem is to find a path from node 1 to node N which visits every node in G exactly once. Our algorithms for solving HPP will be based on:

1. Generate all paths from node 1 to node N.
2. In each path, sort the vertices into increasing order.
3. For each path, check that the result is exactly "$1, 2, 3, \ldots, N$".
4. Output any path which passes the test, if one exists.

In a preparatory step, DNA sequences are designed for the given graph and synthesized. Steps 1-3 will occur as a single self-assembly step, while Step 4 consists of sequencing circular DNA of known length.

For the graph used in [**2**] (shown in Figure 8a), $N = 7$ and we will require a total of 68 DX units of DAE type. Shown in Figure 8b, units 1 through 20 are responsible for Step 1 of the algorithm (the bottom layer in Figure 9a,b); these units are analogous to the oligos in Adleman's solution[10]. Units 19 through 61 are responsible for Step 2 of the algorithm; sorting is accomplished by the Odd-Even Transposition Sort [**8**]. When the symbol ∞ has travelled all the way to the right, the sorting is complete and Step 3 is initiated, using units 62 through 68.

Each terminal complex either (a) encodes a valid Hamiltonian Path, in which case the complex is complete (Figure 9a), and ligation cyclizes the outer ring, but not the inner ring[11]; or (b) encodes an invalid path, in which case the terminal complex contains unfilled, open slots (Figure 9b) and will produce no cyclic strands when ligated[12]. Thus Step 4 can be achieved by separating cyclic from linear DNA strands (e.g. by 2D gel electrophoresis, by exonuclease digestion, or by affinity purification based on the $DONE$ sequence) followed by amplification and sequencing.

Let us briefly compare this molecular algorithm to the one used in [**2**]. To solve a graph with N nodes and E edges, Adleman used roughly $N + E$ oligos and N laboratory steps. We would use roughly $E^2/N + N^2 + N$ DX units (each requiring up to 5 strands) and a constant number of laboratory steps (synthesis, annealing, sequencing)[13].

[10] Adleman's oligos encoded individual edges in the graph, whereas ours encode pairs of edges. Also, knowing that a Hamiltonian path in this graph must visit exactly 7 nodes, our units are devised such that only odd-length paths can form completely.

[11] This can be ensured either by leaving an unmatched base on the sticky ends for interior units, or by phosphorylating only units which occur on the outer edge.

[12] Note that if a path visits a node twice, there will be a gap in the "Step 2" portion of the terminal complex; if a path fails to visit some node, there will be a gap in the "Step 3" portion of the terminal complex.

[13] How feasible these imagined laboratory steps would be is, of course, an open question. However, once the laboratory techniques have been debugged, conceivably our algorithm could be carried out in a single day's work – regardless of the size of the graph (volume permitting). A concern is that, moreso than in Adleman's algorithm, the success of our algorithm is critically dependent upon ligation yields. For example, if ligation is 80% effective, then only $0.8^{30} = 0.1\%$ of the correct terminal complexes will be fully cyclized in our $N = 7$ graph. Also, since each path

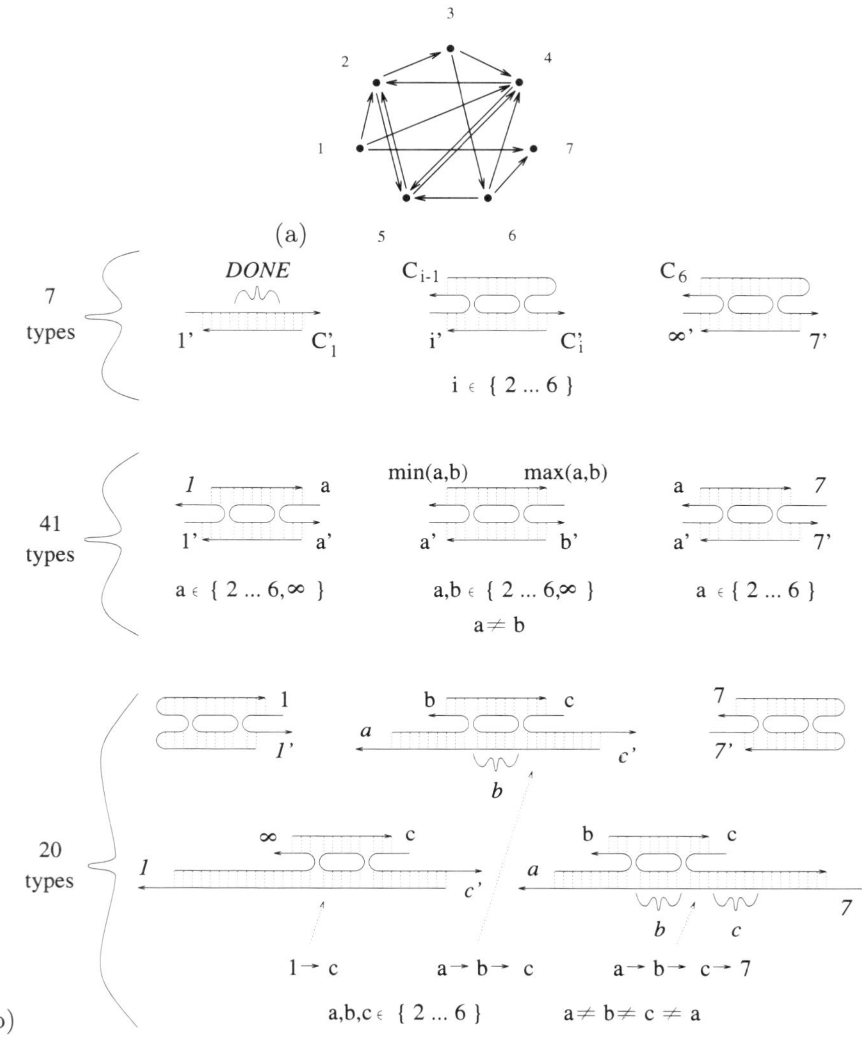

FIGURE 8. (a) The 7 node graph G. (b) Rule molecules (DAE type) for solving the Hamiltonian Path problem. Sticky ends of length $T/2$ are $\{1, 2, 3, 4, 5, 6, 7, \infty, C_1, C_2, C_3, C_4, C_5, C_6\}$ and their complements; sticky ends of length T are $\{\mathit{1}, \mathit{2}, \mathit{3}, \mathit{4}, \mathit{5}, \mathit{6}, \mathit{7}\}$ and their complements. $DONE$ is a special sequence which indicates completion of a lattice. The variables i, a, b, c are used to concisely represent a multiplicity of rule molecules; italicized variables indicate length T sticky ends. For example, in the lower set, 15 units are designed from the central schema, one for each pair of incident paths $a \to b, b \to c$.

Because two dimensional self-assembly can simulate arbitrary cellular automata, similar algorithms can be designed for any computational purpose. For example, an N-variable size s Circuit-SAT problem can be solved using roughly Ns DX units and a constant number of laboratory steps after synthesis.

---

requires a DNA molecule roughly 100 times larger than the DNA used in Adleman's algorithm, greater reaction volumes will be necessary.

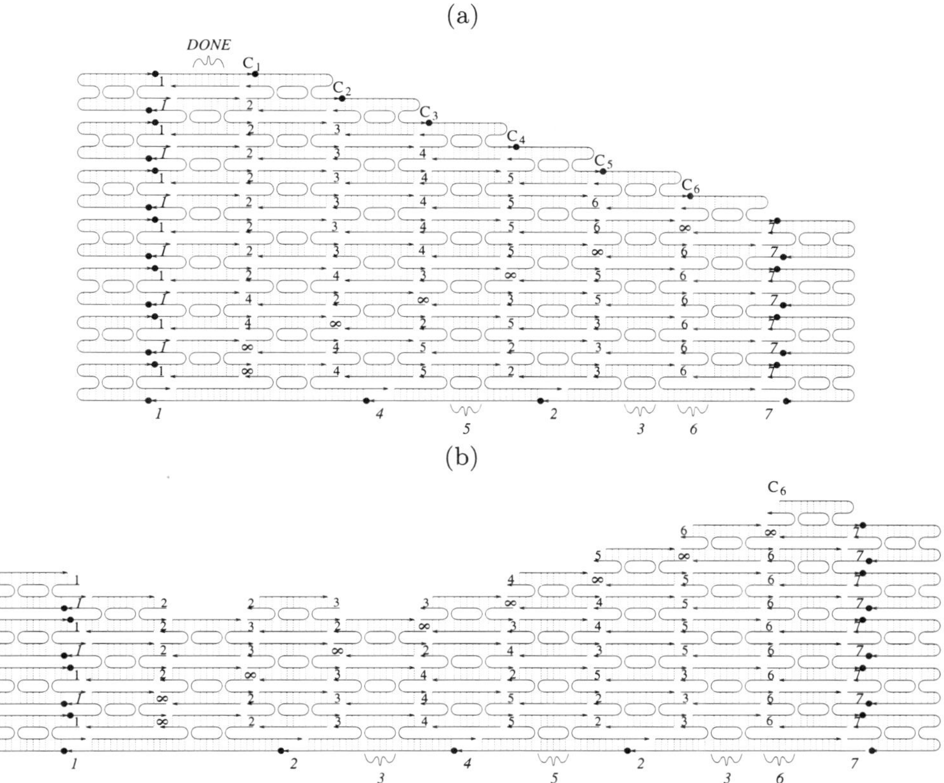

FIGURE 9. Terminal complexes after annealling. Black dots show nicks which will be ligated. (a) The lattice verifying the Hamiltonian path 1452367. (b) The lattice rejecting the invalid path 123452367.

2.7. Augmentation of Computational Power. A trivial corollary of the universality of two-dimensional self-assembly is that if three dimensional structures are allowed, self-assembly is still universal. It is of greater interest if we can exploit all three dimensions to allow for more efficient or more reliable computations. We propose a scheme to do exactly that, again for concreteness using DAO units as our basic building block. In this section, we will present some physical considerations, but we will not formally define R_3^T.

We begin by noting that the solid angle between two adjacent DAO units is determined by the length of the linker arm between them. For the planar lattice, we choose a length such that the angle is approximately 180°. Alternatively, we can choose lengths such that the angle is near 120°, the appropriate value for a "honeycomb lattice" as shown in Figure 10.

As in the case of the two dimensional lattice of DAO units, computation is brought about by judicious choice of the sticky end sequences on several DAO units. The three dimensional lattice thus formed is equivalent to the space-time history of a 2D blocked cellular automata.

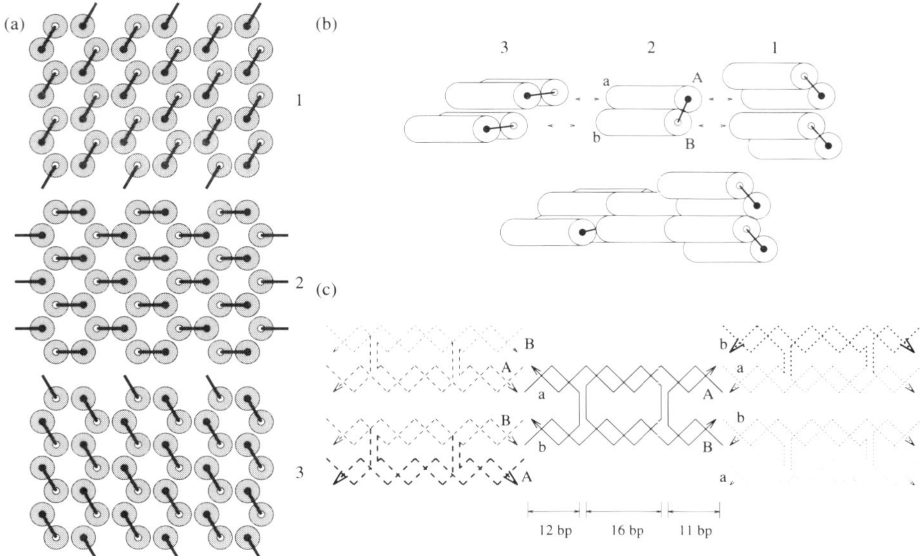

FIGURE 10. Plan for three dimensional lattice. (a) Three cross-sections through the final lattice, corresponding to the three sections indicated in (b). Circles represent cross-sections through a double helix of DNA; bars indicate which helices are part of the same DAO unit. (b) Relative angles of five DAO units are indicated. For perfect 120° angles, helical twist between 31.5 and 35.5 °/ bp is required. (c) Detail of the DAO units. A single DAO unit, with sticky end a complementary to A and b complementary to B, suffices to generate the entire lattice. For computations, the sticky ends are indexed s.t. a_i binds only A_i and b_i binds B_i.

The particular incarnation of three dimensional lattice chosen here is clearly not unique, and it is suggested more as a brain-teaser than as a serious proposal; other geometries are possible, perhaps having preferable practical characteristics.

2.8. Discussion. We have analyzed the computational power of three different regimes of self-assembly in our abstract model, and we have speculated on an extension into the self-assembly of a three dimensional lattice.

The essential construction in the linear case is due to [**2**] who used it to construct paths through graphs. [**4**] and [**12**] observed that linear self-assembly is capable of generating regular languages. Here, we state the result in the context of our formal model, and we show that linear self-assembly of duplexes is *limited* to regular languages. This point requires making the distinction between self-assembly processes with and without hairpins, as shown by the palindromes example. Linear self-assembly has been exploited in many laboratory experiments – both by molecular biologists and by people interested in molecular computation – and although its intricacies are not completely understood, there is a wide foundation of practical experience.

The self-assembly of branched junctions into dendrimer structures seems to be a relatively unexplored idea. For example, in [**10**] it is observed that identical three-armed junctions with two complementary sticky ends can cyclize. If cyclization cannot be prevented, many context-free grammars would be impossible to implement by self-assembly. Another concern comes from geometry: if the desired tree-like structure contains too many branches, steric hindrance may prevent further associations from occurring. Thus it is not known to what extent the technique will be practical.

The self-assembly of DX units into a two-dimensional lattice is also an unconventional idea, yet to be demonstrated in the laboratory. Some first steps in this direction are reported in the next section of the paper, where the slot-filling reaction is explored.

It is interesting to observe that the Chomsky Hierarchy of languages, developed originally for the study of human languages, also arises naturally in the study of self-assembling structures. The progression from regular to context-free to recursively enumerable languages can be seen to parallel both (a) the progression from linear to dendrimer to planar lattice structures, and (b) the progression from "rule molecules" with effectively one input and one output, to those with one input and two outputs, to those with two inputs and two outputs.

One should note that all the previous arguments ignored the kinetic framework implicit in the process of annealing that we originally consider. Specifically, we expect that longer complementary regions hybridize first. Annealing could be represented in our model as recursive computation of languages:

$$\mathcal{L}^{anneal}_{R_1,A,\mathcal{C}} = \mathcal{C}(denature(ligate(\hat{\mathcal{L}}_{R_1^1,\hat{\mathcal{L}}_{R_1^2,\hat{\mathcal{L}}_{R^3,\dots}}})))$$

The kinetic aspects of this model of linear self-assembly may themselves be expoitable for computation. Intermolecular interactions other than the ones considered here might also provide computational advantages. Issues of concentrations and finite supply of DNA must also go into any more practical analysis.

3. Experimental Investigations

The above considerations only point to the possibility of algorithmically-patterned lattices of DNA. There are two major issues to be investigated experimentally. The first is whether homogeneous lattices will form; i.e., whether the geometric structure itself will self-assemble. The second issue is whether, in the presence of multiple units in solution, the logically correct unit will hybridize in each slot. This competitive process for filling each slot is essential for computation, as a single error can propagate throughout the entire computation. Ultimately, error rates will determine the size of lattice in which reliable computation may be performed.

We have begun an investigation of the second issue. We first build a model molecular complex, called `ABC`, which contains a single slot and no other sticky ends. `ABC` is composed of two double crossover molecules, `A` and `C`, and a duplex linker `B`. `ABC` is created by ligating eight oligonucleotides; the final structure contains four hybridized strands. Rather than test the assembly of a double crossover unit into `ABC`'s slot, we model the unit by a linear duplex "linker", called `D`. When `ABC` is properly hybridized to `D`, we call the complex `ABCD`. Completely ligated, `ABCD` is a complex catenane with four interlocked circles. To test the specificity of the hybridization, we also have a mismatched linker `D'`, which is perfectly complementary to only one of the sticky ends in the slot. We expect that `ABCD'` cannot be completely ligated, due to the mismatch, and hence `ABCD'` does not form a catenane. These molecular complexes are diagramed in Figure 11.

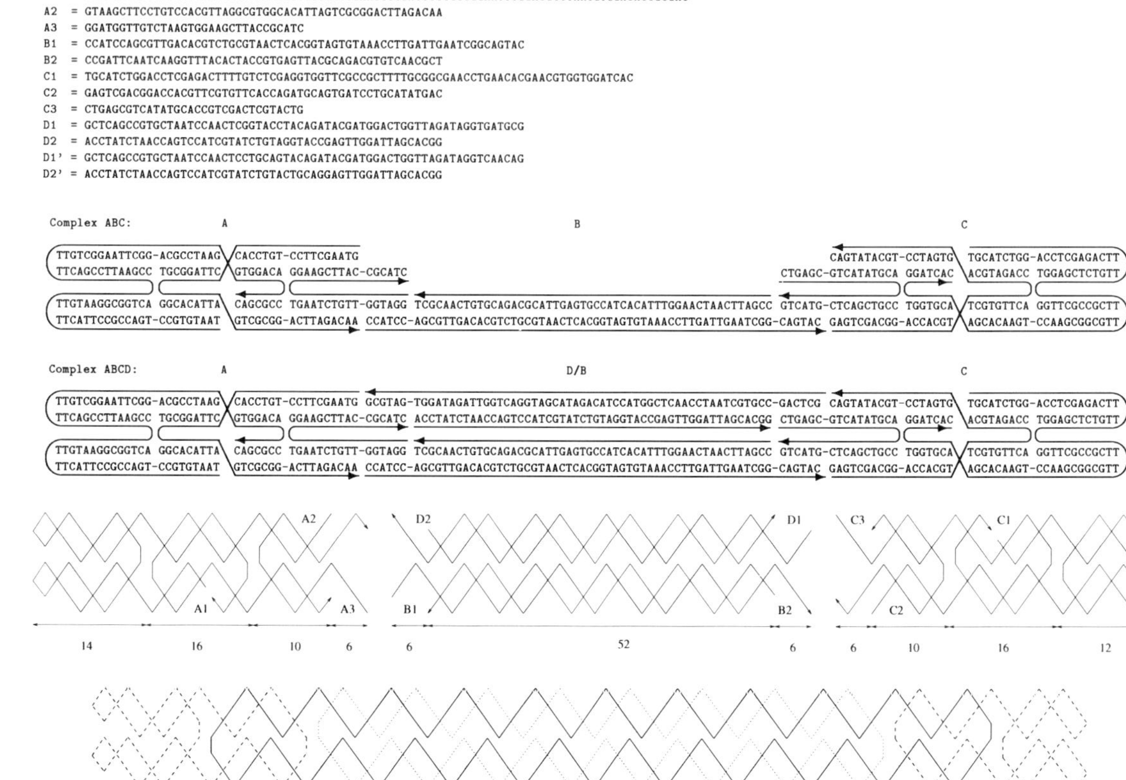

FIGURE 11. (Sequences and Structures for **ABCD**) Sequences are written 5' to 3'. 3' ends are denoted by arrows in the diagrams, and strand labels are near 5' ends. Lengths are measured in nucleotides. Above, sequence details are given along with schematic representations of **ABC** and **ABCD** (not ligated). Below, more geometric detail is sketched for **A**, **B**, **C**, **D**, and **ABCD** (ligated). Diagrams illustrate intended structures only.

Experimentally, we must establish that the double crossover molecules A and C form properly upon annealing their component strands. As developed in [**6**], where the details of hybridization were probed by more extensive structural characterization, a good indication of proper association is a single band of mobility in a non-denaturing gel appropriate for the topology and molecular weight. The ligation products ABC, ABCD, and ABCD' (by which we mean whatever it is we get when we intend to make structures ABC, ABCD, and ABCD' respectively) are examined both in non-denaturing and denaturing gels; in the former we are looking for a single band of approximately the correct apparent molecular weight, while in the latter we are looking for linear strands of the lengths predicted for ligation. Ligation of double crossover molecules has previously been shown to be well-behaved [**9**]. Topologically closed structures, such as ABCD, can be assayed by treating with an exonuclease [**10**]. Although none of these tests is absolutely rigorous, together they may give us confidence that the reactions are proceeding as predicted.

3.1. Materials and Methods. *Sequence Design.* The twelve strands required for A, B, C, D, and D' were designed by applying the principles of sequence symmetry minimization [**11**], where the design process ensures that there are no complementary regions between strands, except as desired. In short, each double crossover molecule is designed by creating sequences appropriate for two asymmetric Holliday junctions, then juxtaposing these sequences as appropriate for a four-stranded DAO, adding hairpin sequences and re-phasing A1 and C1 to put the nick in the central region of the DAO. A1's hairpin regions are longer than C1's to allow A1 and C1 to be distinguished on gels. The lengths of the linkers B and D were chosen such that both DAO units should be nearly coplanar according to an estimated 10.5 base pairs per double helical full turn. Exact sequences are given in Figure 11.

Synthesis and Purification of DNA. All strands were synthesized on an Applied Biosystems 380B automatic DNA synthesizer using routine phosphoramidite procedures [**5**]. DNA strands were purified by denaturing polyacrylamide gel electrophoresis. DNA concentrations were estimated by OD_{260}. All strands were phosphorylated by T4 Polynucleotide Kinase (U.S. Biochemical or Promega), followed by phenol extraction and ethanol precipitation. DNA was not radiolabeled.

Formation of Hydrogen-Bonded Complexes. Complexes A, B, C, and D were formed by mixing stoichiometric quantities of each strand at concentrations near $1\mu M$ in 1x USB T4 DNA Ligase buffer (U.S. Biochemical: 66 mM Tris·HCl (pH 7.6), 6.6 mM $MgCl_2$, 10 mM DTT, 66 μM ATP; or Promega: 30 mM Tris·HCl (pH 7.8), 10 mM $MgCl_2$, 10 mM DTT, 500 μM ATP). These solutions were annealed for two hours from $80°C$ down to room temperature.

Formation of Covalently Bonded Complexes. Complexes AB, BC, ABC, ABCD, and ABCD' were formed by mixing stoichiometric quantities of annealed A, B, C, and D', followed by D after 20 minutes. Up to 50 units of T4 DNA Ligase (U.S. Biochemical or Promega) were added and solutions were incubated in a $16°C$ water bath for 2 or 8 hours. One sample of ABCD was further treated by adding $\frac{1}{10}^{th}$ volume 10x USB Exonuclease III buffer (U.S. Biochemical) and 100 units Exonuclease III (U.S. Biochemical), incubated at $37°C$ for 1 hour. Prior to being loaded in gels, solutions for gels (a) and (b) were heated to $80°C$ and again annealed to room temperature, to denature proteins and re-form hydrogen-bonded complexes. For gels (c), ligation was followed by phenol extraction and ethanol precipitation, then samples were heated to $90°C$ for 5 minutes prior to being loaded.

Denaturing Polyacrylamide Gels. Denaturing gels contain 8.3 M urea and 8% acrylamide (19:1 acrylamide:bisacrylamide). The running buffer is TBE (89 mM Tris·HCl (pH 8.0), 89 mM boric acid, 2 mM EDTA). The sample buffer contains 0.1% bromphenol blue and xylene cyanol FF tracking dyes in 80% formamide with 10 mM EDTA. Samples are heated at $80°C$ for 5 minutes immediately prior to loading. Gels are run at approximately 60 V/cm and 35 Watts, then soaked in StainsAll dye and digitized by DeskScan II on an Apple Macintosh.

Non-denaturing Polyacrylamide Gels. Non-denaturing gels contain 12.5 mM Mg^{++} and 8% acrylamide (19:1 acrylamide:bisacrylamide), 0.75 mm thick. The running buffer is TAE/Mg^{++} (40 mM Tris·HCl (pH 8.0), 20 mM acetic acid, 2 mM EDTA, 12.5 mM magnesium acetate). The loading buffer contains 0.02% bromphenol blue and xylene cyanol FF tracking dyes and 5% glyceroi in ligation buffer. Gels are run at approximately 16 V/cm and 10 Watts at $4°C$, then soaked in StainsAll dye and digitized by DeskScan II on an Apple Macintosh.

3.2. Results. *Formation of Complexes.*

Figures 12a and 12b show several stages in the formation of ABCD. On the non-denaturing gel, the duplexes and double crossover molecules A, B, C, and D form clean bands (a, lanes 1-4) which migrate with approximately the same mobility as equivalent molecular weight duplex DNA. Ligation products AB and BC also show clean bands (a, lanes 9-10). Ligation product ABC appears as the major band in its lane (a, lane 8); another band appears at the level of AB and BC indicating incomplete ligation. Ligation product ABCD also appears, we believe, as the major band in its lane (a, lanes 7 and 5); a slower unidentified band also appears. After exonuclease treatment, the major band of ligation product ABCD is still apparent, though diminished (a, lane 6).

On the denaturing gel, we obtain further evidence of ligation activity by observing the lengths of newly created oligonucleotides. Lanes 1-4 can be used as markers for the lengths of most of the original oligonucleotides: A1 (88), A2 (52), B1 (64), B2 (52), C1 (80), C2 (52), D1 (64), D2 (52). A3 and C3, both 32 nucleotides, ran off the gel. Lane 10, product AB, shows the expected formation of A2B1 (116) and B2A3 (84); lane 9, product BC, likewise shows the formation of B1C2 (116) and C3B2 (84); and lane 8, product ABC, shows the expected formation of A2B1C2 (168) and C3B2A3 (116). Lanes 5 and 7, product ABCD, contain only three significant bands: A1 (88), C1 (80), and a band which migrates slower than a 2000 nucleotide strand, according to the marker (lane 11). This slow band is exonuclease-resistant (lane 6). We therefore conclude that the band contains the catenane A2B1C2D1:A3D2C3B2; i.e., ABCD minus the A1 and C1 loops, which apparently were not ligated. Double crossover molecules with two nicks have been shown to be stable [**13**], suggesting that the nicks in A1 and C1 should not significantly affect the formation or stability of A or C.

Specificity of Reaction.

Figure 12c shows the results of a preliminary experiment investigating the effectiveness of D vs D' in filling the slot created by ABC. Lane 13 contains ABC, and thus has primary bands for A2B1C2 (168) and C3B2A3 (116). Lanes 2 and 12 show ligation of ABCD in 1:1:1:1 stoichiometry. The ligation apparently was not as complete as in 12b, as several bands of "partial products" are observed. The fastest

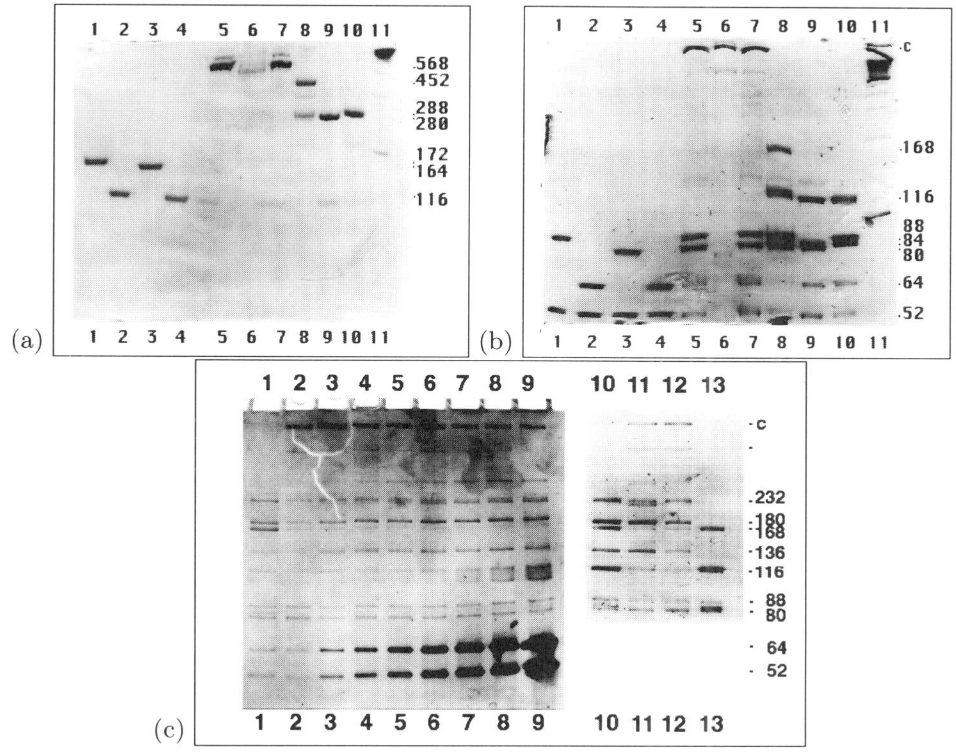

FIGURE 12. . Numbers at right indicate estimated positions for expected products, consistent with the marker lane. (a) Non-denaturing gel electrophoresis, and (b) Denaturing gel electrophoresis. Lane 1: A (172 nucleotides). Lane 2: B (116 nucleotides). Lane 3: C (164 nucleotides). Lane 4: D (116 nucleotides). Lane 5: Product ABCD with ligase. ABCD contains 568 nucleotides. Lane 6: Product ABCD with ligase and exonuclease. Lane 7: product ABCD with ligase. Lane 8: Product ABC with ligase. ABC contains 452 nucleotides. Lane 9: Product BC with ligase. BC contains 280 nucleotides. Lane 10: Product AB with ligase. AB contains 288 nucleotides. Lane 11: 100 base-pair double-stranded ladder. (c) Denaturing gel electrophoresis. All lanes contain A:B:C in 1:1:1 stoichiometry, plus various concentrations of D:D'. Lane 1–9: D:D' = 0:1, 1:0, 1:1, 1:2, 1:4, 1:8, 1:16, 1:32, 1:64. Lanes 10–13: D:D' = 0:20, 1:1, 1:0, 0:0. (Note: the first gel was slightly ripped during staining.)

band is appropriate for linear A3D2C3B2 (168) and cyclical permutations; the next band is appropriate for linear B1C2D1 or D1A2B1 (180); the next major band is appropriate for A2B1C2D1 (232) and cyclical permutations. The band below **c** is known from other gels (not shown) to be exonuclease-resistant, and the two cyclic molecule bands are thought to be an indicator of the formation of ABCD. Lanes 1 and 10 show ligation of ABC with respectively an equimolar amount or a 20-fold excess of D'. We again see the linear bands of lengths 168, 180, and 232, while bands at 136 (D2C3B2) and 116 (C3B2A3 and A3D2C3) become significant. Critically, the slow circular products are missing, suggesting that D' was only ligated on the side where it matches the slot's sticky end. Lanes 11 and 3–9 show ligation of one unit of ABC with an x-fold excess of D' and equimolar D, where x ranges from 1 to 64. In every case, the closed molecule **c** is formed, indicating that ABCD is still formed in the presence of competing D'. Additional bands also appear, possibly due to unexpected interactions involving D1 or D2.

3.3. Discussion. We interpret these results as follows. First, we believe that we are making the ABCD complex, with the caveat that we believe the nicks in A1 and C1 are not being sealed. This suggests that (1) in ABC, the linker B is properly spaced such that double crossover molecules A and C are roughly coplanar, and (2) that in each double crossover molecule, the two helical axes are also roughly coplanar. Second, we observe that D, which matches the sticky ends on both sides of ABC's slot, outcompetes D', which matches only on one side, even when D' is 64-fold more abundant than D. We plan to quantitate the thermodynamics of this reaction in the near future.

The experiments reported here bear on the two-dimensional self-assembly process postulated in Section 2.2. These experiments are meant to model a single slot-filling step during the self-assembly of a two dimensional lattice. In our experiments, this fundamental step can occur, and with some specificity. These results encourage us to examine this step in closer detail in the future, as well as to attempt the self-assembly of an entire sheet. We hope that the self-assembly of an algorithmically patterned sheet of DNA can eventually be verified by TEM or AFM microscopy.

4. Conclusions

The self-assembly of molecules can correspond to several well known computational classes up to and including universal computation. This suggests that external processing is not an intrinsic element of molecular computation; computationally universal "one-pot" reactions seem plausible. We have shown some encouraging but preliminary experimental investigations into the fundamental computational step in our two dimensional self-assembly model. The generality of the approach used here suggests that the potential for universal computation may be widespread among self-assembly processes in nature. In addition to being interesting in its own right as a universal mechanism, it may be worth considering whether the self-assembly processes described here could be useful technologically, perhaps as part of an approach to nanotechnology [**9**].

5. Acknowledgments

Erik Winfree is grateful to the many people who have helped make his foray into the world of molecules possible, enjoyable, and exciting; special thanks go to Len Adleman, Paul Rothemund, Sam Roweis, Dan Abrahams-Gessel, John Hopfield, and John Abelson who generously provided laboratory facilities at Caltech for some of the experiments reported here.

References

[1] Dan Abrahams-Gessel (Dartmouth), personal communication, 1996.

[2] Leonard Adleman. Molecular computation of solutions to combinatorial problems. *Science* 266: 1021–1024, Nov. 11, 1994.

[3] Donald Beaver. A Universal Molecular Computer. In R. J. Lipton and E. B. Baum, editors, *DNA Computers: Proceeding of a DIMACS Workshop,* DIMACS Series in Discrete Mathematics and Theoretical Computer Science, American Mathematical Sociey, 29-36, 1996.

[4] Dan Boneh, Chris Dunworth, Richard Lipton, and Jiri Sgall. On the Computational Power of DNA. Princeton CS Tech-Report CS-TR-499-95. `ftp://ftp.cs.princeton.edu/reports/1995/499.ps.Z`

[5] Marvin H. Caruthers. Gene Synthesis Machines. *Science* 230: 281–285, 1985.

[6] Tsu-Ju Fu, and Nadrian C. Seeman. DNA Double-Crossover Molecules. *Biochemistry* 32: 3211–3220, 1993.

[7] Seymour Ginsburg. *The Mathematical Theory of Context Free Languages*, McGraw-Hill, 1966.

[8] Donald E. Knuth. *The Art of Computer Programming, Volume 3: Sorting and Searching (2nd ed)*, Addison-Wesley, 1973.

[9] Xiaojun Li, Xiaoping Yang, Jing Qi, and Nadrian C. Seeman. Antiparallel DNA Double Crossover Molecules as Components for Nanoconstruction. *J. Am. Chem. Soc.* 118: 6131–6140, 1996.

[10] Rong-Ine Ma, Neville R. Kallenbach, Richard D. Sheardy, Mary L. Petrillo and Nadrian C. Seeman. Three-Arm Nucleic Acid Junctions Are Flexible. *Nucleic Acids Research* 14: 9745–9753, 1986.

[11] Nadrian C. Seeman. Denovo Design of Sequences for Nucleic-acid Structural-Engineering. *Journal of Biomolecular Structure & Dynamics* 8 (3): 573–581, 1990.

[12] Erik Winfree. On the Computational Power of DNA Annealing and Ligation. In R. Lipton and E. Baum, editors, *DNA Based Computers, Volume 27, Proceedings of a DIMACS Workshop, April 4, 1995*, Am. Math. Soc., 1995, 199-215. Also available as `http://dope.caltech.edu-/winfree/Papers/ligation.ps.gz`

[13] Siwei Zhang, and Nadrian C. Seeman. Symmetric Holliday Junction Crossover Isomers. *Journal of Molecular Biology* 238: 658–668, 1994.

(E. Winfree) Computation and Neural Systems Option, California Institute of Technology, Pasadena, CA 91125

E-mail address: `winfree@hope.caltech.edu`

(X. Yang and N. Seeman) Department of Chemistry, New York University, New York, NY 10003

DIMACS Series in Discrete Mathematics
and Theoretical Computer Science
Volume **44**, 1999

The Perils of Polynucleotides: The Experimental Gap Between The Design and Assembly of Unusual DNA Structures

Nadrian C. Seeman, Hui Wang, Bing Liu, Jing Qi, Xiaojun Li, Xiaoping Yang, Furong Liu, Weiqiong Sun, Zhiyong Shen, Ruojie Sha, Chengde Mao, Yinli Wang, Siwei Zhang, Tsu-Ju Fu, Shouming Du, John E. Mueller, Yuwen Zhang and Junghuei Chen

ABSTRACT. DNA computing relies on the successful implementation of physical chemistry techniques involving oligonucleotides of prescribed sequence. Our laboratory has been involved in the assembly and manipulation of designed oligonucleotides in order to pursue studies in genetic recombination and nanofabrication. We have constructed a large number of unusual DNA molecules, including branched DNA molecules, DNA polyhedra, DNA knots, DNA double crossover molecules, and DNA antijunctions and mesojunctions. Our experience with these systems has uncovered a large number of experimental pitfalls that may confront individuals working with DNA computing. We present our experience in this area with the hope that we can help investigators to anticipate the experimental problems that may affect their DNA computing schemes.

1. Introduction

DNA computing, as pioneered by Adleman [**1**], entails the operations of DNA hybridization, DNA ligation [**5**], and separation of target molecules. Lipton's elaboration of Adleman's work also uses hybridization and separation [**28**]. Other schemes for DNA computing involve the use of restriction enzymes [**37**] and more complex DNA structures, such as those containing branch points [**52**]. The techniques used are primarily those of molecular biology, but the nature of the enterprise is chemistry. By this, we mean that the operations require relatively large-scale molecular operations that result in a substantial number of homogeneous target molecules that need to be amplified or purified in the course of the experiment. The answer must be contained in the products in sufficient quantity to be detected by a technique other than genetic selection. Although the correct solution might indeed be 'selected' by its sequence properties, it does not as a rule contain features known from the outset that can result in its selection from a large number of similar

This research has been supported by grants N00014-89-J-3078 from the Office of Naval Research and GM-29554 from the NIH.

molecules by genetic means. Adleman's use of PCR at one step notwithstanding [**1**], successful chemistry with substantial yields appears to be an important component of DNA computing. This is in contrast to molecular biological techniques, in which small yields at successive steps do not prevent a successful cloning operation that selects for known genetic features in the target (*e.g.* [**38**]).

Our laboratory is not involved directly in DNA computing. However, since the early 1980's we have been performing experiments in the construction of DNA molecules with unusual structures and topologies, in the senses of helical branching and strand linkage. Our goals are twofold: 1) understanding the structural equilibria of branched DNA intermediates in the process of genetic recombination, and 2) the assembly of nano-scale objects and designed periodic matter as a way to solve the crystallization problem for biological macromolecules, and for use as potential components in nanofabrication [**40, 41**]. Both of these efforts center on branched DNA molecules called junctions. We have constructed stick-polyhedra, a cube and a truncated octahedron, from junctions that contain fixed branch points. These polyhedra are complex catenanes of cyclic single-stranded DNA molecules. Besides conventional branched junctions, we have also assembled more complex branched structures called double crossovers [**14**], and more general branched structures called antijunctions and mesojunctions [**12, 53**]. In addition, we have constructed DNA trefoil knots of both chiralities, as well as amphicheiral figure-8 knots [**33, 8**]. Thus, we too have been involved in chemical-scale assembly of DNA molecules, through use of the molecular biological techniques of hybridization, ligation, restriction, electrophoresis and affinity separation. Several femtomoles of polyhedra have been assembled, the yields of knots are in the range of picomoles to nanomoles, and the individual small hydrogen bonded complexes are made in larger quantities. The constructions that we have undertaken appear to be relatively simple on paper [**41**]. Nevertheless, each of them has been completed successfully only by virtue of intense and laborious efforts on the parts of the experimentalists who did the laboratory work. We will review this work here, with emphasis on the (often unpublished) problems that have been confronted and overcome in order to obtain the results that were sought. The purpose of this article is to summarize our experience in the experimental treatment of DNA, and to point out pitfalls that may be inobvious to investigators new to DNA chemistry.

2. Branched DNA Molecules

There appear to be no stable DNA branches in nature. If one wishes to build them in the laboratory, and to use them as components for the construction of nanoscale objects, a method must be found to assign sequences to them. We have used sequence symmetry minimization techniques to order to achieve this goal [**41**]. The fundamental assumption in all the DNA sequence assignment we have practiced is that the formation of Watson-Crick-paired double helical DNA is the most favorable state for DNA molecules. It is important to point out that this assumption appears to be true only at appropriate DNA concentrations in conventional neutral aqueous solutions, containing sufficient counterions and moderate salt concentrations, usually at ambient or low temperatures. We will stress the importance of experimental conditions throughout this paper.

Where does the element of design enter into this picture? What problems can be encountered if the sequence is not carefully selected? The antithesis of control over chemical systems is symmetry, in its broadest sense, i.e., multiple outcomes that are equivalent, or nearly so, from the standpoint of the free energy of the system. Thus, one must differentiate unambiguously the target structure from any other possible alternative structures. The alternatives must be made sufficiently unfavorable, relative to the target, that only the target will form in appreciable quantities. This is akin to the strategy familiar to readers of Sherlock Holmes mysteries: "...when you have eliminated the impossible, whatever remains, however improbable, must be the truth" [**7**]. Thus, the goal in DNA structural design is to assign a sequence that will favor the target structure above any alternative structure that the molecule could adopt.

The solution to this problem would be trivial if there were an infinite number of complementary pairs of bases. Benner and his colleagues [**35**] have made headway in increasing the number of base pairs, but the problem remains virtually the same: A small number of bases must be used to assign sequences to DNA molecules containing $10^2 - 10^4$ nucleotides so that the target molecule will form in a unique fashion. The problem has a simple solution for linear duplex DNA, whose length can be millions of nucleotides in the cell; linear duplex is the most favorable structure, and perfect complementarity is the answer. However, branched DNA molecules do not form the most stable structure for DNA [**31**]. Therefore, one is forced to ask whether structures containing a mispair or a looped-out base, or a non-standard base pair are competitive with a perfectly paired sequence in an 'excited state' generated by the presence of a branch point.

The fundamental assumption of the sequence symmetry minimization algorithm [**41**] is that one can treat a series of contiguous nucleotides as a unique cooperative unit. This approach is illustrated in the sequence of the branched junction J1, shown in Figure 1. This molecule contains 4 single strands (Arabic numbers) arranged into 4 double helical 'arms' (Roman numerals). This four-arm junction is made from four hexadecamers. Each hexadecamer is treated as being composed of 13 overlapping tetramers. The first two tetramers (CGCA and GCAA) in strand 1 are boxed in Figure 1. Each tetramer in the entire structure is unique. In addition to tetramer uniqueness, J1 contains no complement to any tetramer spanning the branch point. Thus, no TCAG is present to complement the boxed CTGA at the bend of strand 1. CTGA is one of 12 junction-spanning sequences whose complements have been excluded from the sequence assignment. The ATG sequences in the dashed boxes are examples of redundant sequences, but they are not highly competitive with the target octamers in each arm. The stability of the site of the branch point also requires that it not be flanked by structures whose sequence symmetry is twofold symmetric; otherwise, it can move by an isomerization called branch migration [**17**].

This approach to sequence assignment completely ignores the geometrical and structural factors for which it is a shorthand. Sequence dependent characteristics of secondary structure, such as DNA bends [**22**], the formation of triple helices [**13**], or C-tetrads [**24**] can affect the structures that are formed, but are not now considered in the sequence assignment process. Alternative structures that can be formed from special sequences must also be considered. The propensity of guanines to form G-G pairs (*e.g.*, [**18**]) has led to the exclusion of long stretches of oligo-dG from all structures we have designed.

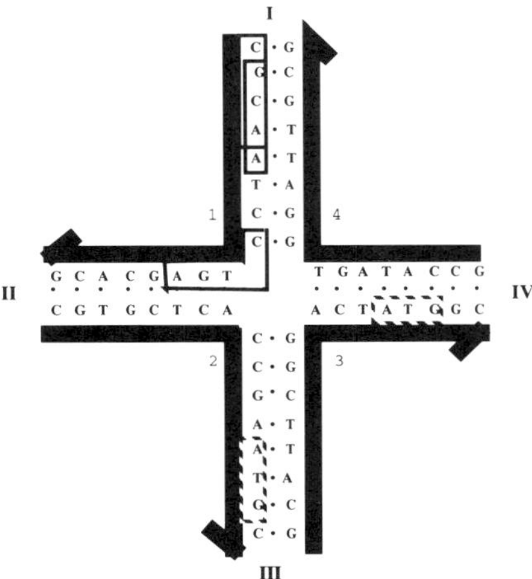

FIGURE 1. J1, A Stable DNA Branched Junction. The junction shown is composed of four strands of DNA, labeled with Arabic numerals. The 3' end of each strand is indicated by the half-arrows. Each strand is paired with two other strands to form a double helical arm. There is no homologous twofold sequence symmetry flanking the central branch point, thereby stabilizing the branch point. At the upper part of arm I, two of the 52 unique tetrameric elements in this complex are boxed. At the corner of strand 1, the sequence CTGA is boxed. The dashed boxes outline redundant ATG sequences.

2.1. Solution Protocols for Assembling Unusual DNA Molecules. It is not sufficient to design a molecule by means of sequence criteria if one wishes to make complex DNA molecules. It is also necessary to choose proper solution conditions. For example, short branched DNA structures, such as J1 (Figure 1) require Mg^{2+} for stability [**42**]: It is possible to visualize the junction as a single band on a non-denaturing gel only if multivalent cations are present in the electrophoresis running buffer in sufficient quantities (ca. $1 - 10\ mM$). When Mg^{2+} is omitted from the running buffer, the complex falls apart. The hybridization protocol is also important: The very first time J1 was formed, a solution containing strands 1 and 2 was mixed with a solution containing strands 3 and 4 at 4 °C, to produce a solution ca. 10 μM in each strand. The branched junction failed to form until the solution was heated and then allowed to cool slowly. We believe the reason for this result is that the 3' half of strand 1 and the 5' half of strand 2 nucleated a stack that permitted weak associations between their respective 5' and 3' ends, while strands 3 and 4 behaved similarly. This result was obtained, even though equimolar mixtures of strands 1 and 3 or 2 and 4 run as single strands on non-denaturing gels [**19**]. Eventually, the junctions would have formed, but the barrier at 4 °C was too high to be overcome quickly just by mixing, followed immediately by spectroscopic examination.

2.2. Structural Considerations in Forming DNA Objects. Structural features of DNA must also be taken into account when a motif is designed. Perhaps the most important of these is the twist of the DNA double helix. The DNA backbone is often represented for convenience as a pair of parallel lines (*e.g.*, Figure 1), but its helical nature cannot be ignored safely in the design of DNA objects.

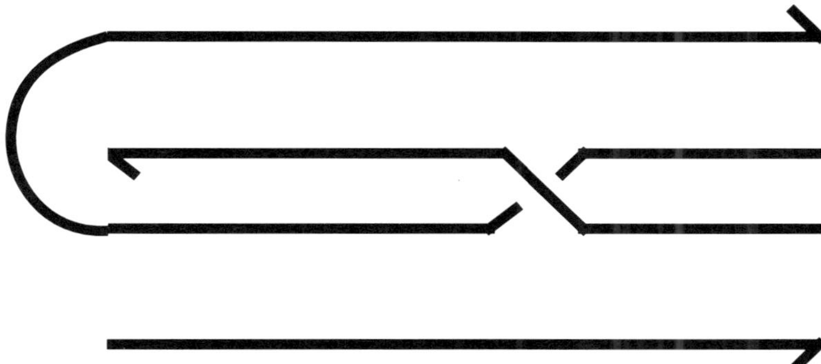

FIGURE 2. A Tethered Junction in which Stacking Domains are Determined by the Length of the Tether. The junction shown is tethered to assume a parallel structure. The arms left of the crossover point are 15 nucleotide pairs long, while those right of the crossover point are 10 nucleotide pairs long. When the tether consists of dT_6, one stacking structure is seen, but different arms stack if the tether is loosened by making it dT_9.

The mathematical and computer-graphical treatments for dealing with this aspect of DNA structure have been worked out [**45, 46**], but it is most often handled experimentally by using an integral number of helical half-turns between the vertices of objects. The cube and truncated octahedron described below contain edges with even numbers of half-turns; we have also constructed a quadrilateral with an odd number of half-turns per edge [**2**]. Other twists per edge are certainly conceivable [**45, 46**], but they have not been used yet in actual construction experiments. A key consequence of the plectonemic nature of the DNA double helix is that the cyclic strands of DNA in complex objects are topologically linked. This can be used to great advantage in the purification of intermediates and products [**3**].

The dimensions of the double helix must also be estimated properly. If the effective radius of the DNA molecule is underestimated, connections between helices may alter the structure. For example, the arms of a branched junction were tethered together (Figure 2), in an attempt to force them to be parallel or antiparallel [**21**]. When the tether between parallel helices was made too short, it was found that the dominant stacking isomer was opposite from that seen for the untethered junction [**4**]. When the tether was lengthened from dT_6 to dT_9, the relaxed stacking isomer was restored.

3. DNA Assemblies

Linear DNA is a construction medium long familiar to those involved in biotechnology [**5**]. The sticky-ended ligation of DNA fragments is arguably the fundamental reaction of genetic engineering; it enables investigators to assemble genetic units with particular functional characteristics. The attachment of particular sticky ends to a branched nucleic acid structure converts it, in principle, into a highly specific nanoscale valence cluster with addressable ends [**41**]. DNA appears to be unique among biological macromolecules, because it is possible to program its intermolecular associations: Complex charge density relationships that determine surface complementarity properties can be reduced to the simple rules of Watson-Crick base pairing. One should beware that other forms of association are possible in the absence of complementary partners, but complementarity appears to prevail at equilibrium, if it is one of the competitive alternatives. Not only is the specificity determined, but so is the local structure: Under conventional conditions, the product of this association is a right-handed double helix approximately 20 Å in diameter with a pitch about 34 Å, containing about 10 nucleotide pairs per turn (B-DNA). By treating them as molecular lego, branched DNA molecules become building blocks that can be assembled into multiply-connected stick-figures and networks. The edges of these figures consist of double helical DNA, and the vertices correspond to the branch points. Thus, it seems possible to design and build complex structures from branched DNA molecules. The basic principles of self-association are shown in Figure 3, where we illustrate the assembly of a 4-arm junction into a quadrilateral that has the potential to form a 2-D lattice.

How good is the specificity of the sticky ends? We have only anecdotal evidence in this regard, but it is enough to raise warning flags. As noted below, we often derive our sticky ends by the restriction of closed DNA hairpin molecules. Once, we restricted with Hinf I a 5-base cutting enzyme (**GANTC**), leaving a sticky end of **ANT**. To our dismay, we found that it ligated improperly to another molecule containing the sticky end **ANT**, in addition to ligating properly to molecules containing the sticky end **AN'T** [**30**] In another study [**26**], two strands in solution contained the same two 3' nucleotides. They were designed to be complexed at different parts of a molecule being ligated, but one strand (designed not to be involved in ligation) was present in slight excess for reasons necessitated by the experiment. That molecule displaced the other one in small quantities, and wound up in some ligation products in place of the intended molecule. Thus, in our hands, T4 DNA ligase is a 'hungry' enzyme: A substrate that is close will often suffice. Schemes that require perfect ligation to transmit information must take this potential source of noise into account.

3.1. Junction and Object Flexibility. When the type of assembly shown in Figure 3 is attempted, one discovers that individual junctions are not rigid at all [**32**]. We can expect that the double helical edges of each object are torsionally and flexurally stiff, because the persistence length of DNA is approximately 500 Å [**16**]: This value leads to a 66.5 Å RMS end-to-end distance for two turns of DNA, nominally 68 Å long. However, the 'valence angles' between the double helical edges of each object are flexible. If a given junction is oligomerized to yield oligolaterals with two turns between branch points, it forms readily a series of cyclic products: trimers, tetramers, pentamers and so on. This is true for 3-arm branched junctions [**26**], 4-arm branched junctions [**34**], and, to a lesser extent, for bulged 3-arm branched junctions [**29**]. Thus, branched junctions are highly flexible, in contrast to valence clusters like the tetrahedral carbon atoms that form the basis of conventional structural chemistry. Junction flexibility is currently the biggest problem in the construction of periodic arrays from branched DNA components. The key lesson here is that one cannot devise schemes that contain sequences that are expected to form periodic patterns in one or more dimensions, without taking steps to ensure that the molecules are incapable of cyclization.

Recently, having determined that bulged 3-arm junctions appear to have somewhat more structural integrity than other species [**29**], we used them to build triangles. Leontis and his colleagues have shown that the presence of a dT_2 bulge in a 3-arm junction of a particular sequence enables the two arms of the junction 3'

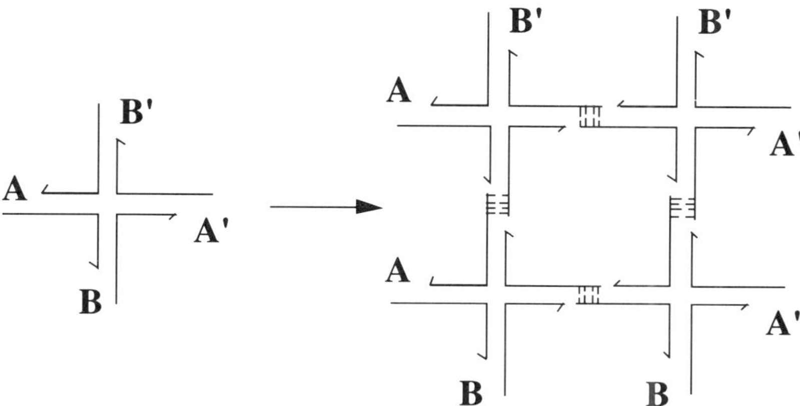

FIGURE 3. Formation of a Two-Dimensional Lattice from a Junction with Sticky Ends. A is a sticky end and A' is its complement. The same relationship exists between B and B'. Four of the monomeric junctions on the left are complexed in parallel orientation to yield the structure on the right. Note that A and B are different from each other, as indicated by the pairing in the complex. DNA ligase can close the gaps left in the complex. Note that the complex has maintained open valences, so that it can be extended by the addition of more monomers.

to the bulge to form a linear stack [**23**]. We built two triangles from bulged 3-arm branched junctions, one triangle with the bulges on the inside and one triangle with bulges on the outside, as shown in Figure 4.

The experiment we wished to conduct entailed ligating these two triangles alternately in solution [**36**]; the presence of a hexamer of triangles as the dominant product would suggest a stiff component, which the triangles were ultimately shown not to be. They were built as the topologically closed molecules shown, because in our experience it is impossible to purify intact target structures from nicked failure structures under non-denaturing conditions. Only topologically closed molecules can withstand the denaturing conditions necessary to effect the purification. Thus, restriction at the PflM I and Bgl I sites leads to exposure of (non-self-complementary) sticky ends, so that we can conduct our ligation experiment.

We encountered two instructive problems in this exercise. First, assembly of the triangle with the bulges on the inside strand resulted in two topoisomers of the triangle. In addition to two strands linked 6 times (as expected for 3 edges two helical turns long), we obtained some triangles that were linked 7 times. Fortunately, it was possible to purify the target material by means of gel electrophoresis. Secondly, restriction was incomplete. As an apparent corollary of Murphy's law, the expense of restriction enzymes appears to be inversely proportional to their activity; restriction endonucleases that expose asymmetric sites (associated with the most control in intermolecular ligation [**55**]) tend to be the most expensive. The presence of unrestricted or partially restricted materials in solution would have acted as a poison in this experiment, so it was necessary to remove them by streptavidin magnetic beads that bound to biotins on the hairpins. In our first experiments, a single biotin on each hairpin (sequence **TTBTT**) was shown to be inadequate,

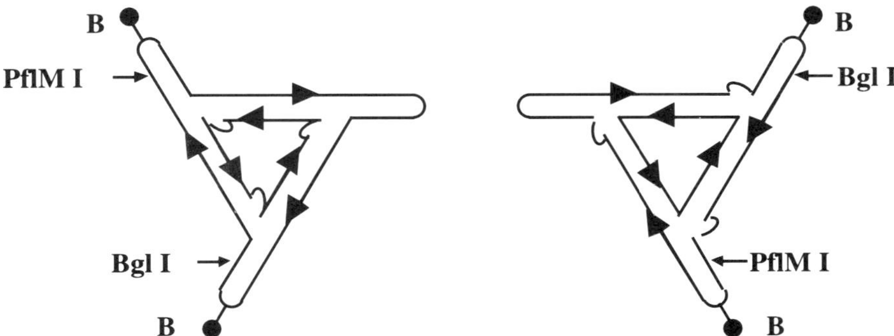

FIGURE 4. The triangle on the left contains the bulges on the inside strand, and the triangle on the right contains the bulges on the outside strand. The arrowheads indicate strand polarities. Each edge of the triangle is designed to contain 20 nucleotide pairs. The restricted ends, when ligated together, also contain 20 pairs, although versions of these triangles have been made to contain 21 and 22 pairs in the inter-triangular helix. The two filled circles labeled 'B' correspond to biotinylated hairpin arms. The restriction sites used to deprotect the two hairpins are labeled as Bgl I and PflM I. The third exocyclic arm and the three edges also contain restriction sites.

leaving about 5% of the incompletely restricted material. The addition of a second biotin to generate hairpins of sequence **TTTBTTTBT** was necessary to remove all the unrestricted material. Thus, procedures requiring selection by affinity measures need to be calibrated in order to determine the accessibility of the affinity probe to its target. In a different application [**25**], even two biotin groups have proved inadequate for complete purification.

3.2. The Construction of DNA Polyhedra. The most exciting prospects for this system entail the assembly of multiply-connected objects in 3-D. We have exploited the flexibility of branched DNA to construct from 3-arm junctions a molecule whose helix axes have the connectivity of a cube [**3**]. A schematic of the 3-dimensional, 3-connected object is shown in Figure 5. It contains 12 edges that consist of double helical DNA. Whereas every edge contains 20 nucleotide pairs of DNA, we expect that their lengths will be about 68 Å. Each of the edges contains a unique site for cleavage by a restriction endonuclease, thereby enabling us to establish the validity of the synthesis. The synthesis of the cube-like object has been demonstrated by breaking it down to standard catenanes, whose electrophoretic properties are known.

There is an integral number of turns in each edge, so every face of the cube corresponds to a single cyclic strand. Therefore, the object in Figure 5 is a complex catenane of 6 single-stranded cyclic molecules, each doubly linked to its four nearest neighbors. The catenated nature of the strands in Figure 5 highlights an important point made earlier about double helical DNA: The braiding of the strands about the helix axis is a critical structural and topological property of any DNA molecule. As noted above, the strands in the cube can be separated from nicked failure products by electrophoresis under denaturing conditions, because the catenane holds together, but nicked molecules separate into individual strands [**3**]. Topologically bonded molecules are much more easily controlled than hydrogen

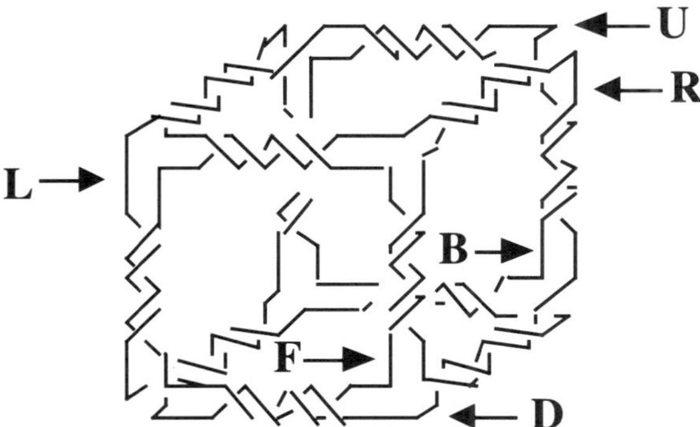

FIGURE 5. A DNA Molecule Whose Helix Axes Have the Connectivity of a Cube. The molecule consists of six cyclic strands that have been catenated together in this particular arrangement. They are labeled by the first letters of their positions, Up, Down, Front, Back, Left and Right.

bonded complexes. The twisting of DNA strands about each other implies that closed structures will be complex catenanes, and infinite constructs can contain polycatenated mesh structures, much like chain mail.

The cube was constructed in solution, and the synthesis is shown in Figure 6. Only sticky ends C and C', D and D' were phosphorylated in the first step. After the ladder-like intermediate was formed, it was necessary to denature the molecule, leaving only the linear triple catenane corresponding to the ultimate left, front and right faces of the cube. The reason for this is that the ladder-like molecule formed only in about 10% yield, and it was impossible to purify it away from the other products under non-denaturing conditions. The two-step ligation entailed both an intermolecular condensation of the two squares (first ligation), and a cyclization of the condensed molecule (second ligation). Condensation is promoted by high concentrations, but cyclization is promoted by low concentrations. Combining the two steps was a grave error that lengthened the synthesis by necessitating the denaturing purification step and then the reconstitution step. This is just one of many examples that demonstrate that the proper concentration of DNA for a given step is one of the key variables in DNA solution chemistry: Too high and cyclizations become condensations; too low, and condensations do not occur.

In response to the problems encountered in constructing the cube, we developed a solid-support based methodology [**55**]. The solid-support permits the convenient removal of reagents and catalysts from the growing product. Each ligation cycle adds a component that creates a covalently-closed, topologically-bonded intermediate. A single edge of an object is formed at a time, thereby eliminating by-products that result from flexible units. Control derives from the restriction endonuclease digestion of hairpin loops that form each side of the new edge. The isolation of growing objects on the support also removes the problem of cross-reactions between growing product molecules. Sequences are chosen to destroy restriction sites when the edge forms. In principle, the component added in condensations can be a junction, a polygon, a group of polygons, a polyhedron, or an array of polyhedra.

In order to test the solid-support methodology, we have constructed a truncated octahedron from DNA [**54**]. This Archimedean solid contains six squares flanking

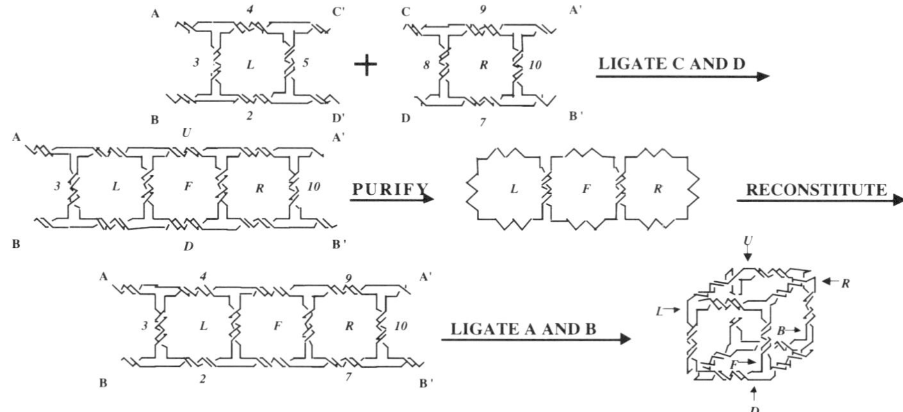

FIGURE 6. The Synthesis of a DNA Cube. The molecule is built from 10 chemically synthesized strands, two 80-mers and eight strands containing about 40 nucleotides. These are hybridized to form two quadrilaterals in the first step. Two ends (C and D) are ligated to form a belt-like molecule that must be denatured and reconstituted in order to purify it from side-products. The belt-like molecule is then cyclized to form the cube-like molecule.

its four-fold symmetry axes, and eight hexagons that surround its 3-fold symmetry axes. There are two turns of DNA per edge, so the final object is a complex 14-catenane. The truncated octahedron is a 3-connected figure, but it has been constructed from 4-arm junctions. The entire molecule contains 2550 nucleotides, and has a molecular weight ca. 790 kd.

In this construction, the objects added to the support are squares and square groupings, as indicated in Figure 7. The structure on the lower left of Figure 7 is a heptacatenane hexasquare complex. The square strands are already intact in this construct, and the hexagons are all formed from the outer strand. The hexagons result from successive intramolecular closures of the sticky ends associated with the restriction enzyme site pairs, S1-S1'...S7-S7'. Whereas the molecules are isolated from each other on the support, it is possible to expose this group of sticky ends successively, by using symmetrically-cleaving restriction enzyme pairs that recognize six nucleotide pairs each. The final step in the synthesis involves releasing the structure from the support and annealing it shut with a hairpin. The synthesis is demonstrated in two stages: First, by showing that all six cyclic strands corresponding to the square molecules are in the heptacatenane, and second, by digesting the final product to the tetracatenanes that flank the squares.

The key problems encountered in this procedure had to do with the failure of restriction enzymes to cleave molecules on the support, even though they had appeared to work in solution. In this case, we were using the most expensive and asymmetric restriction enzymes, type IIs; these recognize an asymmetric site, and cut distal to it. On paper, these are the best enzymes for intermolecular ligations, because their sites can be placed on the hairpins that are then cut off. We had originally planned to add square 1 to the support, square 2 to square 1, then squares 3, 4 and 5 to square 2, and then square 6 to square 5. The tetrasquare complex was a kludge in order to save the synthesis. We assembled it in solution, and then returned to our original plan. The synthesis was done on the 1 picomole

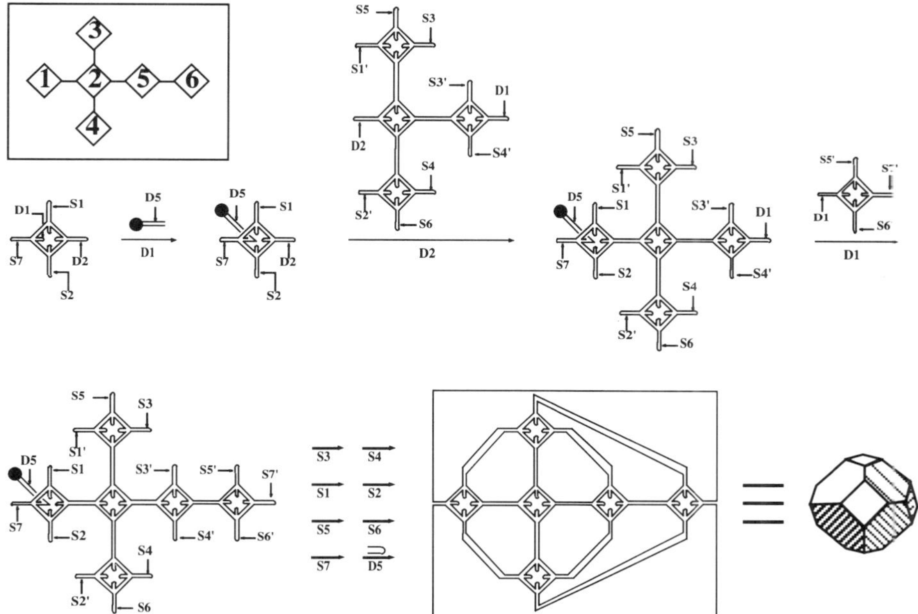

FIGURE 7. The Synthetic Scheme Used to Synthesize the Truncated Octahedron. The boxed diagram in the upper left indicates the numbering of individual squares. Each square in the rest of the diagram is shown with its restriction sites indicated. Symmetric restriction sites are named 'S', indicated in pairs, with one member primed; restriction sites cut distally are named 'D'. A reaction is indicated by a line above a restriction site: This means that the restriction enzyme is added, protecting hairpins are removed and then the two sticky ends are ligated together. The product is shown in two forms. On the left, the S1-S6 closures are shown as triple edges, to emphasize their origins; the two strands of the edge formed by the S7 closure are separated to maintain the symmetry of the picture. On the right, a slightly rotated front view of a polyhedral representation of a truncated octahedron is shown without the exocyclic arms; the 432 cubic symmetry of the ideal object is evident from this view.

scale. Most of the ligations shown above went in nearly quantitative yield, but there was one very bad closure that decreased the final yield to about 10 femtomoles.

4. Tying Knots in DNA

Knotted molecules may serve as a route to the cloning of the DNA polyhedra that we have been constructing [**43**]. The plectonemic nature of DNA makes it an ideal medium for the control of chemical topology. We have demonstrated that a half-turn in DNA can be equated with a node in a knot or catenane [**44, 9**]. Thus, it is quite simple to design target knots from DNA, and to construct them from sequences that embody this relationship between nodes and half-turns. Nodes have a handedness, so there are positive and negative nodes. Negative nodes are produced from conventional right-handed B-DNA, but it requires special sequences and solution conditions to produce positive nodes, because they require the presence of left-handed Z-DNA [**10**]. The biggest problem in the construction of knots is that these molecules typically contain large stretches of specifically unstructured DNA.

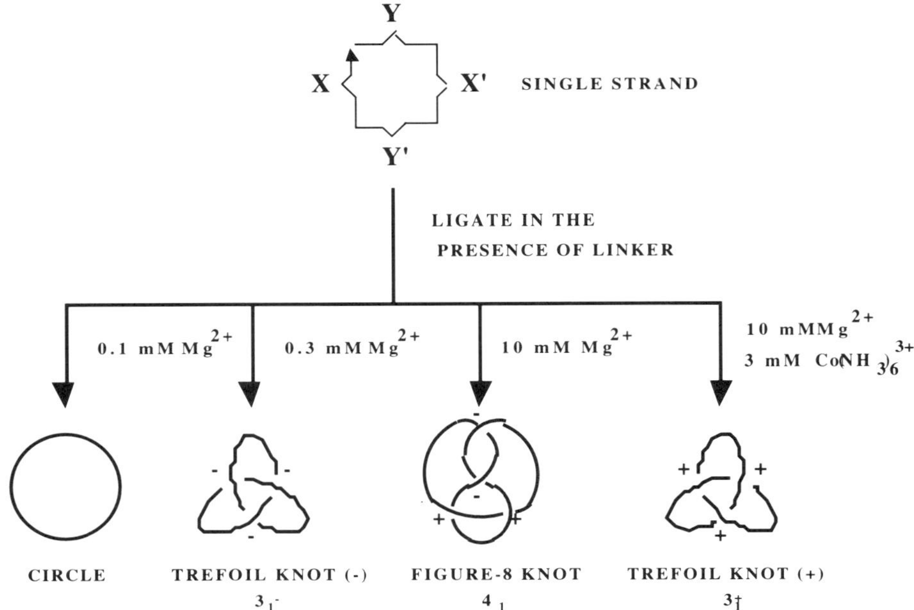

FIGURE 8. The Synthetic Scheme Used to Produce Four Target Topologies. The top of this scheme indicates the molecule from which the target products are produced. The four pairing regions, X and its complement X', Y and its complement Y' are indicated by the bulges from the square.

The sequences of these unstructured regions are usually designated to be oligo-dT regions, and it is often possible for them to form links among themselves that are not intended to be there [**33, 9**]. We have tried to tie them up in a 'topological protection' procedure [**9**], whereby they are complexed into linear duplex with single strands that are later removed from the knot. However, the complement to oligo-dT is oligo-dA, so we wound up linking them even more than in the absence of protection, probably because two oligo-dT strands formed a (dT)-(dA)-(dT) triple helix [**13**]. Nevertheless, topological protection has been successful with random-sequence catenanes, if the protected sequence was two or more turns long [**15**].

Controlling topology has proved to be much easier than controlling structure, at least for simple knots. Relative to the amount of effort we have had to expend, this has been the most tractable of the systems described here. In Figure 8, we show how it is possible to generate four different topologies from a given single strand of DNA [**8**], just by changing solution conditions. The linking is the result of two single-turn double helices, shown as regions X and Y. In this case, the different topologies are intentional, but this is another example of the importance of solution conditions when making molecules from DNA. It is possible to use these molecules as substrates for single-stranded DNA topoisomerases [**11**]. Recently, we have been able to use this system to discover the existence of an RNA topoisomerase [**51**].

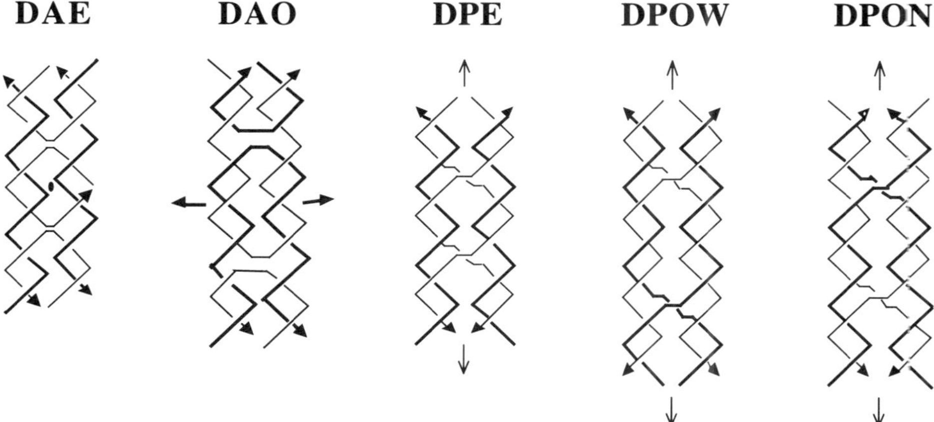

FIGURE 9. Schematic Drawings of the 5 Different Structural Arrangements of Double Crossover Structures. The structures shown are named by the acronym describing their basic characteristics. All names begin with 'D' for double crossover. The second character refers to the relative orientations of their two double helical domains, 'A' for antiparallel and 'P' for parallel. The third character refers to the number (modulus 2) of helical half-turns between crossovers, 'E' for an even number and 'O' for an odd number. A fourth character is needed to describe parallel double crossover molecules with an odd number of helical half-turns between crossovers. The extra half-turn can correspond to a major (wide) groove separation, designated by 'W', or a minor (narrow) groove separation, designated by 'N'. The strands are drawn as zig-zag helical structures, where two consecutive, perpendicular lines correspond to a full helical turn for a strand. The arrowheads at the ends of the strands designate their 3' ends. The lens-shaped figure in the DAE molecule and the pairs of arrowheads outside the other molecules indicate their axes of dyad symmetry.

5. Crowded DNA Molecules

5.1. DNA Double Crossovers. DNA double crossover molecules contain two branched DNA molecules joined at two adjacent arms. These structures have been implicated in recombination involving double strand breaks [**49, 48**]; recently, they have been shown to be intermediates in meiotic recombination [**39**]. Small double crossover molecules can form five different crowded isomers of two-domain structures, as illustrated in Figure 9. The first two isomers (DAE and DAO) contain double helical domains that are antiparallel to each other, and the second three contain parallel helical domains, related by dyad axes vertical on the page. The schematic diagram in Figure 9 illustrates a key source of instability in the parallel molecules, one that was detected when they were constructed [**14**]: The polyanionic strand backbones are brought into close juxtaposition in the parallel molecules, whereas they are interdigitated in the antiparallel molecules. As a consequence, antiparallel molecules appear to be the favored structure for single branched junctions [**27, 47**]. In a similar fashion, antiparallel double crossover molecules are easy to form, and one need not take any more precautions than necessary for single branched junctions. However, the parallel double crossover molecules are very difficult to produce. In order to decrease the apparent repulsion caused by backbone juxtaposition, the strands form stoichiometric multimers of the basic structure (*e.g.*, 2:2:2:2 or 3:3:3:3). These are likely to be structures that permit separation of the charges. It is worth pointing out that there is no protection against the formation of stoichiometric multimers, in any DNA assembly system whose strands can form a cyclic closed structure. One can try to counteract this phenomenon by decreasing the concentration of the strands. It is crucial that all investigators in cyclic systems calibrate their stoichiometry.

5.2. Antijunctions and Mesojunctions. We have shown recently that the DNA branched junctions discussed above represent a special set of the DNA molecules that can flank a polygon [**12, 53**]. Figure 10 illustrates this point for 3-strand and 4-strand molecules. The molecules represented as 3_3 and 4_4 are the conventional branched junctions we have been discussing. Their helix axes point radially at the center of the polygon formed by their strands. The structure labeled 4_0 is an 'antijunction', and its helix axes are circumferential about the polygon, whereas the central dyad axes of its helical domains point radially towards the center of the square. The helical domains in Figure 10 are all shown as single half-turns of DNA, but any integral number of half-turns can be used to construct these molecules. The other three figures illustrated are 'mesojunctions'. These are molecules in which two domains have been switched simultaneously from radial to circumferential. Switching odd numbers of domains leads to topological ambiguities. When the mesojunctions shown here (with two [**53**] or three [**12**] half-turns per domain) are built, they form multimers, like those described for parallel double crossover molecules. The bands are discrete, indicating integral multiples of the target stoichiometry. By using low concentrations, and sometimes higher temperatures, it is possible to isolate and characterize these molecules. This is in contrast to the usual means of stabilizing unusual DNA molecules, where high concentrations and low temperatures are preferred. All of these complexes associate in crowded stacking structures reminiscent of DNA double crossover molecules.

The 4-strand antijunctions (with two or three half-turns per helical domain) are the least tractable molecules we have examined to date. All of the other DNA motifs have formed their target structures, or have at least formed stoichiometric multimers of the targets. These structures form non-stoichiometric assemblies, although with great care it is possible sometimes to isolate discrete antijunctions [**12**]. The reason for this is shown at the bottom of Figure 10, where the antijunction is re-drawn in a stacked structure, rather than the unstacked cyclic representation at the top of the figure. One of the key rules of thumb in nucleic acid structure is that double helices will stack if they possibly can do so. This feature has been noted, among other places, in the structures of transfer RNA [**20**], in structures of 4-arm branched junctions [**4**], and the structures of bulged 3-arm branched junctions [**23**]. It is not surprising that the linear stacking structure of the 4_0 antijunction is preferable to the cyclic unstacked structure drawn at the top of Figure 10. Indeed, characterization of the mesojunctions also indicates that they maximize their stacking. Thus, the four strands of the antijunction form linear complexes of varied extent; their mean length is a function of the concentration of the strands. The relationship between the antijunction shown here and the strand structure used by Adleman [**1**] should be noted.

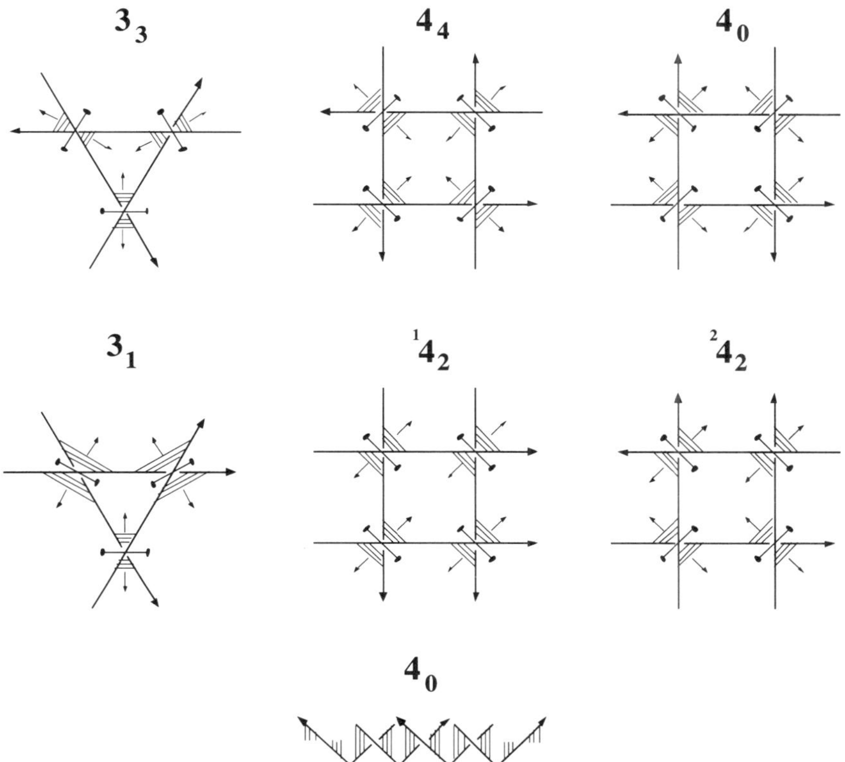

FIGURE 10. Schematics of 3- and 4-Strand Junctions, Antijunctions and Mesojunctions. Each polygon is flanked by strands of DNA that extend beyond the vertices in each direction. The arrowheads indicate the 3' ends of the strands. The vertices correspond to the nodes formed by approximately a half-turn of double helical DNA. Base pairs are represented by lines between antiparallel strands. Thin lines perpendicular to the base pairs represent the axis of each helix. Normal to the helix axes are the central dyad axes of each half-turn. The large number labeling each figure indicates the number of strands, and the subscript indicates the number of helix axes that point to the centers of the polygons formed by the strands. The complexes 3_3 and 4_4 correspond to conventional branched junctions. Their helix axes are all pointing towards the center of the triangle or square along the bisectors of the angles of the polygons ('radial'). The complex 4_0 is a 4-strand antijunction, in which each helix axis ideally is perpendicular to the angular bisectors ('circumferential'). In these cases, the dyad axis is parallel to the angular bisectors of the polygons. The complexes are named by a large number that indicates the number of double helical segments, a subscript which indicates how many of them are pointed along the angular bisectors, and a presuperscript (if necessary) enumerating different members of a class otherwise undifferentiated by the first two numbers. The complexes on the bottom row contain a mix of the two orientations of helix axes. The complex 3_1 contains one radial helix axis and two circumferential axes. The complexes $^14_2$ and $^24_2$, each contain a pair of radial helix axes and a pair of circumferential axes, but the arrangement is different between them: $^14_2$ contains alternating radial and circumferential axes, whereas $^24_2$ contains two radial axes on the right of the square and two circumferential axes on the left. At the bottom of the figure, the 4-strand antijunction, 4_0, is drawn in a linear, rather than in a cyclic representation.

6. Lessons from DNA Construction

This paper has been an excursion through our experiences in DNA construction. We have tried to highlight the most important principles that we have learned in the course of our studies. It is useful to summarize them here:

- *Symmetry is Inimical to Control.* This property is unlikely to need much explication to the readers of this paper, because it does not require experimental failures to appreciate. If one treats symmetry in its broadest sense, the equal, or nearly equal free energies from two possible outcomes of an experiment, it is clear that the most control over molecular associations will be derived from systems that minimize symmetry.
- *Non-Watson-Crick Pairing in Unusual Structures.* Watson-Crick pairing appears to be favored for linear duplex DNA. However, when it is not available, or when it is difficult to achieve, non-Watson Crick pairing can occur. Every base appears capable of pairing with every other base, including itself [**50**].
- *The Importance of Hybridization Protocols.* One must make sure that the pairing structure one want to achieve has an opportunity to overcome kinetic barriers. We routinely heat all mixtures to 90° C for a few minutes, and cool the structure slowly before using it. We also check for the presence of the structure on non-denaturing gels run under the conditions of the experiment, particularly temperature and the DNA and divalent cation concentrations.
- *The Importance of DNA Concentration and Environment.* It is critical to use concentrations of DNA that are appropriate to forming the complex of interest. The significance of appropriate divalent cations, pH, and appropriate temperatures cannot be overestimated.
- *Proper Estimation of DNA Dimensions.* Often a crude model is inadequate to estimate the size of non-canonical DNA features, such as loops between double helical segments. It is useful to vary such parameters experimentally in order to optimize their design.
- *Experimental Determination of DNA Features.* One cannot assume that unusual DNA structures assume the shapes that one feels most comfortable drawing for them. Branched junctions were found not to look like crosses [**4**], tRNA molecules were found not to look like any of their predicted tertiary structures [**20**], and the overall range of structures available to any nucleic acid molecule is often much larger than one is tempted to believe. Furthermore, the structure is often a function of conditions. One should predicate nothing on an unverified DNA tertiary structure.
- *The Fidelity of Ligation is High, but not Perfect.* The examples given here indicate that a 'hungry' ligase molecule may well ligate the wrong molecule to a sticky end, if it bears a close resemblance to the target molecule.
- *DNA Molecules Breathe.* We have implicitly mentioned this point with regard to a strand displacement. However, the bases of DNA molecules do unpair at times, and strands of the same sequence can displace each other. A well-known manifestation of this phenomenon is branch migration, an isomerization that naturally-occurring branched molecules undergo [**17**]: One cannot assume that particular molecules will remain paired non-covalently for very long time periods.

- *Long and Loose Closed DNA Molecules Form Topoisomers.* The plectonemic nature of double helical DNA helices results in catenated closed molecules. There will be a distribution of topological linkage in long molecules [**6**]. Topological impurity can complicate the interpretation of results involving catenated molecules.
- *Affinity Binding is a Filtration Process.* Affinity binding, such as biotin-streptavidin binding, is effective, but often imperfect. Multiple tags may be needed to achieve desired purification, because the tag may not be accessible to the receptor in all conformations.
- *Complex Ligation Mixtures Can Lead to Complex Products.* The ligation of molecules that can form cyclic species can often be complicated by numerous products in addition to the target products.
- *Separation of Hydrogen-Bonded Molecules in Native Conditions is Often Ineffective.* In our experience, purification is most effective under totally denaturing conditions. In order to effect such separations, one must form topologically closed molecules. Nicked and un-nicked hydrogen bonded complexes co-migrate on gels in non-denaturing conditions.
- *Restriction Endonucleases Often Produce Partial Digestion Products.* This is particularly true if the substrate is not simple double-helical DNA. One must test all protocols for the effectiveness of the restriction enzymes involved, and it is often necessary to find means to remove undigested material.
- *Multimerization of Cyclic Structures Can Occur.* All the discrete structures described here involve cyclic pairing schemes, *e.g.*, $1 \rightarrow 2 \rightarrow 3 \rightarrow 4 \rightarrow 1$. One hopes that the final molecule '1' and the first molecule '1' are the same molecule, but cyclic schemes that do not take steps to ensure the identity of the two '1' molecules are in danger of multimerization if this phenomenon will promote favorable interactions or decrease unfavorable ones.
- *Base Stacking is Often the Determining Interaction.* It is tempting to build up hydrogen bonding schemes when one is designing DNA molecules, but base stacking may well turn out to be the dominant physical interaction in the system.
- *Treat DNA as a Physical Chemical System.* In order to deal with DNA molecules, it is necessary to examine them under the conditions in which they will be used, even if that is sometimes inconvenient. It is useful to vary the temperature, concentration of DNA and the solution environment, and if using electrophoresis, the gel concentration. In order to develop a system that works well for any purpose, one must vary the parameters that define the system, to understand its strengths and limitations. There are many 'bibles' available to provide molecular biological protocols (*e.g.*, [**38**]). These sources are invaluable, but we suggest that levels of chemical efficiency that lead to successful molecular biology are often inadequate for successful chemistry, and that each protocol be examined carefully and not followed blindly. Likewise, it can be useful to ignore enzyme optima, and concentrate on conditions that ensure that the desired DNA complex is present in the experiment.

This article has been written in the hope that the experience that we have gathered in the course of more than ten years of work will be of value to investigators in the new field of DNA computing. The senior author came to this work with experience only in the computational aspects of crystallography, and with minimal hands-on laboratory experience. It is useful to collaborate with an experienced experimentalist when working with DNA for the first time; our collaboration with Neville Kallenbach during the 1980's certainly accelerated our progress. We hope that the readers of this paper will be able to avoid the traps into which we have fallen. We offer the comments partly as a cautionary tale, but our primary goal is to encourage workers in DNA computing that it is possible to over come the pitfalls and to achieve successful results with DNA physical chemistry.

References

[1] Adleman, L.M., Molecular Computation of Solutions to Combinatorial Problems. *Science*, 266:1021-1024, 1994.

[2] Chen, J.H., Kallenbach, N.R. and Seeman, N. C., *J. Am. Chem. Soc.*, 111:6402-6407, 1989.

[3] Chen, J. and Seeman, N.C., *Nature (London)*, 350:631-633, 1991.

[4] Churchill, M.E.A., Tullius, T.D., Kallenbach, N.R. and Seeman, N.C., *Proc. Nat. Acad. Sci. (USA)*, 85:4653-4656, 1988.

[5] Cohen, S.N., Chang, A.C.Y., Boyer, H.W. and Helling, R.B., *Proc. Nat. Acad. Sci. (USA)*, 70:3240-3244, 1973.

[6] Depew, R.E. and Wang, J., *Proc. Nat. Acad. Sci. (USA)*, 72:4275-4279, 1975.

[7] Doyle, A.C., The Sign of Four. *The Complete Sherlock Holmes*, The Literary Guild, New York, 1936, p. 118.

[8] Du, S.M., Stollar, B.D. and Seeman, N.C., *J. Am. Chem. Soc.*, 117:1194-1200, 1995.

[9] Du, S.M. and Seeman, N.C., *Biopolymers*, 34:31-37, 1994.

[10] Du, S.M. and Seeman, N.C., *J. Am. Chem. Soc.*, 114:9652-9655, 1992.

[11] Du, S.M., Wang, H., Tse-Dinh,Y.-C. and Seeman, N.C., *Biochemistry*, 34:673-682, 1995.

[12] Du, S.M., Zhang, S. and Seeman, N.C., *Biochemistry*, 31:10955-10963, 1992.

[13] Felsenfeld, G., Davies, D.R. and Rich, A., *J. Am. Chem. Soc.*, 79:2023-2024, 1957.

[14] Fu, T.-J., and Seeman, N.C., *Biochemistry*, 32:3211-3220, 1993.

[15] Fu, T.-J., Tse-Dinh, Y.-C. and Seeman, N.C., *J. Mol. Biol.*, 236:91-105, 1994.

[16] Hagerman, P.J., Ann. Rev. Biophys. and Biophys. Chem. 17, 265-286 (1988).

[17] Hsieh, P. and Panyutin, I.G., *Nucleic Acids and Molecular Biology 9*, F. Eckstein and D.M.J. Lilley eds, Springer-Verlag, Berlin, pp 42-65, 1995.

[18] Jin, R., Breslauer, K.J., Jones, R.A. and Gaffney, B.L., *Science*, 250:543-546, 1990.

[19] Kallenbach, N.R., Ma, R.-I. and Seeman, N.C., *Nature* 305:829-831, 1983.

[20] Kim, S.H., Suddath, F.L., Quigley, G.J., McPherson, A., Sussman, J.L., Wang, A.H.-J., Seeman, N.C. and Rich, A., *Science*, 185:435-440, 1974.

[21] Kimball, A., Guo, Q., Lu, M., Kallenbach, N. R., Cunningham, R. P., Seeman, N.C. and Tullius, T. D., *J. Biol. Chem.*, 265:6544-6547, 1990.

[22] Koo, H.-S., Wu, H.-M., and Crothers, D. M., *Nature*, 320:501-506, 1986.

[23] Leontis, N.B., Hills, M.T., Piotto, M., Malhotra, A., Nussbaum, J. and Gorenstein, D.G., *J. Biomol. Str. and Dyns.* 11:215-223, 1993.

[24] Leroy, J.-L. and Gueron, M., em Structure 3:101-120, 1995.

[25] Li, X., Wang, H. and N. C. Seeman, *Biochemistry*, 36:4240-4249 (1997).

[26] Li, X.,Yang, X., Qi, J. and Seeman, N.C., *J. Am. Chem. Soc.*, 118:6131-6140, 1996.

[27] Lilley, D.M.J. and Clegg, R.M., *Ann. Rev. Biophys. Biomol. Str.*, 22:299-328, 1993.

[28] Lipton, R.J., DNA Solution of Hard Computational Problems. *Science*, 268:542-545, 1995.

[29] Liu, B., Leontis, N.B. and Seeman, N.C., *Nanobiol.* 3:177-188, 1994.

[30] Liu, F., H. Wang and N. C. Seeman, *Nanobiology*, in press (1998).

[31] Lu, M., Guo, Q., Marky, L.A., Seeman, N.C. and Kallenbach,N.R., *J. Mol. Biol.*, 223:781-789, 1992.

[32] Ma, R.-I., Kallenbach, N.R., Sheardy, R.D., Petrillo, M.L. and Seeman, N.C., *Nucl. Acids Res.* 14:9745-9753, 1986.

[33] Mueller, J.E., Du, S.M. and Seeman, N.C., *J. Am. Chem. Soc.*, 113:6306-6308, 1991.

[34] Petrillo, M.L., Newton, C.J., Cunningham, R.P., R.-I. Ma, Kallenbach, N.R.and Seeman, N.C., Biopolymers 27, 1337-1352, 1988.

[35] Piccirilli, J.A., Krauch, T., Moroney, S.E. and Benner, S.A., *Nature*, 343:33-37, 1990.

[36] Qi, J., Li, X.,Yang, X. and Seeman, N.C., *J. Am. Chem. Soc.*, 118:6121-6130, 1996.
[37] Rothemund, P.W.K., A DNA and Restriction Enzyme Implementation of Turing Machines, In R. J. Lipton and E. B. Baum, editors, *DNA Computers: Proceeding of a DIMACS Workshop,* DIMACS Series in Discrete Mathematics and Theoretical Computer Science, American Mathematical Sociey, 29-36, 1996.
[38] Sambrook, J., Fritsch, E.F., and Maniatis, T., *Molecular Cloning, second edition*, Cold Spring Harbor Laboratory Press, 1989.
[39] Schwacha, A. and Kleckner, N., *Cell* 83:783-791, 1995.
[40] Seeman, N.C., *The Chemical Intelligencer* 1(3):38-47, 1995.
[41] Seeman, N.C., *J. Theor. Biol.*, 99:237-247, 1982.
[42] Seeman, N.C., Maestre, M.F., Ma, R.-I. and Kallenbach, N.R., *Progress in Clinical and Biological Research, 172A: The Molecular Basis of Cancer*, R. Rein edit., Alan R. Liss Inc., New York, pp. 99-108, 1985.
[43] Seeman, N.C., *DNA and Cell Biology*, 10:475-486, 1991.
[44] Seeman, N.C., *Mol. Engineering*, 2:297-307, 1992.
[45] Seeman, N.C., *J. Biomol. Str. and Dyns.*, 3:11-34, 1985.
[46] Seeman, N.C., *J. Mol. Graphics*, 3:34-39, 1985.
[47] Seeman, N.C. and Kallenbach, N.R., *Ann. Rev. Biophys. Biomol. Str.*, 23:53-86, 1994.
[48] Sun, H., Treco, D. and Szostak, J.W., *Cell*, 64:1155-1161, 1991.
[49] Thaler, D.S. and Stahl, F.W., *Ann. Rev. Genet.*, 22:169-197, 1988.
[50] Voet, D. and Rich, A., *Prog. Nucl. Acid Res. and Mol. Biol.*, 10:183-265, 1970.
[51] Wang, H., Di Gate, R.J. and Seeman, N.C., *Proc. Nat. Acad. Sci. (USA)*, in press (1996).
[52] Winfree, E., On the Computational Power of DNA Annealing and Ligation. In R. J. Lipton and E. B. Baum, editors, *DNA Computers: Proceeding of a DIMACS Workshop,* DIMACS Series in Discrete Mathematics and Theoretical Computer Science, American Mathematical Sociey, 75-119, 1996.
[53] Wang, H. and Seeman, N.C., *Biochemistry*, 34:920-929, 1995.
[54] Zhang, Y. and Seeman, N.C., *J. Am. Chem. Soc.*, 116:1661-1669, 1994.
[55] Zhang, Y. and Seeman, N.C., *J. Am. Chem. Soc.*, 114:2656-2663, 1992.

Department of Chemistry, New York University, New York, NY 10003 USA
E-mail address: ned.seeman@nyu.edu

Current address, S. Zhang: Department of Cell Biology, Fairchild Center, Stanford University, Stanford, CA 94305.

Current address, T. J. Fu: Department of Biology, University of California, M-0634, La Jolla, CA 92093-0634.

Current address, S. Du: Tularik, Inc, 270 E. Grand Ave, South San Francisco, CA 94080.

Current address, J. E. Mueller: Axlerod Institute, NY State Dept. of Health, 120 New Scotland Ave., Albany, NY 12201

Current address, Y. Zhang: Microbiological Associates, 9900 Blackwell Road, Rockville, MD 20580

Current address, J. Chen: Department of Chemistry, University of Delaware, Newark, DE 19716

DIMACS Series in Discrete Mathematics
and Theoretical Computer Science
Volume **44**, 1999

DNA Sequences Useful for Computation

Eric B. Baum

ABSTRACT. Recent proposals for DNA based computing [**1, 2, 3**] encode Boolean vector component values with sequences of DNA. It has previously been assumed that sufficient length random subsequences could be used to encode component values. However use of such subsequences will inadvertently result in long complementary subsequences. Complementary subsequences of sufficient length would stick to each other and cause mistakes or delays in computation. We suggest some constraints on DNA subsequences to be used in encodings, and describe maximal sets of subsequences satisfying these constraints. We also propose that by use of exotic materials, the problem of unintended complementarity can be greatly alleviated, potentially improving the chances for practical molecular computing.

1. Introduction

A seminal paper of Adleman [**1**] recently sparked growing interest in the use of DNA and the methods of molecular biology to do computation. Adleman's approach required the encoding of computer science problems into DNA sequences, and relied heavily on "extraction" of sequences containing a particular subsequence by use of a complementary subsequence. Subsequently Lipton [**2**] proposed an approach for using DNA to solve Satisfiability and other problems in the computational class NP. Lipton proposed a particular encoding of Boolean vectors and relied on similar extraction operations using complementary subsequences. Baum proposed to use a similar encoding, and some variants, for Content Addressable memories [**3**]. Several other proposed computer algorithms rely on similar encodings and methods [**4, 5**].

For each of these proposals, a set of DNA subsequences must be chosen. Practical considerations will force these subsequences to satisfy certain requirements. Lipton and Adleman suggested using random subsequences. However, practical requirements impose constraints that can not be met by random sequences, and it is not *a priori* obvious that they can be satisfied at all. This paper describes such constraints and proposes sets of subsequences which satisfy them. We also propose that use of exotic materials may greatly alleviate problems of unintended complementarity, potentially improving greatly the chances for practical molecular computing.

For concreteness we briefly review the encoding of Lipton [**2**]. One wishes to encode the Boolean vectors $\{0,1\}^n$, where n is about 60. For each $i = 1, ..., n$, one chooses two subsequences of DNA X_i and Y_i, corresponding respectively to a 0 or a 1 in the ith component. A vector in $\{0,1\}^n$ is then encoded by the concatenation of the appropriate subsequences, perhaps spaced by a fixed subsequence, or a subsequence corresponding to the number of the component[1]. This evidently requires at least 120 suitable subsequences. In some of the algorithms proposed [**4, 5**], the initial vectors are extended by appending additional subsequences in a similar fashion (say corresponding to a tag that the vector encoded satisfies some Boolean circuit) so that the number of subsequences needed will grow substantially, and may reach thousands or more. If enough suitable subsequences are not available, this will constrain algorithmic possibilities.

Let Z be a sequence of DNA. Then let $\bar{Z}$ denote the sequence of DNA which is Watson-Crick complementary to Z. The Watson Crick complement of the sequence Z is the sequence gotten by replacing A by T and C by G and vice versa, and then taking the sequence in the reverse order. For example if $Z = AGTCC$, then $\bar{Z} == GGACT$.

Now given an encoding as above, a key operation in many of the algorithms proposed is an "extract". Here one produces a subsequence $\bar{X}_i$ or $\bar{Y}_i$, and affixes it to a magnetic bead. Placing these magnetic beads into one's test-tube computer, the subsequences introduced bind[2] to any molecules already present containing the complementary subsequence. These bound molecules can now be extracted magnetically. This process allows one to search the testtube for vectors having particular component values.

Now we must impose some constraints on our molecules to avoid two difficulties. First, when we come to extract X_i, say, we want $\bar{X}_i$ to stick only to molecules containing X_i. If there is a sufficiently long subsequence in common to X_i and some molecule not containing X_i, we may have erroneous extracts. To avoid this we will require there be no long subsequence common to X_i and either any one of the other X_i or Y_i, or to the overlap region between some X_i or Y_i and whatever sequence may follow it on a molecule. Second, we don't want our molecule to stick to itself, or one of the other molecules in our soup. If this happened, it would gum up our computer and slow it down or make it unreliable. Thus we will require that there is no long subsequence such that both it and its complement may appear.

We will assume in this paper that a subsequence of length k is long enough to present a possibility of sticking, and seek not to have accidental collisions as long as k. It is not clear how large a k may be permissable in practice. It is possible that conditions may enforce k as low as 5, and seems likely that our computer will work more accurately and more rapidly the lower a k we can work with. Achieving low k values is a difficult constraint. If suitable subsequences can not be found,

[1]Given a directed graph, Adleman [**1**] described a procedure for constructing an associated set of DNA molecules by mixing in subsequences corresponding to all nodes in his graph, and subsequences corresponding to all edges, which served to splint the node sequences together. Lipton [**2**] views the encoding of Boolean vectors as a version of Adleman's directed graph construction for a certain fixed graph. In this view, each spacer subsequence must be distinct, or Lipton's graph would collapse, with all those nodes of degree four assigned the identical sequence being identified. We remark that, if instead of mixing in all DNA molecules simultaneously, as Adleman did, those corresponding to the distinct nodes in Lipton's graph are poured in sequentially, so that molecules are formed coding first for 2 bit strings, then for 3 bit words, etc., then molecules can potentially be built up coding for all n bit strings but using a single, identical spacer.

[2]This is an idealization. In practice, of course, some molecules may bind incorrectly or fail to bind where they should.

the entire enterprise of DNA based computing along the lines discussed to date is potentially in jeopardy.

Previous authors (e.g. [**1, 2**]) had noted that by using sufficiently long random sequences of bases as component subsequences X_i, the base itself would not appear. However if complementary subsequences of a finite length $k+1$ may stick, it is no help that the whole sequence is not present.

2. Useful Sequences

Thus we are motivated to ask the following question.

QUESTION 1. *Find a spacer sequence S_0 and a set of sequences $\{Z_j\}$ as large as possible with the property that there is no subsequence x of length k, which (a) occurs in any two sequences SZ_jS or (b) x occurs in some $S_0Z_jS_0$ and $\bar{x}$ occurs in some $S_0Z_iS_0$. We will also require that the Z_j are themselves at least of length k, so that they can be extracted on.*

As DNA computing evolves, related questions will be asked. For example, we might also ask for a set of spacer sequences, which we will not extract on, and a set of Z_j which we will. We hope analysis similar to that presented here can be applied to such generalizations.

We assert first that without loss of generality the sequences Z_j can all be taken to be of length k. If they are longer, this merely induces more constraints. Using longer sequences will thus require a smaller set, but may have practical advantages in the extract operation. We discuss this briefly at the end.

Next we observe that the set of sequences of length k are divided into equivalence classes by the operation Complement. There are 4^k strings of length k. If $k = 2m+1$ (for m an integer) then no sequence is equal to its complement, so there are $4^k/2$ equivalence classes each of size 2. If $k = 2m$, then there are 4^m sequences equal their complements (since each of the first m components can be chosen in any of 4 ways, thus determining a sequence if we demand it is equal its complement). So for even k we have $(4^k - 4^m)/2$ equivalence classes of size 2 and 4^m of size 1.

For future reference, we note that if we restrict to sequences neither beginning nor ending with A nor T, then if $k = 2m+1$ we find analagously that there are $4^{k-1}/2$ equivalence classes each of size 2, and if $k = 2m$ there are $2 \times 4^{m-1}$ classes of size 1 and $2 \times 4^{k-2} - 4^{m-1}$ classes of size 2.

Say we choose as spacer sequence k X's, for X some base. Then it is apparent that we can form a large set satisfying our constraints by choosing one element from each equivalence class except those which either begin or end with X or $\bar{X}$.

However we can build a larger set. For concreteness, let the spacer be k A's. So that the above rule would not have allowed sequences beginning or ending with T's or A's. Assume for the moment $k = 2m+1$ so there are no equivalence classes of size 1. For each string S of length $k-1$, not beginning nor ending with A or T, and chosen such that $S \neq \bar{S}$, consider its complement $\bar{S}$. We may choose ST and TS, provided we do not choose $\bar{S}T$ nor $T\bar{S}$. Having accumulated each pair ST and TS we must then strike out of consideration, from every equivalence class in the set of length k strings, any that either begin with or end with $\bar{S}$. What we hope is that this will not remove from consideration *both* elements of any equivalence classes. When does this happen? For both elements of an equivalence class to be removed, we must remove both a string SX and the string $\bar{X}\bar{S}$. This would happen when we have added two strings S_1T and TS_2 such that $\overline{\bar{S}_1 X} \equiv \bar{X}S_1 = Y\bar{S}_2$. But this implies $\bar{S}_1 = S_2$, which we have forbidden.

Note that, for those $k-1$ strings which are self complementary, we can not both add ST, and TS, and any representative of the class XS, for $X = \{C, G\}$. So we exclude ST and TS of this form.

Thus for each size 2 equivalence class of $k-1$ length strings not beginning nor ending with A nor T, we may choose a representative S, and add both TS and ST. We additionally may add a representative from every length k class which does not end or begin with A or T, selecting a member still allowed after the necessary cross offs. Following the count of number of equivalence classes from above, this simple algorithm allows us to select a set of size $2 \times (2 \times 4^{k-3} - 4^{m-1}) + 2 \times 4^{k-2} = 3 \times 4^{k-2} - 2 \times 4^{m-1}$.

Can we build an even larger set? If you add ST, then you can't add $\bar{S}A$, nor $A\bar{S}$. If you add SA, then you would not be able to add XS, and you would also be unable to add $\bar{S}\bar{X}$. Likewise AS is ruled out. So there is no possible gain in terms of adding strings beginning or ending with A.

However it is possible to create a larger set, as described below, by using length $k-1$ sequences which are permitted to end in A and then appending a T.

We denote by R_i^j a string of length i whose first element is neither A nor T. The superscript here serves solely to denote different such strings and will be omitted where no confusion is possible. Let Y^j denote a single base either C or G. Again the superscript is only to distinguish different elements. So Y is simply an alternative name for R_1.

If $S = R_{k-2}A$, then $\bar{S}$ begins with T. We already know that if we add ST, we can not add $\bar{S}T$. So we can either choose to consider adding strings ending in AT, or ending and beginning in T, but not both. We consider the set of strings of form $R_{k-2}AT$. First note that we can safely add all strings of form $R_{k-3}AAT$ or $R_{k-3}TAT$, precisely because[3] we have excluded strings which both end and begin with T.

Next we seek to add strings of form $R_{k-3}YAT$, for $Y \in \{C, G\}$. It will be clearer if we denote these Q^iAT, where the i indexes different strings, the Q are of length $k-2$, and they neither begin nor end with A nor T. When we add a string of form QAT, then we can not add a string of form $T\bar{Q}X$. So we will seek instead to include $T\bar{X}Q$. This will be fine unless

$$\bar{X}Q^i = \bar{Q}^j Y \tag{1}$$

for any Y, where we have included both Q^i and Q^j. To insure this does not happen, each time we add a string Q^iAT to our set, we must cross off conflicting $Q's$. That is, each time we add a Q, we must construct two others by $CQ = \bar{Q}'Y$, and $GQ = \bar{Q}''Y$ and cross Q' and Q'' off from consideration. However, each time we add $Q = R_{k-3}C$, we may add also $R_{k-3}G$, with no further cross offs. So for each two Q's we add, we must cross off two. Now note, equation 1 will never cross off any potential Q which has an A or a T in its second to last position. That is: because Q^i in equation 1 never begins with A nor T, by definition, any Q^j satisfying this equation on the right hand side will not have an A nor a T in its second to last position.

[3] That is, if we call $R_{k-3}AA$ string S, adding ST would conflict with $\bar{S}T$ but we are not adding this since it both begins and ends with T, so there is no conflict.

In summary, this procedure gets the following strings. All strings of form $R_{k-3}AAT$ and $R_{k-3}TAT$ and of form $R_{k-4}ACAT$, $R_{k-4}TCAT$, $R_{k-4}AGAT$, and $R_{k-4}TGAT$. Half of the strings of form $R_{k-4}CCAT$, $R_{k-4}GCAT$, $R_{k-4}CGAT$, $R_{k-4}GGAT$. Half of the strings of form ST for $S \neq \bar{S}$ and $S = R_{k-2}C$ or $R_{k-2}G$, and for each of these the corresponding strings TS. A representative of each class neither ending nor beginning with A nor T. Assuming $k = 2m + 1 \geq 5$, this works out to be $3 \times 4^{k-2} + 5 \times 4^{k-4} - 2 \times 4^{m-1}$.

If $k = 2m$, virtually identical analysis yields a set consisting of:(1) all strings of form $R_{k-3}AAT$ and $R_{k-3}TAT$, (2) all strings of form $R_{k-4}ACAT$, $R_{k-4}TCAT$, $R_{k-4}AGAT$, $R_{k-4}TGAT$; (3) half of the strings of form $R_{k-4}CCAT$, $R_{k-4}GCAT$, $R_{k-4}CGAT$, $R_{k-4}GGAT$; (4) half of the strings of form $R_{k-2}CT$ and $R_{k-2}GT$ and for each the corresponding string of form $TR_{k-2}C$ and $TR_{k-2}G$, and a representative of each length k class of size 2 neither beginning nor ending with A nor T. This works out to be also $3 \times 4^{k-2} + 5 \times 4^{k-4} - 2 \times 4^{m-1}$.

Is there any possibility of a still larger set? We have representatives of each size 2 class not beginning nor ending with A nor T. As discussed, considering adding classes beginning or ending with A will exclude more additional strings than in this set. We have a representative of every possible class of form ST or TS where S neither begins nor ends with A nor T. As discussed, considering strings both beginning with and ending with T will do worse than the construction given. So the only additional freedom is to consider strings of form ST where S ends in T or A. The reader can convince himself that using strings ending in T will cost more than it gains. We have analyzed the set of strings ending in AT in order to get this maximal set. Thus we see that if any string ommitted here is included, at least an equal number of strings included here would have to be ommitted.

The reader should, with thought, be able to satisfy himself that using a different sequence as spacer will not expand the set. By using k A's, we have been able to rule out all translations into the spacer with only a requirement that neither the first nor last letter of any sequence be A. If we used a more complex spacer sequence, we acquire additional constraints. If we used a spacer sequence beginning with k A's, and ending with k copies of some other letter, we could generalize our construction, but would not achieve a larger one.

We have:

THEOREM 1. *The maximum size set of strings satisfying question 1, for* $k = 2m + 1$ *or* $k = 2m$, $k \geq 5$, *is* $3 \times 4^{k-2} + 5 \times 4^{k-4} - 2 \times 4^{m-1}$.

Note, one could imagine modifying question 1 by using different spacers S^i to come after different sequences. We do not believe a larger set could be produced in this way, but have not investigated the question in exhaustive detail.

The constraints are quite restrictive for small k. The number of sequences we can have in our set is given below for various values of k.

k= 4	5	6	7	8	9
N(k)= 50	208	816	3232	13440	54144

Clearly k must be at least five for any hope of interesting applications. k as low as five may however be consistent with many DNA computing applications, potentially speeding up and avoiding errors in retrieval. If it is practically possible to work with large k, say as large as 9 for example, we might consider sacrificing the number of distinct sequences achievable in order to work with longer sequences. If we work with length L sequences, for $L > k$, then in an extract step we would have a stronger fit between the retrieval sequence and its mate. The number of such sequences achievable is certainly no larger than $N(k)/(L-k)$. We have not yet investigated how large a set may actually be constructed.

3. Exotic Materials

At the workshop, Nadrian Seeman remarked in a question period that one need not be restricted to the bases used by biology. In particular, [**8**] incorporated an exotic new base. Use of this would of course allow greater numbers of sequences satisfying Question 1. Theorem 1 above could be readily generalized to this case. One objection (raised in personal communication by H. Rubin) is that DNA made with these exotic bases may not be very stable under PCR, which might limit the flexibility of algorithms if they are incorporated.

Another interesting possibility is to use as a spacer some inert strand rather than single strand DNA. What we really want for DNA computing are active regions where we can write information, and on which we can do extract operations. Spacers have been proposed here as a method to alleviate unintended complementarity, particularly arising at the overlap of two coding regions. But as such, the optimal spacer would probably be fairly rigid (preventing the molecule from curling up and binding to itself) and inert. Quite plausibly some exotic material would serve much better than, e.g. the strand of Adenines discussed previously in this paper.

When I asked him about this, Nadrian Seeman suggested that branching phosphoramidites[**7**] might be an example of what I am looking for. These are commercially available[**6**]. These allow one to use as a spacer a sequence of double stranded DNA which is strongly bound, and so will not separate into single strands if one is careful.

If one uses novel inert spacers of this type, then the analysis of the paper is superfluous, but one can possibly achieve much better results.

4. Conclusion

In conclusion we have discusses constraints on encodings to be used in DNA based computers arising from the possibility of undesired complementarity. Two types of constraints were noted. One arose to try to minimize errors in extraction, when a string complementary to an encoding region is introduced, one doesn't want it to stick other regions. The other constraint arose from the possibility of one's coding sequences sticking to themselves or other coding sequences, turning the computer into jello. We gave an encoding which hopefully will minimize these types of errors. We also discussed some novel materials which may greatly alleviate these problems, potentially making DNA Computing more practically realizable.

I thank Peter Kaplan and Dan Boneh for conversations, and Nadrian Seeman for noting at the workshop that one might use more than the four bases provided by nature, and calling my attention to branching phosphoramidites.

References

[1] L. M. Adleman, Molecular computation of solutions to combinatorial problems. *Science*, 266: 1021-1024, 1994.

[2] R. J. Lipton, DNA Solution of Hard Computational Problems. *Science*, 268:542-544, 1995.

[3] E. B. Baum, Building an associative memory vastly larger than the brain. *Science*, 268:583-585, 1995.
[4] D. Boneh, C. Dunworth, R. Lipton, J. Sgall, *On the computational power of DNA*, preprint, available at http://www.CS.Princeton.EDU/ dabo/biocomp.html.
[5] D. Boneh, C. Dunworth, R. Lipton, *Breaking DES using a molecular computer*, In R. J. Lipton and E. B. Baum, editors, *DNA Computers: Proceeding of a DIMACS Workshop*. DIMACS Series in Discrete Mathematics and Theoretical Computer Science, American Mathematical Sociey, 37-65, 1996.
[6] Clontech Laboratories Inc. Catalog, Palo Alto CA, Product number 5251 in 1995-96 catalog.
[7] T. Horn, M. S. Urdea, Forks and combs and DNA. *Nucleic Acids Res.*, 17:6959-67, 1989.
[8] J. A. Piccirilli, T. Krauch, S. E. Moroney, and S. A. Benner, Enzymatic Incorporation of a New Base Pair into DNA and RNA Extends the Genetic Alphabet. *Nature*, 343:33-37, 1990.

NEC Research Institute, 4 Independence Way, Princeton, NJ 08540
E-mail address: eric@research.nj.nec.com

DIMACS Series in Discrete Mathematics
and Theoretical Computer Science
Volume **44**, 1999

A Restricted Genetic Alphabet For DNA Computing

Kalim U. Mir

1. Introduction

Since Adleman [**1**] demonstrated his original groundbreaking scheme, a simpler approach suggested by Lipton [**7**] has widened the range of problems that can be addressed by DNA computing. A single molecular operation, DNA annealing, is required for Lipton's scheme. This also forms the basis of Baum's proposal for a content-addressable DNA memory [**3**]. In both cases an extractor or cue oligonucleotide, most likely to be in the solid-phase, attached to beads, would anneal to a longer single-stranded target present in the graph or memory. So far, most work on DNA computing has rightly concentrated on what is theoretically possible. Here however, I will discuss some practical issues and offer a means to overcome some practical obstacles.

2. Sequencing the Solution

For a real world application of DNA computing it is not enough only to know that a solution exists; we need to know what it is. Hence, a means for reading the molecular output of the computation or retrieval is required. However, conventional means of sequencing bases require enzymatic reactions and gel chromotography and are slow, expensive, laborius, and would not be easy to integrate into a DNA Computer. An attractive alternative approach that is currently being developed at Oxford and elsewhere, called Sequencing by Hybridisation (SbH), would offer a one-step automated process which could easily be integrated into a future DNA computer. This method takes the concept of sequence-specific recognition of DNA by short oligonucleotides to its logical conclusion: target strands are hybridised to a complete set of oligonucleotides (for instance, every possible 8mer) and the sequence of the target is reconstructed from the hybridisation data obtained [**2**]. In the most common implementation [**13**] a target strand is hybridised to an array of oligonucleotide probes, synthesised on a flat solid surface.* Of course if more than one solution is likely to be obtained from the DNA computation then the strands encoding the solution need to be separated before SbH can be performed on each of them individually. Conventionally, this would be done by cloning the

*In another implementation [**6**] oligonucleotides are serially hybridised to an immobilised target.

solutions. However, an alternative *in vitro* approach [**5**] which may be quicker should be considered.

It should be pointed out that SbH is experimentally related to DNA computing in that it also requires annealing of an oligonucleotide to a longer single-stranded target. As with DNA computing, much effort has been expended in developing the theoretical potential of SbH. In fact, highly sophisticated methods for fabrication of high-density oligonucleotide arrays called "DNA Chips" have also been developed [**11**]. However so far, although it has been demonstrated with a model DNA consisting entirely of pyrimidines [**13**], the approach has floundered when the full four base genetic alphabet is retained in each strand [**9**].

3. The Effect of Intramolecular Structure on DNA Annealing

The failure of SbH in application to targets containing all four bases can be attributed to intramolecular folding within the target molecule which hinders the intermolecular interaction between target and probe. This results in the occurrence of false negatives, that is, oligonucleotides on the array that fail to hybridise to complementary sequences which are present in the target. For oligonucleotide annealing to achieve sufficient discrimination of mismatches the length of oligonucleotide should be short. However, under the conditions required for a short oligonucleotides to achieve stable duplex with a long single-stranded target, intramolecular interactions within the target strand become a dominant feature [**10, 14**]. This would be as much of a problem for DNA computing as it is for DNA sequencing and would be a major source of errors. Using longer oligonucleotides coupled with higher annealing temperatures may reduce the problem but would concomitantly lead to a higher rate of mismatch. Folding of the target strand, is of course not a problem when either purines or pyrimidines – nor when combinations of adenine (A) and cytosine (C) or thymine (T) and guanine (G) – are only permitted in the strands. These strands have no potential for internal Watson-Crick interactions. Hybridisation of a polypyrimidine target to an array of complementary oligonucleotides has shown a uniform level of yield from all oligonucleotide complements. This is in marked contrast to the large variation in hybridisation yields obtained when a natural target is hybridised to a complementary array [**12**].

4. A Restricted Alphabet

The above puts forward a strong case for employing a restricted genetic alphabet for coding in DNA computing. A two base alphabet could give 4096 12mer and 32,768 15mer variables; the test tube contents would be resistant to intramolecular as well as intermolecular interactions based on canonical base-pairing. If relevant applications can be found, a mixed pool of strands each composed of either {A, G}, {T, C}, {A, C} or {T, G} combinations would quadruple the number of possible variables; the test tube contents would be resistant to intramolecular interactions but not intermolecular interactions. As the latter are bimolecular they would not pose as much of a problem as the unimolecular intramolecular associations. A dividend of an all purine or all pyrimidine encoding would be that the most stable end mismatch, G.A, would not occur between oligonucleotide and target; G and A would be on the same strand. A comprehensive set of purine-pyrimidine hybridisation data are already available in the literature which would faciliate selection of suitable oligonucleotides. It has been found that the use of 4.4M Tetramethylammonium chloride (TMACl) as the hybridisation solvent optimally reduces the effect of base composition on hybridisation as well as dramatically increasing yield [**8**]. The use of TMACl would delimit one potential drawback of a restricted alphabet as explained in the following.

Reducing the number of letters in the genetic alphabet reduces the number of variables of a certain length that are available. Hence, as variables would be chosen from a smaller pool the sequence difference between variables would be less and so there would be a greater chance of annealing between incomplete matches. However, if the effects of base composition are minimised by using TMACl, duplex stability would become more length dependent and it may be possible to discriminate between partial and complete matches by employing a temperature at which only the full match is stable. So, although there would be an increase in the occurrence of partial matches, these would be transient reversible associations. Any slowing down effects that this may have would be offset by the speed-up in finding the true match that is gained by such partial matches effectively increasing the concentration of oligonucleotides around the target strand and, the reduction in dimensionality in the diffusion process [**4**]. Hopping of the oligonucleotide from one partially matched site to another would occur until the stable – defined by the conditions used – perfect match is found. Much of the search would be reduced from a 3-D process to a faster 2-D process.

5. Conclusion

Adoption of a restricted genetic alphabet may offer advantages for DNA Computing. Moreover, such a restricted two base coding would open the way for the use of SbH for reading the output of the computation or retrieval.

6. Acknowledgements

Special thanks to Gareth Williams. Thanks also to Len Adleman, Paul Rothemund, Uwe Maskos, Ed Southern, K. Ohkuma, N. Gemma and K. Ando.

References

[1] L. Adleman, Molecular Computation of Solutions to Combinatorial Problems. *Science*, 266:1021-1024, (Nov. 11) 1994.

[2] M. Baringa, Will DNA Chip Speed Genome Initiative? *Science*, 253: 1489, (Sept. 27) 1991.

[3] E.B. Baum, Building an Associative Memory Vastly Larger Than the Brain, *Science*, 268: 583-585, (Apr. 28) 1995.

[4] V. Chan, D. J. Graves and S. E. McKenzie, The Biophysics of DNA Hybridization with Immobilized Oligonucleotide Probes, *Biophysical Journal*, 69:2243-2255, (Dec.) 1995.

[5] H.V. Chetverina and A. B. Chetverin, Cloning of RNA Molecules in vitro. *Nucleic Acids Research*, 21: 2349-2353, 1993.

[6] R. Dramanac, Z. Dramanac, Z. Strezoska et al, DNA Sequence Determination by Hybridization: A Strategy for Efficient Large-Scale Sequencing. *Science*, 260: 1649-1652, (Jun. 11) 1993.

[7] R. Lipton, DNA Solution of Hard Computational Problems. *Science*, 268:542-545, (Apr. 28) 1995.

[8] U. Maskos and E. M. Southern, A study of oligonucleotide reassociation using large arrays of oligonucleotides synthesised on a glass support. *Nucleic Acids Research*, 21: 4663-4669, (Oct. 11) 1993.

[9] U. Maskos; unpublished data.

[10] K.U. Mir and E.M. Southern; manuscript in preparation.

[11] A.C. Pease, D. Solas, E.J. Sullivan, M.T. Cornin, C.P. Holmes and S.P.A. Fodor, Light-generated oligonucleotide arrays for rapid DNA sequence analysis. *Proc. Natl. Acad. Sci. USA.*, 91: 5022-5026.

[12] E. M. Southern, S. C. Case-Green, J.K. Elder, M. Johnson, K.U. Mir, L.Wang, and J.C. Williams, Arrays of complementary oligonucleotides for analysing hybridisation behaviour of nucleic acids. *Nucleic Acids Research*, 22: 1368-1373, (Apr. 25) 1995.

[13] E.M. Southern, U. Maskos and J.K. Elder, Analyzing and comparing nucleic acid sequences by hybridization to arrays of oligonucleotides: evaluation using experimental models. *Genomics*, 13: 1008-1017, (Aug) 1992.

[14] R. Wu, Nucleotide Sequence Analysis of DNA, *Nature: New Biol*, 236: 198-200, (Apr. 19) 1972.

Current address: Advanced Research Laboratory, Toshiba Research and Development Center, 1, Komukai Toshiba-cho, Saiwai-ku, Kawasaki 210, Japan

E-mail address: `kalim@arl.rdc.toshiba.co.jp`

DEPARTMENT OF BIOCHEMISTRY, UNIVERSITY OF OXFORD, SOUTH PARKS RD, OXFORD OX1 3QU, UK.

E-mail address: `kalim@bioch.ox.ac.uk`

DIMACS Series in Discrete Mathematics
and Theoretical Computer Science
Volume **44**, 1999

Good Encodings for DNA-Based Solutions to Combinatorial Problems

R. Deaton, R. C. Murphy, M. Garzon, D. R. Franceschetti, and S. E. Stevens, Jr.

ABSTRACT. Adleman has solved the Hamiltonian path problem by encoding the vertices and edges of the graph in oligonucleotides of DNA, hybridizing the oligonucleotides to produce potential answers, and extracting any DNA which corresponds to the Hamiltonian path. Depending on the conditions under which the DNA reactions occur, two oligonucleotides can hybridize without exact matching between their base pairs. This possibility was verified by experiment. For DNA-based solutions to combinatorial problems to become a viable and practical technology, the possibility of false positives must be eliminated. The primary mechanism for the production of false positives is hybridization stringency that depends on the reaction conditions, of which the most important is temperature. Evidence is provided that encoding the vertices and edges of the graph in DNA oligonucleotides that are a minimum distance apart results in reliable encodings that virtually eliminate the risk of false positives. A genetic algorithm was shown to be useful to search the space of possible codewords. The Hamming bound is shown to be an upper bound on the number of reliable encodings. Laboratory results confirmed that the choice of good encodings is very dependent on the reaction conditions.

1. Introduction

Using techniques from molecular biology, Adleman [**1**] has discovered a method for solving hard, combinatorial problems. His example problem was the directed Hamiltonian path problem, which is NP-complete [**4**]. The vertices and edges of the graph were encoded in oligonucleotides of DNA from which the Hamiltonian path was produced through the processes of hybridization, ligation, and amplification. Later, Lipton [**8**] modified and extended Adleman's technique to SATISFIABILITY and other NP-complete problems. For both efficiency and error tolerance, Baum [**2**] has studied constraints on DNA subsequences. Specifically, he bounds the size of a set of spacer sequences and information sequences such that no sequences of length k stick.

Both Adleman and Lipton used random encodings for their respective problems. The assumption was that a random encoding was adequate. Adleman, however, does point out that as the size of the problem grows, particular attention must be paid to errors and the formation of "pseudopaths," *i.e.* false positives. In this paper, the goal is to study the types of errors that lead to false positives in Adleman's original technique. Efficiency of the encoding is a related issue. In addition, given a large enough batch of DNA-encoded vertices and edges, false negatives, or answers to the problem which exist, but which the DNA search cannot find, are assumed not to occur.

Computing based on DNA has the potential to exploit massive parallelism in solving large, combinatorial search problems. Its success, however, depends on its reliability. A reliable technique does not produce false positive solutions, or encodings that produce sequences which appear to be valid Hamiltonian paths, but actually are not. In this paper, encodings of problems in DNA that produce reliable solutions are investigated, and a bound on the size of problems that can be solved reliably is given. A genetic algorithm (GA) was used to generate encodings for DNA search. The actual occurrence of false positives, and the reliability of encodings, suggested by a GA, were verified experimentally.

The efficiency of the encoding relates to the fraction of DNA oligonucleotides that actually contribute to the result. Depending on reaction conditions, oligonucleotides can hybridize in ways that prevent their contribution to the DNA solution. Hybridizations that affect efficiency were simulated in the GA which, then, was able to produce efficient encodings for DNA search.

Hybridization stringency refers to the number of complement base pairs that have to match for DNA oligonucleotides to bind. Hybridization stringency depends on reaction conditions (such as sodium concentration, temperature, and relative percentage of $A+T$'s to $G+C$'s), and it increases with temperature [**3, 7**]. Temperature is the most significant variable, and attention will be focused on it. Therefore, sequences which at one temperature might have to match exactly to hybridize, at another temperature, might only have to match at some fraction of the total length of the string. Since the oligonucleotides bind with less than exact or perfect matching of complementary base pairs, there is a potential source of error. The quality of the encoding is a function of temperature, and therefore, reliability of the results is dependent on the experimental conditions for hybridization.

Good encodings are both reliable and efficient. In this paper, the analysis concentrates on the properties and generation of good codes. The Hamming bound is shown to be an upper bound on the size of a problem (*i.e.* number of vertices in the graph) that can be solved by Adleman's technique without the possibility of false positives. A genetic algorithm (GA) was used to search for good encodings as the percentage of oligonucleotide matching required for hybridization, and error-mechanisms changed. The genetic algorithm proved to be a good tool for searching for encodings. The GA was also used to investigate the efficiency of the encodings and to verify the Hamming bound. In this work, good encodings must satisfy a minimum distance property, or must all be a certain distance from each other. Laboratory results showed the possibility of false positives, and the goodness of GA-generated encodings.

2. Methodology

The vertices and edges of the graph are encoded in oligonucleotides, or single strands, of DNA [**11**]. The encoding alphabet consists of the nucleic acid bases A, T, G, C, which bind according to the Watson-Crick complement condition, $\overline{A} \equiv T$

$$< q_j| < p_j|p_i > |q_i >= \left\{ \begin{array}{cc} {}^{5'}p_i & q_i^{3'} \\ & \| \\ {}^{3'}q_j & p_j^{5'} \end{array} \right. \qquad |p_i >< q_j|q_i >< p_j| = \left\{ \begin{array}{cc} {}^{5'}p_i & q_i^{3'} \\ & \| \\ {}^{3'}q_j & p_j^{5'} \end{array} \right.$$

(a) (b)

FIGURE 1. Splice operation of Definition 2.

and $\overline{G} \equiv C$. The chemically distinct ends, $5'$ and $3'$, of complementary DNA sequences are bound together by hybridization. The enzyme ligase seals the double stranded DNA in a process called ligation. These are the essential processes considered in this work. In addition, Adleman used polymerase chain reaction (PCR) to extract the answer from knowledge of the beginning and ending vertices, and from the length of the path.

A $5'$ to $3'$ oligonucleotide is denoted as $|p_i >$, and a $3'$ to $5'$ oligonucleotide as $< p_i|$, where p_i is an arbitrary sequence of nucleotide bases. The Watson-Crick complement of an oligonucleotide, $|p_i >$, is $< \overline{p}_i|$. A hybridization and ligation reaction between two arbitrary oligonucleotides, $< p_i|$ and $|p_j >$, is $< p_i|p_j >$, where perfect complement base pair matching occurs if $< p_i| =< \overline{p}_j|$. In Adleman's approach, each vertex is associated with a random n-mer, $|O_i >$.

DEFINITION 1 (Join). *Let $|p_i >$ and $|q_i >$ be two DNA oligonucleotides. The operation $\vee$ ("join") is defined as $|p_i > \vee |q_i >= |p_i q_i >$, which corresponds to the $3'$ to $5'$ phosphodiester link from p_i to q_i. Likewise, $< q_i| \vee < p_i| =< q_i p_i| = |p_i q_i >$.*

Each n-mer, $|O_i >$, consists of the join of the $\frac{n}{2}$-mers, $|p_i > \vee |q_i >= |p_i q_i >$. The sets of $\frac{n}{2}$-mers, $|q_i > \in Q$ and $|p_i > \in P$, are two distinct sets of vertex codewords, a context which will become clear in the next section. The edges, $i \to j$, in the graph are formed by joining the Watson-Crick complement of $|q_i >$ ($< \overline{q}_i|$) to the complement of $|p_j >$ ($< \overline{p}_j|$) which results in the edge, $< \overline{q}_i| \vee < \overline{p}_j| =< \overline{q}_i \overline{p}_j|$. To determine a path (vertex-vertex ...), the operation of splicing is defined as follows.

DEFINITION 2 (Splice). *Let $|p_i q_i >$ and $|p_j q_j >$ be DNA oligonucleotides. The operation $\wedge$ ("splice") is defined as the production of a potentially double stranded overlap between $|p_i q_i >$ and $|p_j q_j >$, and can result in two possibilities (Figure 1) that could lead to false positives,*

$$< q_j p_j| \wedge |p_i q_i >= \left\{ \begin{array}{c} < q_j| < p_j|p_i > |q_i > \\ or \\ |p_i >< q_j|q_i >< p_j| \end{array} \right. .$$

A path through a graph is formed through hybridization and ligation by the edge oligonucleotides, $< \overline{q}_i \overline{p}_j|$, splicing together the sequence of vertices,

$$|p_i q_i >, \ |p_j q_j >, \dots .$$

Therefore, the formation of a vertex-edge-vertex combination is

$$\big(|p_i q_i > \wedge < \overline{q}_i \overline{p}_j|\big) \wedge |p_j q_j > .$$

For this work, a match was defined as the hybridization of an $\frac{n}{2}$-mer to another $\frac{n}{2}$-mer. To produce false positives, $\frac{n}{2}$-mers must hybridize with a number of complementary base pairs that are in sequence, but not necessarily consecutive. The number of complementary base pairs needed for hybridization is a function of the reaction conditions, including temperature. An *error* was defined as a non-complementary base pair in a hybridized strand ($< A\underline{G}CTC\underline{G}|T\underline{A}GA\underline{T}C >$, where the non-complementary base pairs are underlined), and a *mismatching hybridization* as a bound pair of oligonucleotides that contain at least one error.

When a hybridization looks like a valid vertex–edge splice, a mismatch potentially produces a false positive to the Hamiltonian path problem. Mismatches that might produce false positives involve hybridizations among the p_i's, and among the q_j's, but not between p_i's and q_j's. They are a vertex–reverse vertex splice,

$$\begin{array}{c} < O_j| \ \wedge \ |O_i > \\ \text{or} \\ |p_i >< q_j|q_i >< p_j|, \text{ where } < q_j| \neq < \overline{q}_i|, \end{array} \tag{1}$$

an edge–reverse edge splice,

$$\begin{array}{c} < O_{k\to l}| \ \wedge \ |O_{i\to j} > \\ \text{or} \\ |\overline{q}_i >< \overline{p}_k|\overline{p}_j >< \overline{q}_l|, \text{ where } < p_k| \neq < p_j|, \end{array} \tag{2}$$

and a wrong vertex–edge splice

$$\begin{array}{c} < O_{k\to l}| \ \wedge \ |O_i > \\ \text{or} \\ |p_i >< \overline{q}_k|q_i >< \overline{p}_l|, \text{ where } < \overline{q}_k| \neq < \overline{q}_i|, \end{array} \tag{3}$$

Other hybridizations, which will affect the efficiency of the encoding, but will not produce false positives, are a hairpin match,

$$< q_i|p_i >, \tag{4}$$

and a non-splicing match,

$$\begin{array}{c} < O_j| \ \wedge \ |O_i > \\ \text{or} \\ < p_j| < q_j|p_i > |q_i > . \end{array} \tag{5}$$

A shareware genetic algorithm (GA) [**5**], written in C and run on a Sun SPARCstation LX, was used to search for good encodings. Genetic algorithms conduct many simultaneous, parallel searches of a solution space [**6**]. Exhaustive search for good encodings produces a combinatorial explosion as the size of the n-mer to represent each vertex, and the number of vertices in the graph increases. Therefore, the GA was chosen as a more practical method to search the space of n-mers for good encodings. In addition, the GA can incorporate mismatch mechanisms that can occur, but that do not lend themselves to analytical or laboratory results. For the GA, the crossover probability was 60%, and the mutation rate was 0.1%.

Initially, the graph from Adleman's paper was used. The vertex encodings were represented with strings of binary digits which were interpreted two at a time as a specific base. The encoding of the edges was determined from that of the vertices. A fitness function was written to evaluate the encoding for errors of the type shown in Equations 1 - 5. The fitness function generated the edges for the specific graph from the vertex encoding, and checked for mismatches between vertices and vertices,

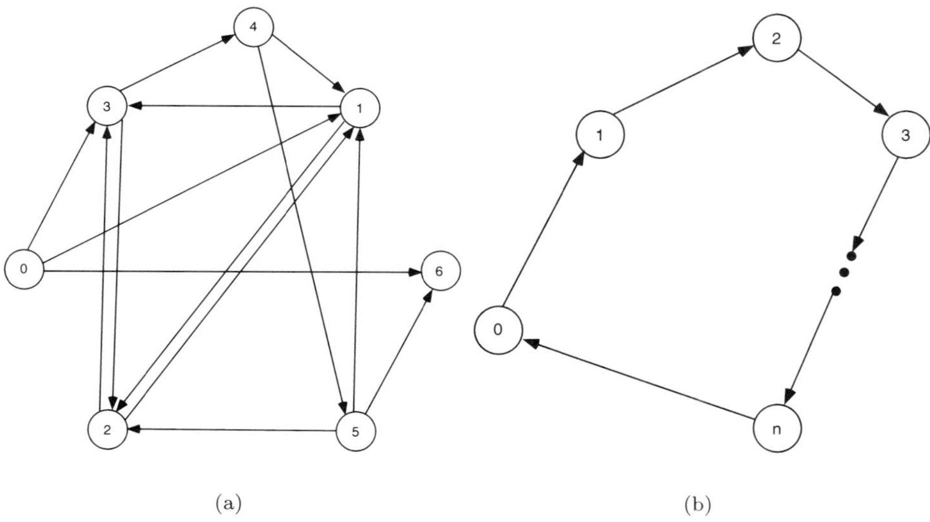

FIGURE 2. (a) Graph from Adleman. The Hamiltonian path is $0 \to 1 \to 2 \to 3 \to 4 \to 5 \to 6$. (b) Graph for verification of Hamming Bound and experimental trials. For experiments, $n = 6$.

edges and edges, and vertices and edges. The fitness, which the GA minimized, was the total number of mismatches. In addition, functions were written to convert the binary strings to nucleic acid base representation, to analyze the distribution of bases in the encoding, and to calculate the distance properties of the vertex codes.

Using the graph of Figure 2b with $n = 6$, molecular biology techniques were used to show the possibility of false positives and the goodness of the GA generated codes. The GA was used to generate 20-mers to represent the vertices of the graph. In the GA, strands were allowed to match with 5/10 complementary base pairs. All mismatch mechanisms (Equations 1-5), were incorporated into the GA. With the specified mismatch mechanisms, only 18 vertices could be encoded without the possibility of mismatches, even though the Hamming bound (described next, Eq. 6) for the 10-mer codewords of sets Q and P is 24. Each vertex codeword was at least a distance of 6 from any other 10-mer codeword that was a member of the same set, Q or P. Out of the 18 possible vertex oligonucleotides, $|O_i >$, a random 7 were chosen for laboratory work (Table 1). Following Adleman [**1**], the edges were generated from the vertices. While Adleman did his experiments at room temperature, the temperature was controlled for these experiments at 20°C. Hybridization, ligation, and agarose gel electrophoresis were done to determine that a path of the correct length was generated with the GA generated encoding. In addition, an experiment was done in which no Hamiltonian path existed in order to verify the event of false positives.

3. Theoretical Result

Recall that the Hamming distance between two strings is the number of corresponding places where two characters differ.

<table>
<tr><th>Vertex</th><th>Oligonucleotide Encoding</th></tr>
<tr><td>0</td><td>|TACTCATATGGGGTTATACG></td></tr>
<tr><td>1</td><td>|CTCCGCCTGGGCTTAGCTTA></td></tr>
<tr><td>2</td><td>|GATCCTCTGTTTCCTCAGCT></td></tr>
<tr><td>3</td><td>|GGCTCCACTTACTCTCTTGT></td></tr>
<tr><td>4</td><td>|TATGGGCTAGCGGTCCGGTT></td></tr>
<tr><td>5</td><td>|GCCCTTGTAGTCTCGGGTCC></td></tr>
<tr><td>6</td><td>|TTCCTGTAACTTGCCTCTAA></td></tr>
</table>

TABLE 1. Oligonucleotide encodings of vertices for experimental trials.

THEOREM 1. *For Adleman's solution to the Hamiltonian path problem, an upper bound on the number of vertices, $|C|$, that can be encoded in oligonucleotides of length n without producing mismatched splices between vertex and edge encodings is given by the Hamming bound* [**9**],

$$|C| \sum_{i=0}^{t} \binom{n/2}{i} (q-1)^i \leq q^{\frac{n}{2}}, \tag{6}$$

where t is related to the minimum distance, $d_{min} = 2t+1$, between encodings, and q is number of symbols in the alphabet ($q = 4$ for DNA).

PROOF. Consideration is restricted to vertex–edge mismatches, Equation 3. Each valid vertex–edge splice is in one of the two forms given in Definition 2, with $q_j = \overline{q}_i$ or $p_j = \overline{p}_i$, respectively. From now on, attention will be restricted to the second case in Definition 2 (Figure 1b), with the understanding that the argument also applies in an independent way to the first case (Figure 1a). A mismatched splice is one in which $q_j \neq \overline{q}_i$ in a hybridized strand. According to the reaction conditions of the hybridization, a number, d_{hyb}, of errors is allowed between oligonucleotides for hybridization, or splicing, to occur. The possibilities for matches between vertex and edge encodings involves matches between the q_i's and $\overline{q}_i$'s. If we allow d_{hyb} errors for hybridization, then, the vertex encodings, q_i's, must be a Hamming distance, d_{min} greater than d_{hyb} apart to insure against false vertex–edge splices. If the distance between q_i and q_j, $d(q_i, q_j) \leq d_{\text{hyb}}$, for i and j two distinct vertices, then, q_i can hybridize to $\overline{q}_j$, and vice versa, and produce false vertex–edge splices. If no false vertex–edge splices occur, then, no false positives will be produced. The same argument applies to the p_i's, independently from the q_i's. Therefore, we can view each of the two sets of $\frac{n}{2}$-mers, Q and P, as a code [**9**], C, that must satisfy a minimum distance property, $d(q_i, q_j) \geq d_{\text{min}}$, $\forall q_i, q_j \in Q, i \neq j$, and $d(p_i, p_j) \geq d_{\text{min}}$, $\forall p_i, p_j \in P, i \neq j$. The Hamming bound is an upper bound for the number of words such that every codeword is at least a distance d_{min} from any other codeword. Therefore, the sets of $\frac{n}{2}$-mers, Q and P, must each satisfy the Hamming bound to insure that no false positives occur from vertex–edge mismatches. □

Vertex	Good Codes	Bad Codes
0	CTTGTGACCGCTTCTGGGGA	ATCAGCTGGATTCATCTGAA
1	CATTGGCGGCGCGTAGGCTT	ATCAACAGAAATCCGCGGAA
2	ATAGAGTGGATAGTTCTGGG	ATCAGCTGAGGTCTGGTGAG
3	GATGGTGCTTAGAGAAGTGG	GTCCGCTGTATTCTCGTGAT
4	TGTATCTCGTTTTAACATCC	TTCAACTGTTTTCAGCTGTG
5	GAAAAAGGACCAAAAGAGAG	TTCACCTTTATTGAGCCGAA
6	TTGTAAGCCTACTGCGTGAC	TTCAGCCGATTTGCGGAGAA

TABLE 2. Good and bad encodings of Adleman's graph with $d_{\text{min}} = 6$.

4. Genetic Algorithm Results

The GA was used to search for error-free (as defined above) encodings for Adleman's graph, Figure 2a. All the mismatch and error mechanisms in Equations 1 - 5 were incorporated. The number of errors allowed for hybridization was varied, simulating different hybridization temperatures. Using 20-mers to encode the vertices, mismatch-free encodings were found, as were encodings that produced many mismatches. Examples are given in Table 2 for $d_{\text{min}} = 6$ in Eq. 6. The good encodings all satisfied the minimum distance properties dictated by the Hamming bound, and the bad encodings did not. The bad encodings were obtained by having the GA maximize the fitness, or number of mismatches. The distance properties of good and bad encodings are given in Table 3. In addition, the bad encodings in Table 2 had a potential false positive consisting of the path $0 \to 2 \to 3 \to 1 \to 4 \to 5 \to 6$. The false positive path is

$$|p_0 >< \overline{q_3 p_2 q_2 p_3 q_3 p_4 q_5 p_6 q_4 p_5 q_5 p_6}| q_0 p_2 q_2 p_3 q_3 p_1 q_1 p_4 q_4 p_5 q_5 p_6 > |q_6 > .$$

This path began with vertex 0 and ended with vertex 6, and was of the same length as the correct path, $0 \to 1 \to 2 \to 3 \to 4 \to 5 \to 6$. The false path is possible because oligonucleotides that are a Hamming distance < 6 apart can match.

The GA was used to verify that the Hamming bound is a true bound for good encodings of the vertices. The graph used to verify this is shown in Figure 2b. In this graph, one and only one path existed. The testing was done for vertex encodings with 6-mers. For the case of 6-mers, the Hamming bounds (equality in Eq. 1) correspond to 64, 9, and 4 vertices with 0, 1, and 2 errors respectively. When only vertex–edge matches were considered (Equation 3), good encodings were found for all bounds. When the additional error mechanisms were introduced (Equations 1, 2, 4, 5), the GA was unable to find a good encoding corresponding to the Hamming bound, but was able to find good encodings below the Hamming bound.

5. Lab Results

The potential for false positive solutions using Adleman's technique was explored by searching for the Hamilotonian path for a single-solution directed Hamiltonion graph (Figure 2b). Figure 3 demonstrated that the anticipated correct 140 base pair path length solution was obtained when the oligonucleotides of Table 1 are

Vertex	p_0	q_0	p_1	q_1	p_2	q_2	p_3	q_3	p_4	q_4	p_5	q_5	p_6	q_6
p_0	0	5	7	8	7	8	7	8	7	6	10	6	7	7
q_0		0	5	8	5	7	8	8	7	8	8	7	8	7
p_1			0	8	7	6	7	9	8	8	7	6	9	6
q_1				0	8	8	5	8	8	9	7	8	4	8
p_2					0	7	9	5	8	8	7	6	8	5
q_2						0	9	8	5	8	8	7	7	8
p_3							0	8	7	9	7	9	7	8
q_3								0	8	8	6	7	7	8
p_4									0	6	9	9	7	8
q_4										0	7	7	6	8
p_5											0	4	7	7
q_5												0	9	7
p_6													0	10
q_6														0

a) Good Encoding.

Vertex	p_0	q_0	p_1	q_1	p_2	q_2	p_3	q_3	p_4	q_4	p_5	q_5	p_6	q_6
p_0	0	3	3	3	2	5	3	6	4	3	4	4	4	6
q_0		0	3	4	3	5	4	4	3	3	3	3	3	5
p_1			0	3	3	6	5	6	4	5	5	4	4	5
q_1				0	3	5	3	6	6	5	6	4	4	4
p_2					0	3	4	5	4	2	5	4	3	6
q_2						0	4	3	6	4	7	6	5	5
p_3							0	6	4	3	4	5	5	5
q_3								0	4	5	5	6	4	5
p_4									0	2	3	5	3	7
q_4										0	3	4	3	6
p_5											0	5	5	7
q_5												0	2	3
p_6													0	5
q_6														0

b) Bad Encoding.

TABLE 3. Distance properties for good and bad encodings to Adleman's graph with $d_{\text{min}} = 6$.

employed in the hybridization/ligation reaction (lanes 2). The importance of carefully choosing sequence encodings for the vertices and their corresponding edges was next examined by repeating this experiment under conditions in which no correct solution existed. The edge sequence

$$< \overline{q}_3\overline{p}_4| =< TGAGAGAACAATACCCGATC|$$

was replaced with a similar sequence ($< TGAGAGAACAATACTCGATC|$) that differed in one base, forcing a base-pair mismatch to occur during the hybridization/ligation step to provide a 140 base pair path. Figure 3 showed that a false positive 140 base pair band (lanes 3) equal in intensity to the true positive signal band was observed.

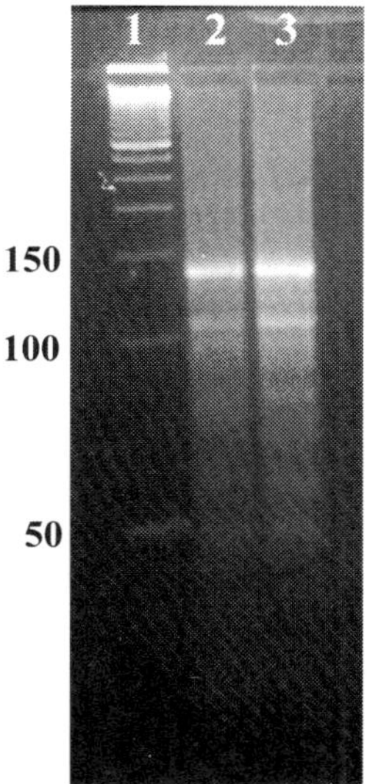

FIGURE 3. Ethidium bromide-stained agarose gel following electrophoresis of ligation reactions carried out as described by Adlelman. Ligations were carried out under controlled temperature (20° C) conditions. Lane 1: 50kbp DNA ladder employed as molecular size standard (BRL). Lane 2 and 3: Products of ligation reactions employing vertices 0-6 plus (lane 2) six oligonucleotide sequences representing exactly matching edges, and (lane 3) five oligonucleotide sequences representing exactly matching edges plus one oligomer containing a single mismatch. Equal amounts (20μl) of ligation product were loaded in lanes 2 and 3.

The hybridization stringency is a function of sodium molar concentration, temperature, percentage of $(G+C)$ bases, and overall length of the strand. The temperature at which half the population of perfectly matched double stranded DNA of length N will have separated into single stranded DNA is called the melting temperature, $T_m(°\mathrm{C})$. T_m describes both the melting and hybridzation processes (*i.e.* there is no hysteresis in the percent melted versus temperature curve). An empirical formula for oligonucleotides and DNA molecules from approximately 14 to 70 base pairs long is given by,

$$(7) \qquad T_m = 81.5 - 16.6\log_{10}[\mathrm{Na}^+] + 0.41\%(G+C) - 600/n,$$

where Na^+ is the molar sodium concentration, and $\%(G+C)$ is the percentage of G's and C's in the long oligonucleotide of length n. As a rough guide, the T_m decreases approximately 1°C per 1% mismatch between hybridizing oligonucleotides [**10**], resulting in a melting temperature for the mismatched hybridization, T_{mm}. Let T_r be a specified reaction temperature. For each degree difference between the perfectly matched T_m, and the T_r, hybridization may occur with a 1% mismatch between probe and target sequence. This occurs until the melting temperature for the mismatched hybrid, T_{mm}, is below the reaction temperature, T_r. Therefore, the total number of errors allowed is, roughly,

$$(8) \qquad d_{\mathrm{hyb}} = (T_m - T_r)(0.01)n,$$

where n is the length of the sequence. From Eq. 8, the Hamming bound (Eq. 6 with $t = (d_{\text{min}} - 1)/2$), or total number of codewords at most a distance $d_{\text{min}} = d_{\text{hyb}} + 1$ apart, was calculated, and plotted in Figure 4 as a function of T_r for $n = 15$, and as a function of sequence length for $T_r = 20°$C. In both cases, $\%(G+C) = 50\%$, and $\text{Na}^+ = 50$ mmoles/liter. Figure 4a shows the approximate effect of temperature on the number of codewords that are a distance d_{min} apart. As the reaction temperature increases, the stringency for hybridization increases, and there are less possibilities of mismatch. At a given temperature (Figure 4b), the stringency decreases as the length of the strand increases, producing more possibilities for mismatch and keeping the number of encodings from increasing at an exponential rate. Therefore, increasing the sequence length does not have as much effect on the number of encodings as might be expected. It must be emphasized that this analysis applies to a population of strands, and that T_m specifies a 50% melted or hybridized population. Therefore, significant variation in the number of codewords and errors at a given temperature is possible. More experiments and analysis are needed in this area.

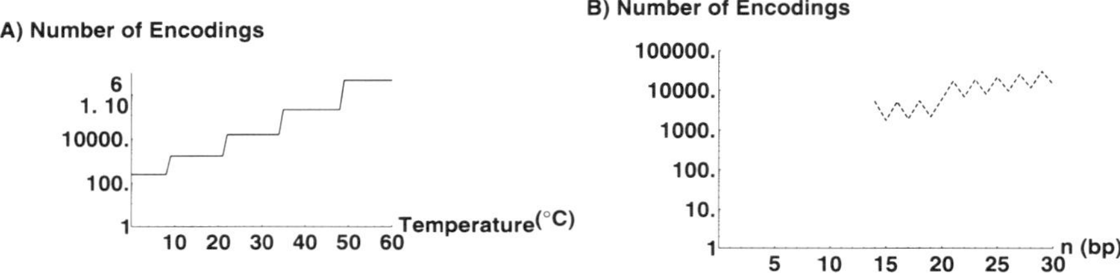

FIGURE 4. (A) Number of good encodings as a function of reaction temperature (T_r) for $n = 15$, and (B) sequence length (n) at 20°C, with $\%(G + C) = 50\%$, $\text{Na}^+ = 50$ mmoles/liter, and T_m calculated with Eq. 7.

6. Discussion

False positives are a function of the reaction conditions (primarily, temperature) under which hybridization occurs. If strings are allowed to hybridize with d_{hyb} or less errors, then, the complements of all oligonucleotides within a Hamming distance d_{hyb} of a given vertex codeword, potentially, will hybridize with that vertex codeword. Therefore, to produce an encoding of the vertices that excludes possibilities of false positives, all words within a distance d_{hyb} of any vertex codeword must be excluded from the encoding. This is the property that the Hamming bound satisfies.

The types of errors that produce false positives are seemingly valid vertex–edge matches. Suppose our hybridization conditions are such that two $\frac{n}{2}$-mers will hybridize if they match in 4 out of 6 complementary bases. These matching bases must be in sequence, but do not have to be consecutive, (Figure 5a). They must be in sequence to substitute for a valid vertex-edge splice in a path and produce a valid length path. Matchings that are not in sequence either will not splice easily into a path, or will produce paths that are of invalid length. Matchings between the sets of vertex codewords, Q and P, will not produce false positives, as the do not

$$\begin{array}{cccccccccccccccccc}
A & G & C & T & C & G & C & A & T & T & T & G & & & & & & \\
 & & & & & & \| & & \| & & \| & \| & & & & & & \\
 & & & & & & G & A & A & C & A & C & T & C & G & A & G & C
\end{array}$$

a) Form of valid vertex–edge splice.

$$\begin{array}{cccccccccccc}
A & G & C & T & C & T & C & T & T & G & T & G \\
\| & \| & & & \| & \| & & & & & & \\
G & C & G & C & G & A & T & C & G & A & G & C
\end{array}$$

b) Invalid form for vertex–edge splice.

FIGURE 5

have the proper form of a valid vertex–edge match, Figure 5b, though they would reduce the efficiency. Therefore, the determination of codewords for Q and P is independent, and both should satisfy the Hamming bound. In general, it became increasingly difficult for the GA to find good encodings when the number of vertices approached the Hamming bound. The results showed that the search took longer the closer the number of vertices was to the Hamming bound, that good codes existed when the bound said that they should, and that no good codes were found in over a million generations when the bound said they should not exist.

Besides vertex–edge, other mismatch possibilities that could result in false positives are matches between the vertices and the reverse of the vertices, and the edges and the reverse of the edges. When these were introduced, the GA was still able to find good encodings. The upper limit established by the Hamming bound, however, overestimates the number of vertices when these mismatch mechanisms are included. This was verified by GA search. Other mismatches between oligonucleotides are possible which affect the efficiency of Adleman's technique. For instance, hairpins (Eq. 4), and matches between codewords that could not produce false positives (Eq. 5) reduce the number of oligonucleotides available to search for the correct path. These mechanisms were also incorporated into the GA, and good encodings were found. These additional mechanisms, however, further reduce the upper bound on the number of vertices for which good encodings exist, as verified by GA search. For these additional mismatch mechanisms, and for efficiency, the n-mers, $|p_i q_i>$, which are used to represent the vertices, should also satisfy a minimum distance property.

If vertex codewords are used that are a minimum distance apart, which depends on reaction conditions, then, false positives are prevented. The experimental results verified the possibility of false positive solutions. In Adleman [**1**], the gel photographs of the solution are smeared, possibly indicating mismatched strands of numerous lengths. In this work, when the GA-generated encoding was used, distinct bands at multiples of 20 base pairs (bp) (20-mers were used in the encoding), and no bands higher than 140 bp were observed (Figure 3). This distinct banding and the absence of smearing indicates that only valid vertex–edge splices were produced from the GA-generated encoding. This supports the conclusion that the encoding of the problem matters for the reliability of the result.

7. Conclusion

This work examined the effect of hybridization reaction conditions on the reliability of Adleman's DNA-based technique for the solution of hard, combinatorial optimization problems. With random encodings, the existence of false positives was verified in laboratory conditions similar to Adleman's experiment. The main result establishes a criterion for good encodings, and a quantitative framework for their search and analysis. The analytically derived Hamming bound is a practical upper bound on the size of good encodings for fault-tolerant DNA searches. This bound was based on encodings that preclude the possibility of producing false positives in the DNA search. Other mechanisms for mismatch between oligonucleotides reduce this bound. The actual number of good encodings is dependent on the reaction conditions, primarily temperature, under which hybridization occurs. False negatives have not been considered since they can be controlled by using the proper amount of oligonucleotides in the test tube.

A second result is that more conventional methods (still biologically inspired, such as genetic algorithms) are useful for searching for good encodings of the Hamiltonian path problem in DNA. In this work, a GA was able to generate good encodings capable of ruling out errors from various sources that affect the efficiency and reliability of DNA-based techniques.

The laboratory results showed that the encoding of the problem and the experimental conditions may affect substantially the reliability, and therefore, the validity of DNA results. Our experiments verified the existence of false positive, and in the photograph of our gel (Figure 3), the distinct banding and the absence of smearing indicates that only valid vertex–edge splices were produced from the GA-generated encoding. There is enough evidence to conclude that the encoding of the problems matters in the reliability of the result.

References

[1] L. M. Adleman. Molecular computation of solutions to combinatorial problems. *Science*, 266:1021–1024, 1994.

[2] E. B. Baum. DNA sequences for useful computation. In L. F. Landweber and E. B. Baum, editors, *DNA Based Computers : Proceeding of the Second DIMACS Workshop*, DIMACS Series in Mathematics and Theoretical Computer Science, pages 235–241. American Mathematical Society, 1998.

[3] E. T. Bolton and B. J. McCarthy. A general method for the isolation of RNA complementary to DNA. *Proc. Nat. Acad. Sci.*, 48:1390–1397, 1962.

[4] Michael R. Garey and David S. Johnson. *Computers and Intractability.* Freeman, New York, 1979.

[5] John J. Grefenstette. Genesis (genetic search implementation system) , v. 5.0, 1990.

[6] John H. Holland. *Adaptation in Natural and Artificial Systems.* MIT Press, Cambridge, MA, 1992.

[7] K. Itakura, J. J. Rossi, and R. B. Wallace. Synthesis and use of synthetic oligonucleotides. *Ann. Rev. Biochem.*, 53:323, 1984.

[8] R. J. Lipton. DNA solution of hard computational problems. *Science*, 268:542–545, 1995.

[9] F. J. MacWilliams and N. J. A. Sloane. *The Theory of Error-Correcting Codes.* Elsevier Science Publishers B.V., Amsterdam, 1977.

[10] J. Sambrook, E. F. Fritsch, and T. Maniatis. *Molecular Cloning: A Laboratory Manual.* Cold Spring Harbor Laboratory Press, second edition, 1989.

[11] James D. Watson, Nancy H. Hopkins, Jeffrey W. Roberts, Joan A. Steitz, and Alan M. Weiner. *Molecular Biology of the Gene.* The Benjamin/Cummings Publishing Co., Inc, Menlo Park, CA, fourth edition, 1987.

The Molecular Computing Group, The University of Memphis, Memphis, TN 38152
E-mail address: r-deaton@memphis.edu

DIMACS Series in Discrete Mathematics
and Theoretical Computer Science
Volume **44**, 1999

DNA Computations Can Have Global Memory

Richard J. Lipton

1. Introduction

Ever since Adleman's seminal paper [**1**] there has been a flood of ideas on how one could use DNA to compute. I first showed that DNA could be used to solve more than just a variation of TSP [**8**]. Since then there have been many other papers on using DNA to solve various computational problems. [**2, 4, 3, 5, 6, 10**]

At the top-most level all these papers use DNA in the same way. Each strand of DNA encodes the state of a processor. Each processor operates *independently* : there is no communication from one processor to another. Processors each search their own part of a large space. The papers differ in the details of how they encode the state into the DNA. They all, however, perform independent searches.

Our new result is that we can allow, for the first time, *global memory.* That is, we show that the individual DNA strands can communication with each other. This changes everything. The point is that DNA can do much more general parallel computations than previously realized.

The class of computations that allow global memory are much more powerful than independent searches. For example, one of the main ways to search large spaces is the so called "branch-and-bound" method. In this method the search in one part of the space is "pruned" by values found in other parts of the space. Previously, it was not possible to even imagine how DNA computations could do this: now we know how.

Our method is based on the same bio-technology operations that we have used previously. We use no new bio-operations. Thus, our addition of global memory to DNA computations has all the promise and difficulties that previous DNA computations had. The key is that we use nothing new: we just exploit the power of the existing methods in a new way.

We will now outline the basic idea. The details are the subject of the rest of the paper. Imagine that each processor has a bit b_i. Further, suppose that the processors want the following global value:

$$b_1 \vee \cdots \vee b_N$$

where there are N processors, i.e. there are N strands. That is the processors want to know the value of the "OR" of the bits $b_1, \dots, b_N$. (Note, our method can be

Supported in part by NSF CCR-9304718.

made to handle other global memory functions. The case of "OR" is just to make the explaination easier.)

How can the processors get this value? It's simple. Perform an extraction on the bit b. This creates two test tubes. One has all the processor (i.e. strands) that have $b_i = 0$ and one where $b_i = 1$. Now assume that we can *non-destructively* test the latter test tube to see if it is empty. There are two cases:

1. The test tube is empty. In this case pour the two together and set the answer bit in all the processors to 0.
2. The test tube is not empty. In this case pour the two test tubes together and now set the answer bit to 1.

That's it. The way that we get global information to processors is to use detection to externally compute information. Then, we write it *back* to the processors.

There is a problem, of course. What if the detection of strands destroys the contents of the test tube? It turns out there is a simple method that we can use to handle this case. This will all be presented shortly.

2. Model of Parallel Computation

In this section we will present our model of parallel computation. We assume that there are N processors: each has a local memory. We use $m_i^{(p)}$ to denote the ith bit of the memory of the pth processor. Initially, processor p starts with its memory set to p, i.e. processors know their identity.

We assume that the processors operate synchronously. At each time step they either:

1. Do a *local step*. In this case, for some memory locations i, j, k and all processors p,
$$m_i^{(p)} \leftarrow m_j^{(p)} \circ m_k^{(p)}$$
where $\circ$ is a boolean operation. Thus, all processors update the same locations in the same way. Note, that while processors do the same step, the value they compute may be different since it depends on local state information. Thus, some will set $m_i^{(p)}$ to 0 and some to 1.
2. Do a *global step*. In this case, for some memory location i,
$$z \leftarrow \bigvee_{p=1}^{N} m_i^{(p)}.$$
Then, for all processors p,
$$m_i^{(p)} \leftarrow z.$$
Note, that all the processors get the same common value z.

Clearly, global steps allow processors to share information. This is the new type of operation that we are able to support. Previously, all DNA computations could only perform local steps.

A global step is powerful, but there is a more powerful kind of step that we also can support. We call it a *group step*. A group step is a kind of "parallel" global step. The processors divide themselves into disjoint groups. Essentially, each of the groups does a global operation. Since this happens at the same time it is clearly more powerful than doing several global steps. Formally, a group step is:

1. Partition the processor names into groups $G_1, \ldots, G_m$. Set for each group j,
$$z_j \leftarrow \bigvee_{p \in G_j} m_{i_j}^{(p)}.$$
Then, for all groups j and all processors $p \in G_j$,
$$m_{i_j}^{(p)} \leftarrow z_j.$$

Note, that each processor gets the group's value.

In all the above we only allow "global" memory operations that take the "OR" of bits. Clearly, this is a powerful function. However, it may be interesting to implement other functions. We will consider the following standard functions.

1. The "AND" function.
2. The "MAJORITY" function.
3. The "THRESHOLD" function.
4. The "PARITY" function.

Note, (1) is trivial: clearly "AND" can be computed by using DeMorgan's Law. However, the other operations are more interesting. Essentially, we are able to handle "MAJORITY" and "THRESHOLD" in restricted cases. The case of "PARITY" is open. We do not know if it is possible to handle this function with DNA computations.

3. Model of DNA

The beauty of our addition of global memory is that it works for many methods of doing DNA computations. It is quite general an idea. We need only make a few minimal assumptions.

First, we assume that strands encode the state of each processor. We can use the methods outlined in [**8**] or other new methods such as the "sticker" model [**2**]. The key is only that we assume that strands encode a bit pattern.

Second, we assume that it is possible to do *extraction.* That is we assume that we can remove from a test tube of strands all strands that have a bit set to a given value. Most current suggested models of DNA computing support this operation. Note, we do not address the issue of errors. The methods outlined here will work provided the error rates are low enough.

However, there is one point that we wish to make about errors. It seems possible that we can arrange the parallel algorithms so that they can tolerate small numbers of errors. For example, we may be able to arrange that whenever an "OR" computes a "1," there are several "1"'s present. It is an interesting problem to answer how much this will cost in performance.

Third, we assume that we can *set* bits of strands. Most of the methods suggested to date allow this. For example, the sticker model [**2**] is attractive because this operation is so inexpensive.

Finally, we assume that it is possible to do either *non-destructive* detection or *destructive* detection. In the former we assume that given a test tube we can determine whether or not the test contains *any* strands at all. We assume that the test does not change the contents of the strands.

In the latter case the detection is assumed to be destructive. We assume that it determines the sequence of the strands but destroys the test tube otherwise. Note, this is what would happen if, for example, one needed to use PCR to amplify the test tube in order to sequence it. Also we asssume that the cost of this operation is *linear* in the number of strands: if a test tube has k distinct strands, then we assume that the bio-operation takes time $O(k)$.

4. Simulation of Parallel Computations by DNA

We first will consider the cases of non-destructive testing. The memory operations are all assumed to be "OR"/"AND".

THEOREM 4.1. *A parallel computation with p processors and t local/global time steps, can be simulated by p strands of DNA in $O(t\log(m))$ bio-operations, where no group is larger than m.*

In the case of destructive testing the time of the simulation increases by a factor of $\log(p)$:

THEOREM 4.2. *A parallel computation with p processors and t local/global time steps, can be simulated by p strands of DNA in $O(t\log(p)\log(m))$ bio-operations, where no group is larger than m.*

We now turn to sketch the proofs of the above theorems. There are two key issues: First, how do we handle the group operations? Second, how do we handle destructive testing?

Proof of Theorem 4.1: The method outlined in the introduction is sufficient to handle the non-group case. Clearly, it takes one extraction per global operation. Since the extraction is non-destructive, the cost is linear in t. The interesting point is that group operations only slows down the simulation by a factor of at most $\log(m)$ where m is the number of groups.

Suppose that there are m groups. We claim that in $\log(m)$ parallel extractions we can form m test tubes. Each contains exactly those strands that correspond to each group. (We assume that strands have small $\log(m)$ sized "tags" that allow the extraction to work.) Further, we assume that we can do bio-operations in *parallel.* That is, we require that we can perform extractions on different test tubes at the same time. Then, we do the detections on the final test tubes and determine the individual bits that correspond to each group. We then set the bits. □

Proof of Theorem 4.2: The key issue now is what to do about a destructive test. We inductively assume that there two kinds of strands: *normal* and *damaged.* All normal strands are in the test tube. All damaged strands are *not* in any test tube; however, we assume that we know their complete sequence. Here is how we perform a computation step. A local step is easy. We do the bio-operations on the test tube; we perform the same operations formally on the damaged strands. Now suppose that we want to perform some global operation. Assume for easy of discussion that it is not a group step.

Compute the "OR" of the damaged strands. If it is a "1," then set all the answer bits. This is the easy case. The hard case is when this bit is "0". Then, perform the extraction on the desired bit as before on the test tube that contains all the normal strands. Let there be k strands in this test tube. We must determine whether or not $k > 0$.

We operate as follows. Guess that the test tube has $k = p$ strands. By repeated dilution form two test tubes: Let test tube A contain approximately $1/p$ of the strands and let B contain the rest. Test A to see if it has any strands. There are two cases:

1. There are strands in A. In this case we determine their sequence. We continue the simulation with B. All the strands from A are added to the current collection of damaged ones.
2. There are no strands in A. In this case dilute B again by a factor or $2/p$. Continue in this way with $4/p$ and so on. Clearly, in at most $\log(p)$ steps we are in case 1 or there are no strands at all.

□

Finally, there is the question of handling other memory operations such as "MAJORITY." We can easily extend our theorems to this case provided the cardinality is always bounded away from $p/2$ where there are p strands. That is we can handle the case where the computation either has 49% or 51% of the bits equal to 1. We cannot yet determine what happens with other functions such as parity.

5. Comparison to "Electronic" Computations

The reason that DNA computations are exciting is that a liter of water can contain in solution about 10^{21} bases. This is at the reasonable rate of 5 grams of DNA per one liter of water. Of course DNA operations are slow: some take as long as 1 hour; others take minutes. Also strands need to be about 1,000 long to encode a reasonable sized processor state. Thus, the overall potential for DNA computation is that they can do about 10^{15} operations per second: this is 1,000 tera-ops.

This is a large number. However, electronic computations continue to growth at about 50% per year. If this rate continues, electronic ones will be at this level in a few years. Thus, there is a window for DNA computations, but it is not forever.

However, the addition of global memory to DNA computations changes this simple calculation. There is an argument that shows that DNA computations could potentially gain an additional factor of 1,000 over electronic computations. The key observation is that its "hard" for electronic computations to simulate parallel computations that have global memory.

In order to see this let us consider how electronic computers would simulate a parallel computation. If the computation has *no* global memory, the simulation is simple. The electronic computer will simply simulate each processor one at a time. Clearly, the following is trivally true:

THEOREM 5.1. *A parallel computation with p processors and t local time steps, can be simulated on a serial machine in pt time and t space.*

However, the addition of global memory changes this. Now the best naiive simulation seems to be the following:

THEOREM 5.2. *A parallel computation with p processors and t local/global time steps, can be simulated on a serial machine in pt^2 time and t space.*

The critical point is that now the global memory requires the serial machine to recompute the states of each processor. We do not know if this simulation is best. If it is, then this is how we pick up a factor of t over electronic computations. Thus, if t is 1,000 this is a large additional advantage of DNA computations over electronic ones.

We can prove the following:

THEOREM 5.3. *Any general simulation of a parallel computation with p processors and t local/global time steps, requires $\Omega(pt^2)$ steps on a serial machine that has only $p/2$ storage.*

By general simulation we mean that the "pebbling" model is used. In this model the simulation is restricted to operate as follows. The simulation can record at most $p/2$ states of the processors. At each step it can compute from a state to a next state. It can keep this state as long as there is room. The simulation can also "throw away" states. If these states are needed later on, the simulation must recompute them.

6. Conclusions

We have shown that DNA computations can have global operations. In particular, we have shown that they can share global memory. A critical next question is: Can they exploit this global memory to solve important problems? We call this the "search for a killer app." Even if DNA computations can be made to work, it is unclear what we would use them for. Thus, it is important for us to find some application that has two properties:

1. It can be mapped onto DNA computations.
2. It is important to solve.

The trouble with the current examples of DNA computations is that none of the applications are compelling yet. None is clearly so important that they by themselves justify the construction of DNA computers. We hope that the addition of global memory to the capabilities of DNA computations with help us discover the "killer app."

References

[1] Leonard M. Adleman. Molecular Computation of Solution to Combinatorial Problems. *Science*, 266:1021–1024, 1994.

[2] S. Roweis, E. Winfree, R. Burgoyne, N. Chelyapov, M. Goodman, P. Rothemund, L. Adleman, A Sticker Based Architecture for DNA Computation. In L. Landweber and E. Baum, editors, *DNA Computers: Proceeding of the Second DIMACS Workshop,* DIMACS Series in Discrete Mathematics and Theoretical Computer Science, American Mathematical Sociey, 1-29, 1998.

[3] E. Bach, A. Condon, E. Glaser and C. Tanguay, DNA Models and Algorithms for NP-Complete Problems. In *Proceedings of the 11th Annual IEEE Conference on Computational Complexity*, 1996.

[4] D. Beaver, A Universal Molecular Computer. In R. J. Lipton and E. B. Baum, editors, *DNA Computers: Proceeding of a DIMACS Workshop,* DIMACS Series in Discrete Mathematics and Theoretical Computer Science, American Mathematical Sociey, 29-36, 1996.

[5] D. Boneh, C. Dunworth, R. Lipton, Breaking DES Using a Molecular Computer. In R. J. Lipton and E. B. Baum, editors, *DNA Computers: Proceeding of a DIMACS Workshop,* DIMACS Series in Discrete Mathematics and Theoretical Computer Science, American Mathematical Sociey, 37-65, 1996.

[6] D. Boneh, C. Dunworth, R. Lipton, J. Sgall, On the Computational Power of DNA, *CS-TR-499-95.*

[7] D. Boneh and R. Lipton, Making DNA Computers Error Resistant. In L. Landweber and E. Baum, editors, *DNA Computers: Proceeding of the Second DIMACS Workshop,* DIMACS Series in Discrete Mathematics and Theoretical Computer Science, American Mathematical Sociey, 31-44, 1998.

[8] Richard J. Lipton. DNA Solution of Hard Computational Problems. *Science*, 268:542-545, 1995.

[9] C. Papadimitrou, private communication, 1994.

[10] J. Reif, Parallel Molecular Computation: Models and Simulations. *Seventh Annual ACM Symposium on Parallel Algorithms and Architectures*, 1995.

DEPARTMENT OF COMPUTER SCIENCE, PRINCETON UNIVERSITY, PRINCETON, NJ 08544
E-mail address: rjl@princeton.edu

DIMACS Series in Discrete Mathematics
and Theoretical Computer Science
Volume **44**, 1999

Exascale Computer Algebra Problems Interconnect with Molecular Reactions and Complexity Theory

Robert M. Williams and David Harlan Wood

ABSTRACT. In discussing exascale (exa $= 10^{18}$) computer algebra problems we interconnect three themes. First, computer algebra is similar to DNA laboratory reactions. Both rearrange identical subunits. Second, DNA is an attractive medium for computation because of its density and parallelism. Third, determinant and/or permanent expansions exemplify many levels of complexity.

These three issues are combined in a proposed experiment using a DNA algorithm to evaluate or approximate the permanent of a matrix of zeros and ones, a well-known problem in the class #P-complete. Such problems are harder than those previously addressed by DNA techniques in the pioneering articles of Adleman and Lipton.

Problems of fairly interesting size can be done because scale-up difficulties are reduced by alternating intermediate steps of building and filtering. The example algorithm suggests directions for the general problem of expanding symbolic determinants and permanents given their zero entries.

1. Introduction

Exascale (exa $= 10^{18}$) computer algebra problems interconnect three themes.
(1) Symbolic computations, such as are found in computer algebra, are analogous to DNA laboratory reactions, in that both change the arrangement of finite numbers of identical subunits.
(2) Determinant and/or permanent expansions for different patterns of zero entries exemplify many levels of complexity. Some of these expansions are as hard as problems in the class #P-Complete, which turn are widely regarded to be harder than problems in the complexity classes previously addressed by DNA techniques in the pioneering articles of Adleman and Lipton.
(3) DNA is an attractive medium for computation because of its potential parallelism from having up to 10^{18} bytes reacting in a single test tube of DNA solution. Also attractive is its high information storage density of 10^{23} bytes per kilogram. This number dwarfs the human genome, 10^9 bytes, and even the 1995 USA hard disk production of 2×10^{17} bytes.

1991 *Mathematics Subject Classification.* Primary 68Q40, 68U99, 68Q15.
Both authors were supported in part by the Office of Naval Research.

In Section 7, an example algorithm is given combining all three of the above issues. Our algorithm exploits DNA laboratory procedures to evaluate or approximate the permanent of an $n \times n$ matrix of zeros and ones, a well-known problem in the complexity class #P-Complete. Scale-up difficulties are reduced in our approach by alternating intermediate steps of building and filtering. This is less wasteful than existing approaches which generally depend on filtering their answers from about e^n times as much DNA.

Our computations of permanents are of low cost because they exploit a once-and-for-all laboratory construction that suffices for even the worst case problem output size. (Which makes these computations somewhat wasteful in their use of DNA.) In a related companion paper **[13]** submitted to this workshop, algorithms are given for expanding symbolic determinants and permanents given their patterns of zero entries. The algorithms in the companion paper use more expensive one-at-a-time constructions, but they are limited only by the output size of the individual problem. (Which makes these computations efficient in their use of DNA.)

2. Molecular Computing

By the late 1980s molecular computing was a well established field. A comprehensive 90 page survey **[10]** appeared in 1990. In November 1992, a special issue of *Computer* **[9]** was devoted to molecular computing.

Two years after this special issue, molecular computing was no longer hypothetical. In a major breakthrough, Adleman **[2]** was able to say, "This experiment demonstrates the feasibility of carrying out computations at the molecular level." Out of the fantastic range of biological and non-biological molecular reactions, Adleman choose to exploit the simplest: DNA and its attendant laboratory technology.

Within a few months, a landmark paper **[14]** by Lipton showed how DNA laboratory techniques could be use to solve the versatile **3SAT** problem in NP-Complete. This was particularly intriguing because so many NP-Complete problems can be reduced to **3SAT** in a natural way.

Shortly thereafter, the first Workshop on DNA Based Computers was held **[15]**. The topics included NP-Complete problems, integer factoring, universal and reprogrammable DNA computers including various Turing machines, biological paradigms, and the possible necessity of annealing and ligation.

However, Adleman and Lipton and other researchers generally use techniques that extract a (small) answer from the set of all conceivable answers. This last set can be impractically large for even moderately challenging problems. It is hard to decide which fact seems more amazing—that DNA can offer 10^{18} parallel processors, or that 10^{18} DNA processors is not enough to compete with conventional computers!

This apparent dilemma is of course due to the different algorithms used. A successful algorithm must be efficient and must also suit the means of computation. DNA computation seems to require algorithms that are quite different from those used on conventional computers (one notable exception is found in **[3]**). Clearly, much is yet to be learned about DNA algorithms. Furthermore, we need not limit ourselves to DNA among the possible biological computing media **[10][18][7]**. Thinking of computing as the pairing of computing media with algorithms, we are left with the question, "What is a problem for which molecular computing is the preferred technique?"

We present one part of a possible answer to this question. We give an algorithm for evaluating or approximating a 0-1 permanent by using DNA laboratory techniques. More generally, we find DNA laboratory techniques are well suited to the problem of expanding a determinant or permanent having symbolic entries. We denote this problem **ESD0**. In this problem, given a pattern of zeros in a matrix, we produce the set of nonzero terms that form the expansion.

To even find the *number of terms* in this expansion is equivalent to the evaluation of a 0-1 permanent, a problem denoted **P0-1** in the complexity class #P-Complete. While **3SAT** and the other problems in NP-Complete are widely assumed to require an effort at least proportional to 2^n, where n is a measure of the size of the input, the problems in #P-Complete are harder, being widely assumed to require an effort proportional to $n\,2^n$.

An important implication is that a liter of DNA solution will hold all 2^n possible outcomes of a **3SAT** problem of size $n = 70$ (nearly big enough to be interesting) or it can hold all $n!$ possible outcomes of a **ESD0** or **P0-1** problem of size $n = 20$ (begins to be big enough to be interesting). In a companion paper [**13**] the **ESD0** algorithms are less restricted: only enough DNA is needed to represent the answer.

3. Computer Algebra

The computer algebra subfield of computer science produces exact symbolic results by finite computations on exact symbolic inputs. To give a basis for this observation, we now review some of the history of computers and computation. Early computations (1940s) were based on exact arithmetic over a finite sub-set of integers or rational numbers. Extreme care was needed to ensure that results remained within a fixed range after arithmetic operations. Next, computation was extended (1950s) to approximate (floating point) numerical arithmetic computations. Also, more advanced, but still approximate, mathematical operations, such as root finding and linear algebra, were possible. Finally, symbolic computation developed (1960s) based on exact computations with non-numeric objects, such as polynomials and other functions. Furthermore, non-arithmetic operations such as ordinary and partial differentiation of these functions can be computed exactly. A key feature of computer algebra is obtaining exact, non-numeric results using only finite computer resources. The noteworthy feature of computer algebra that we find echoed in DNA reactions is the transformation of exact, non-numeric inputs into exact, non-numeric outputs.

In this section we only present an analogy between computer algebra and DNA reactions. Here we wish to emphasize that our point of view is that DNA reactions are natural candidates for performing computer algebra computations of high complexity (exceeding the purported 2^n complexity of NP-Complete problems). In other words, DNA processes do not appear, from our point of view, to be suited to numeric computations. Rather, they seem suited to computer algebra.

4. Determinant Ruler

In Table 1 below we interconnect computer algebra and complexity theory by providing classes of **EDS0** problems resulting from various zero patterns. Common space/time complexity levels, up to and beyond the level of #P-Complete, are spanned by this one paradigm: symbolic determinants with various zero patterns. This is fact is completely independent of whether or not P = NP.

TABLE 1. Special cases of **ESD0** (Expansion of Symbolic Determinants given their zero patterns) span complexity classes up to and including $n!$. The first column reflects both the space and time complexity of the **ESD0** problems cited in the second column.

In the first column the number of nonzero terms in **EDS0** is shown in big O notation; this reflects space complexity. This in turn gives a lower bound on time complexity. This underestimates time complexity because for each term μ_n as much time is required for performing the n multiplications within each term. The time cost μ_n depends on the nature of the matrix elements. Usually, μ_n is n or greater, and often much greater, as for example when the matrix entries are polynomials.

The last column reflects the currently known time complexities of the numeric and decision problems cited in the third column. The time complexity upper bounds shown in the last column are roughly aligned with the time complexities shown in the first column.

A horizontal line in this table divides exponential complexity from mere polynomial complexity. This division is definite for **ESD0** problems, but ambiguous for decision problems because it is not known whether NP-Complete problems are only Polynomial or are 'Exponential' (slightly abusing this term to mean 'more difficult than Polynomial').

ESD0 Time	ESD0 and Related Problems		Decision Problems & Numeric Problems	Time Complexity
$O(n!)\ \mu_n$	**ESD0** det with no zeros		Generate permutations of n data	$O(n!)$
	[1]**ESD0** find count of terms	$\Longleftrightarrow$	**P0-1** in #P-Complete	$O(n\,2^n)$
$O(2^n)\ \mu_n$	**ESD0** det(Hessenberg)		Generate all subsets of n data	$O(2^n)$
			NP-Complete problems, if P≠NP	Exponential
———	———		**TSP** & **3SAT** in NP-Complete	$O(2^n)$
			P and all NP problems, if P=NP	Polynomial
$O(n^6)\ \mu_n$	**ESD0** det($\bigotimes^3$Arrow)		Linear programming by ellipsoids	$O(n^6)$
	[1]**ESD0** have nonzero terms?	$\Longleftrightarrow$	Full bipartite network flow? [**12**]	$O(n^{5/2})$
$O(n^2)\ \mu_n$	**ESD0** det(BrGi)		Matrix-vector multiply [**5**]	$O(n^2)$
$O(n)\ \mu_n$	**ESD0** for det(Arrow)		Search n unordered data	$O(n)$
$O(1)\ \mu_n$	**ESD0** det(Triangular)		Select one datum from a set	$O(1)$

5. Facts about Permanents

Here we treat permanents (and determinants) from the point of view of permutations.

DEFINITION 5.1. The permanent of a square matrix A is given by

$$\mathrm{per}(A) = \sum \sigma \cdot a_{1,p_1} \cdot a_{2,p_2} \cdots a_{n,p_n}. \tag{5.1}$$

In each term, $p_1, p_2, \ldots, p_n$ is a permutation of $1, 2, \ldots, n$. The sum is taken over all permutations. In each term the factor σ is taken to be 1.

The definition of the determinant of a square matrix A is the same as that of the permanent except the factor σ in each term is $+1$ if p is an even permutation, and -1 if it is odd.

Matrices with a fixed pattern of zero entries occur very frequently in applications. These zeros can greatly reduce the number of terms. For example, one 3×3 matrix having two zero entries has

$$\mathrm{per}\begin{pmatrix} a_{1,1} & 0 & a_{1,3} \\ 0 & a_{2,2} & a_{2,3} \\ a_{3,1} & a_{3,2} & a_{3,3} \end{pmatrix} = a_{1,1}\, a_{2,2}\, a_{3,3} + a_{1,1}\, a_{2,3}\, a_{3,2} + a_{1,3}\, a_{2,2}\, a_{3,1}. \tag{5.2}$$

Our objective in finding the expansion of a permanent (or determinant) is to obtain the set of the permutations that correspond to the nonzero terms. In this case, the set is

$$\{1,2,3\}, \{1,3,2\}, \{3,2,1\}. \tag{5.3}$$

A different pattern having two zeros, but in different locations, happens to give fewer terms,

$$\mathrm{per}\begin{pmatrix} a_{1,1} & 0 & a_{1,3} \\ a_{2,1} & a_{2,2} & a_{2,3} \\ a_{3,1} & 0 & a_{3,3} \end{pmatrix} = a_{1,1}\, a_{2,2}\, a_{3,3} + a_{1,3}\, a_{2,2}\, a_{3,1}. \tag{5.4}$$

Here the desired set permutations is

$$\{1,2,3\}, \{3,2,1\}. \tag{5.5}$$

The following **[8]** is easily observed in examples such as Eqs 5.2 and 5.4.

FACT 5.2. Let A have independent symbols for entries, aside from a given zero pattern. Then the number of nonzero terms in the expansion of $\det(A)$ or $\mathrm{per}(A)$ is equal to the value of $\mathrm{per}(A^{'})$, where the matrix $A^{'}$ is formed by replacing each nonzero entry of A by 1.

The solution of **ESD0**, given a pattern of zeros, is a set of permutations, as in Eqs 5.3, and 5.5. Fact 5.2 tells us that to merely find the *size* of this set is equivalent to the evaluation of an 0-1 permanent. This evaluation is known **[11]** to be an intractable problem.

FACT 5.3. The problem **P0-1**, computing the permanent of an $n \times n$ matrix of zeros and ones, is in the complexity class #P-Complete.

The problems in the class #P-Complete can be thought of as decision problems (often from the complexity class NP-Complete) with the additional requirement of determining not just the existence of solutions, but determining the exact number of solutions. For the full details, see **[11]**. We know **[16]** an algorithm for **P0-1**.

FACT 5.4. A 0-1 permanent of size $n \times n$ can be computed with $O(n\,2^n)$ arithmetic operations.

6. Bilingual Dictionary of DNA Terms and Computer Concepts

Many common DNA reactions (changing the arrangement of finite numbers of identical subunits) are exact symbolic computations. When we consider DNA from the viewpoint of computation, we recall the well-known list data structure of (static) DNA. Next, we consider the question: "What class of dynamic computational processes or algorithms could correspond to DNA reactions?" Some standard laboratory processes have natural counterparts in computing. These are shown in Table 2. These identifications have been used by several researchers **[1]** **[2]** **[4]** **[6]** **[14]**. Of course, there are also many other laboratory processes **[17]** that are not included here, some of which have also been proposed for DNA Computing.

TABLE 2. Some standard laboratory processes have natural counterparts in computing.

COMPUTING PRIMITIVES	LABORATORY PROCESSES
CPU Clock Cycle	Laboratory Steps; Reaction Times
Write List or Tape	Static DNA Synthesis
Read List or Tape	Static DNA Sequencing
Append Lists	DNA Reaction: Annealing and Ligation
Sublist Selection	DNA Reaction: Restriction Enzymes
List Sorting and Selection	Electrophoresis or Affinity Purification

7. A DNA Algorithm for Evaluating a 0-1 Permanent

We now give an algorithm for DNA computation for the problem **P0-1**, namely, evaluating the permanent of a 0-1 matrix, given its pattern of zeros.

- **Requires:** (a) Off-the-shelf mixture of DNA representations of all permutations of n indices
  (b) Off-the-shelf set of n^2 precoded complementary indices
- **Algorithm**
  - **Input:** A $n \times n$ pattern of zeros selected from complementary indices
  - **Output:** Ratio of (value of permanent) / $n!$
  - **Body:**
    1. To a standard mixture of all possible permutations of indices, add complementary DNA coded indices of all entries which are zero
    2. These "zeros" stick only to those permutations that cancel out of answer
    3. Filter out the permutations having adhering zeros
    4. Output: Measure the ratio of remaining permutations to original permutations

Modified Adleman Algorithm for Permutation Mixture. We now give an algorithm for the construction of the required off-the-shelf mixture of permutations. We modify Adleman's approach in a way that scale-up problems are reduced by alternating the intermediate steps of building and filtering.

- **Datatype:** Partial permutations of length k, namely double strands of DNA encoding $|i_1|i_2|i_3|\dots|i_k|$, subsequences of permutations of integers.
- **Input:** Integer n.
- **Requires:** DNA strands encoding each of the integers $1, \dots, n$.
- **Output:** Mixture of DNA representations of permutations of size n
- **Algorithm**
  - **Initialize:** All strands of length 1.
  - **Body:**
    1. FOR K=2 through N DO
       (a) Extend lengths by appending random integer in $1, \dots, n$
       (b) Filter out strands containing duplicate integers

The mixture of permutations could also be done with the techniques of [**13**].

8. Discussion of the P0-1 Algorithm

For interesting size problems, current technology limits our ability to carry out step 4 of the first of the above algorithm. Our algorithm is best regarded as a practical way to *approximate* the value of a 0-1 permanent. For some selected NP-Complete problems it has been shown, roughly speaking, that approximating the solution is about as hard as actually obtaining it. However, it appears to be an open question whether the problem **P0-1** has this property or not. In particular, no polynomial time approximation algorithm is known for this problem.

The main virtue of our algorithm is that it always involves only the same small number of laboratory steps involving only mass-produced materials. This is the case regardless of the value of the permanent and regardless of the size of the matrix (up to the limit of the standard mixture of permutations).

Recall that the value of an $n \times n$ permanent having zeros and ones for elements is an integer between 0 and $n!$. The most interesting cases are when the value is large. This is because it is known [**19**] from rather general considerations that if the value of the permanent is t, then it is possible to find it in time proportional to a polynomial in t—which could be an efficient method for conventional computers when t is small. There is also an easy test for $t = 0$ using a $O(n^{5/2})$ algorithm [**12**].

The main disadvantage of our algorithm is that it is impractical for matrices larger than about 20×20. This is simply because the standard mixture would have to contain all $20! \approx 10^{18}$ permutations of length 20. This would require roughly a liter of DNA solution. A 21×21 matrix would need 21 times as much solution, and so forth.

9. Toward Solving ESD0: The Full Expansion

Much more information is available from the above algorithm. The result of the first three steps is a mixture of all permutations corresponding to nonzero terms in the expansion of the permanent (or determinant) of the matrix with the given zero pattern. From these permutations, the full expansion is determined (taking into account the parity of the permutations in the case of the determinant). For a simple example, we see that from Eq 5.3 we can recover the right hand side of Eq 5.2.

Presumably, a symbolic expansion with a billion or more terms will not be comprehensible without further processing. Ideally this would be done by another DNA algorithm that accepts input in the form of the output produced by our algorithm. Otherwise, step 4 has to become "Output: Decode the remaining permutations." This is likely to be a formidable job.

A companion paper [**13**] gives a readout algorithm. Also, algorithms for **ESD0** are given there that are restricted only by the size of the intermediate results, not by the size of the input matrix. However, since they can not use a standard mixture of permutations, they must construct the nonzero terms anew for every input zero pattern.

References

[1] L. M. Adleman, *On constructing a molecular computer*, University of Southern California Department of Computer Science Draft.

[2] Leonard M. Adleman, *Molecular computation of solutions to combinatorial problems.*, Science **266** (1994), 1021–1024.

[3] Eric B. Baum and Dan Boneh, *Running dynamic programming algorithms on a DNA computer*, Proceedings of the Second Annual Meeting on DNA Based Computers, held at Princeton University, June 10-12, 1996, to appear.

[4] Donald Beaver, *Factoring: The DNA solution*, J. of Computational Biology **1** (1995).

[5] Dario Bini and Victor Pan, *Polynomial and matrix computations*, Birkhäuser Verlag, Boston, 1994.

[6] Dan Boneh, Christopher Dunworth, and Richard J. Lipton, *Breaking DES using a molecular computer*, Tech. Report CS-TR-489-95, Princeton University, May 1995.

[7] Dennis Bray, *Protein molecules as computational elements in living cells*, Nature **376** (1995), 307–312.

[8] Richard A. Brualdi and Herbert J. Ryser, *Combinatorial matrix theory*, Cambridge University Press, Cambridge, 1991, Encyclopedia of Mathematics and Its Applications, Volume 39, Edited by G.-C. Rota.

[9] *Special issue on molecular computer paradigms*, Computer (Michael Conrad, ed.), IEEE Computer Society, Los Alamitos, CA, November 1992.

[10] Michael Conrad, *Molecular Computing*, Advances in Computers, Volume 31 (Marshall C. Yovits, ed.), Academic Press, New York, 1990, pp. 235–324.

[11] Michael R. Garey and David S. Johnson, *Computers and intractability, a guide to the theory of NP-completeness*, W. H. Freeman, New York, 1979.

[12] John E. Hopcroft and Richard M. Karp, *An* $n^{5/2}$ *algorithm for maximum matchings in bipartite graphs*, SIAM Journal on Computing **2** (1973), no. 4, 225–231.

[13] Thomas H. Leete, Matthew D. Schwartz, Robert M. Williams, David Harlan Wood, Jerome S. Salem, and Harvey Rubin, *Massively parallel DNA computation: Expansion of symbolic determinants*, DNA Based Computers: Proceedings of a DIMACS Workshop, June 10-12, 1996, Princeton University, American Mathematical Society, to appear.

[14] Richard J. Lipton, *Using DNA to solve NP-complete problems*, Science **268** (1995), 542–545.

[15] Richard J. Lipton and Eric B. Baum (eds.), *DNA based computers: Proceedings of a DIMACS workshop, April 4, 1995, Princeton University*, DIMACS series in discrete mathematics and theoretical computer science, vol. 27, Providence, American Mathematical Society, 1996.

[16] Albert Nijenhuis and Herbert S. Wilf, *Combinatorial algorithms*, Academic Press, New York, 1975.

[17] J. Sambrook, E. F. Fritsch, and T. Maniatis, *Molecular cloning, A laboratory manual*, second ed., Cold Spring Harbor Laboratory Press, 1989.

[18] Warren D. Smith and Allan Schweitzer, *DNA computers in vitro and vivo*, Tech. report, NEC Research Institute, 4 Independence Way, Princeton, NJ 08544, March 1995.

[19] Leslie G. Valiant, *The complexity of enumeration and reliability problems*, SIAM Journal on Computing **8** (1979), no. 3, 410–421.

(R. M. Williams) Naval Air Warfare Center, Warminster, PA
Current address: Naval Air Warfare Center, Patuxent River, MD
E-mail address: bobw@henry.nawcad.navy.mil

(D. Wood) Department of Computer Science, University of Delaware Newark, DE
E-mail address: wood@cis.udel.edu